CMOS and Beyond

Get up to speed with the future of logic switch design with this indispensable overview of the most promising successors to modern CMOS transistors.

Learn how to overcome existing design challenges using novel device concepts, presented using an in-depth, accessible, tutorial-style approach. Drawing on the expertise of leading researchers from both industry and academia, and including insightful contributions from the developers of many of these alternative logic devices, new concepts are introduced and discussed from a range of different viewpoints, covering all the necessary theoretical background and developmental context.

Covering cutting-edge developments with the potential to overcome existing limitations on transistor performance, such as tunneling field-effect transistors (TFETs), alternative charge-based devices, spin-based devices, and more exotic approaches, this is essential reading for academic researchers, professional engineers, and graduate students working with semiconductor devices and technology.

Tsu-Jae King Liu is the TSMC Distinguished Professor in Microelectronics, Department of Electrical Engineering and Computer Sciences at the University of California, Berkeley. A co-inventor of the FinFET, she has been awarded the IEEE Kiyo Tomiyasu Award (2010), the Electrochemical Society Thomas D. Callinan Award (2011), and the SIA University Researcher Award (2014). She is a Fellow of the IEEE.

Kelin Kuhn is the Mary Upson Visiting Professor at Cornell University. She retired from Intel Corporation in 2014, where she was an Intel Fellow and Director of Advanced Device Technology in the Technology and Manufacturing Group at Intel Corporation. At Intel she was responsible for research on new device architectures and has been awarded two Intel Achievement Awards, one for high-k metal-gate (2006) and one for Trigate (2008), as well as the IEEE Paul Rappaport Award (2013). She is a Fellow of the IEEE.

CMOS and Beyond

Logic Switches for Terascale Integrated Circuits

TSU-JAE KING LIU
University of California, Berkeley

KELIN KUHN
Intel Corporation

CAMBRIDGE
UNIVERSITY PRESS

Shaftesbury Road, Cambridge CB2 8EA, United Kingdom

One Liberty Plaza, 20th Floor, New York, NY 10006, USA

477 Williamstown Road, Port Melbourne, VIC 3207, Australia

314–321, 3rd Floor, Plot 3, Splendor Forum, Jasola District Centre, New Delhi – 110025, India

103 Penang Road, #05–06/07, Visioncrest Commercial, Singapore 238467

Cambridge University Press is part of Cambridge University Press & Assessment,
a department of the University of Cambridge.

We share the University's mission to contribute to society through the pursuit of
education, learning and research at the highest international levels of excellence.

www.cambridge.org
Information on this title: www.cambridge.org/9781107043183

© Cambridge University Press & Assessment 2015

First published 2015

A catalogue record for this publication is available from the British Library

Library of Congress Cataloging-in-Publication data
Liu, Tsu-Jae King.
CMOS and beyond : logic switches for terascale integrated circuits / Tsu-Jac King Liu, University of
California, Berkeley ; Kelin Kuhn, Intel Corporation.
    pages   cm.
ISBN 978-1-107-04318-3 (Hardback)
1. Metal oxide semiconductors, Complementary.   2. Integrated circuits.   I. Kuhn, Kelin.   II. Title.
TK7871.99.M44L63 2014
621.39′5–dc23   2014020938

ISBN   978-1-107-04318-3   Hardback

Contents

7   Graphene and 2D crystal tunnel transistors   144

Qin Zhang, Pei Zhao, Nan Ma, Grace (Huili) Xing, and Debdeep Jena

8   Bilayer pseudospin field effect transistor   175

Dharmendar Reddy, Leonard F. Register, and Sanjay K. Bannerjee

Section III Alternative field effect devices

**9   Computation and learning with metal–insulator transitions and emergent
          phases in correlated oxides   209**

You Zhou, Sieu D. Ha, and Shriram Ramanathan

10   The piezoelectronic transistor   236

Paul M. Solomon, Bruce G. Elmegreen, Matt Copel, Marcelo A. Kuroda,
Susan Trolier-McKinstry, Glenn J. Martyna, and Dennis M. Newns

Section IV Spin-based devices

Contributors

Sapan Agarwal
University of California, Berkeley

Elad Alon
University of California, Berkeley

Sanjay K. Banerjee
University of Texas at Austin

Gary H. Bernstein
University of Notre Dame

George I. Bourianoff
Intel Corporation

Ahmet Ceyhan
Georgia Institute of Technology

Matt Copel
IBM

György Csaba
University of Notre Dame

Bruce G. Elmegreen
IBM

Sieu D. Ha
Harvard University

X. Sharon Hu
University of Notre Dame

Zachery A. Jacobson
University of California, Berkeley

Debdeep Jena
University of Notre Dame

Zhengping Jiang
Purdue University

Asif Islam Khan
University of California, Berkeley

Kelin J. Kuhn
Intel Corporation, Cornell University

Alexander Khitun
University of California, Riverside

Tsu-Jae King Liu
University of California, Berkeley

Gerhard Klimeck
Purdue University

Kelin J. Kuhn
Intel Corporation

Marcelo A. Kuroda
IBM

Nan Ma
University of Notre Dame

Glenn J. Martyna
IBM

Azad Naeemi
Georgia Institute of Technology

Rhesa Nathanael
University of California, Berkeley

Dennis M. Newns
IBM

Michael T. Niemier
University of Notre Dame

Dmitri E. Nikonov
Intel Corporation

Alexei Orlov
University of Notre Dame

Wolfgang Porod
University of Notre Dame

Shaloo Rakheja
Massachusetts Institute of Technology

Shriram Ramanathan
Harvard University

Dharmendar Reddy
University of Texas at Austin

Leonard F. Register
University of Texas at Austin

Sayeef Salahuddin
University of California, Berkeley

Alan Seabaugh
University of Notre Dame

Paul M. Solomon
IBM

Susan Trolier-McKinstry
Pennsylvania State University

Grace (Huili) Xing
University of Notre Dame

Eli Yablonovitch
University of California, Berkeley

Qin Zhang
University of Notre Dame

Pei Zhao
University of Notre Dame

You Zhou
Harvard University

Preface

Steady miniaturization of CMOS (complementary metal–oxide–semiconductor) transistors – the predominant type of electrical switches used in digital integrated circuit "chips" – has yielded continual improvements in the performance and cost-per-function of electronic devices over the past four decades. This relentless miniaturization has resulted in ubiquitous information technology with dramatic global impact on virtually every aspect of life in modern society.

CMOS technology is reaching a state of maturity wherein continued transistor scaling will not be as straightforward in the future as it has been in the past. This is already apparent from the slowdown in certain aspects of scaling (e.g., chip supply voltage scaling, transistor off-state leakage current scaling, and so on). Clearly, improved switch designs will be needed to sustain the growth of the electronics industry beyond the next decade. A wide variety of alternative switch designs are being discussed in the research community, many of which use operating principles dramatically different from those of conventional CMOS transistors. Unfortunately, papers published by the research community in rapidly developing fields are rarely tutorial. Thus, much of this important new information is not readily comprehensible to the mainstream electronics community.

To help address this communication gap, we approached recognized experts in the research community with requests to create tutorial essays in their area of speciality. This book organizes these essays into sections, beginning with background information on the power–performance trade-off (motivating steep sub-threshold swing devices), continuing with tunneling-based devices, alternative field effect devices, and spin-based (magnetic) devices. It closes by reviewing the challenges of interconnects for these evolving new switch designs.

The first section reviews chip design considerations and benchmarks various alternative switching devices, with particular emphasis on devices with steeper sub-threshold swing. In Chapter 1, Elad Alon introduces the concepts underlying historical transistor scaling and analyses the key trade-offs between density, power, and performance which drive modern CMOS chip designs. Circuit design techniques such as power gating and parallelism are reviewed in context of constraints for continued dimensional scaling. The energy-efficiency limit for CMOS technology due to the 60 mV/dec sub-threshold swing limitation is discussed in relation to the potential benefit of beyond-CMOS devices with steeper sub-threshold swings. In Chapters 2 and 3, Zachery A. Jacobson and Kelin Kuhn review and benchmark a broad range of alternative devices being explored in the research community. These chapters focus on electronic (as opposed to magnetic) devices that either incorporate new materials or mechanisms for enhanced switching behaviour. Chapter 2 briefly reviews the history and operating principles of these devices, while Chapter 3 benchmarks them with respect to drive current, energy efficiency, fabrication cost, complexity, and memory cell area. In Chapter 4, Asif Islam Khan and Sayeef Salahuddin explore the idea of overcoming the 60 mV/dec sub-threshold swing limit by incorporating a ferroelectric layer within the gate stack of a CMOS transistor. They discuss both the theory and recent experiments which support the possibility of achieving CMOS transistors with negative small-signal capacitance.

The second section covers device designs which utilize quantum mechanical tunneling as the switching mechanism to achieve steeper sub-threshold swing. In Chapter 5, Sapan Agarwal and Eli Yablonovitch assess the promise of tunnel field effect transistors (TFETs) in light of the requirements for simultaneous steep sub-threshold swing, large on/off-current ratio, and high on-state conductance. The authors explore the impact of p-n junction dimensionality and discuss various design trade-offs and the advantages of lateral, vertical, and bilayer implementations. Recent experimental data is evaluated in light of the various design requirements. In Chapter 6, Alan Seabaugh, Zhengping Jiang, and Gerhard Klimeck continue the discussion on TFETs, but with a focus on III-V semiconductor material systems. Design trade-offs for homojunction versus heterojunction III-V systems and challenges for achieving high performance with p-channel TFETs are discussed, along with non-idealities of particular relevance to III-V systems (such as traps, interface roughness, and alloy disorder). In Chapter 7, Qin Zhang, Pei Zhao, Nan Ma, Grace (Huili) Xing, and Debdeep Jena further extend the TFET discussion by evaluating the potential for TFETs built with graphene and two-dimensional semiconductor materials. In-plane tunneling and inter-layer tunneling devices are reviewed, and recent experimental results are assessed in relation to theoretical understanding. In Chapter 8, Dharmendar Reddy, Leonard F. Register, and Sanjay K. Banerjee review a novel tunneling device, the bilayer pseudospin field effect transistor (BiSFET). The BiSFET relies on the possibility of room temperature excitonic (electron-hole) superfluid condensation in two dielectrically separated graphene layers. Formation of a room temperature condensate is essential for BiSFET operation, and the authors discuss the key physics and challenges of creating such a condensate. BiSFET compact models and circuit designs are also discussed and used to project performance benefits over CMOS.

The third section covers devices that employ alternative approaches to achieving steeper switching behaviour. In Chapter 9, You Zhou, Sieu D. Ha, and Shriram Ramanathan discuss the possibility of making devices from electron correlated materials, which can transition between insulator and metal phases. The authors discuss the physics of the metal–insulator transition, with particular emphasis on the vanadium dioxide (VO_2) system. Mott FET devices, solid-state VO_2 FETs and ionic liquid-gated VO_2 FETs are reviewed and circuit architectures which exploit these devices are discussed. In Chapter 10, Paul M. Solomon, Bruce G. Elmegreen, Matt Copel, Marcelo A. Kuroda, Susan Trolier-McKinstry, Glenn. J. Martyna, and Dennis M. Newns introduce piezoelectronic transistor (PET) devices. The PET is essentially a solid-state relay in which a piezoelectric element provides the mechanical force and a piezoresistive element transduces the mechanical force to electrical switching. The basic physics of piezoelectric and piezoresistive materials are discussed, along with process integration challenges. PET dynamics, compact models, and circuit designs are also discussed and used to project performance benefits over CMOS. In Chapter 11, Rhesa Nathanael and Tsu-Jae King Liu discuss nanoscale electromechanical relays as logic switches. Relays use mechanical movement to physically short or open an electrical connection between two contacts, and have the ideal characteristics of zero off-state leakage current, abrupt sub-threshold swing, and low gate leakage. The authors review materials requirements and process integration challenges specific to nanorelays, describe a variety of relay designs that provide for more compact implementation of complex logic circuits, and discuss scaling methodologies.

The fourth section covers devices that use magnetic effects or electronic spin to carry information. These can be used to implement nanomagnetic logic (wherein small magnets are used to construct circuits), spin torque logic, or spinwave logic (wherein electron spin is the information token). In Chapter 12, György Csaba, Gary H. Bernstein, Alexei Orlov, Michael T. Niemier, X. Sharon Hu, and Wolfgang Porod discuss the possibility of making circuits out of small single-domain magnets. The switching properties of single-domain nanomagnets are introduced and various clocking schemes are discussed. A full-adder structure is benchmarked against CMOS and design issues in nanomagnetic logic are reviewed. In Chapter 13, Dmitri E. Nikonov and George I. Bourianoff introduce the possibility of making majority gate logic circuits using the spin torque effect. In these devices the combined action of spin torques from the various inputs transfers enough torque to switch the magnetization of the output. Detailed simulations of in-plane and perpendicular spin torque switching are reviewed. An adder circuit is discussed and benchmarked against CMOS. In Chapter 14, Alexander Khitun analyses the possibility of using spin waves for logic functions. A spin wave is a collective oscillation of spins in a spin lattice around the direction of magnetization. The physics of spin wave devices is introduced and experimental results discussed. Various spin wave circuits and architectures are reviewed and benchmarked against CMOS.

A critical (but frequently neglected) issue in assessment of beyond CMOS devices is the interconnect architecture. Creating a fabulous new switch is valueless if it cannot be connected to other active or passive devices! This is particularly relevant for magnetic and spin-based devices as they do not (typically) interface directly to conventional electronic devices. Thus, in Chapter 15, Shaloo Rakheja, Ahmet Ceyhan, and Azad Naeemi close this book with a comprehensive evaluation of the interconnect considerations for advanced logic devices. This includes both interconnect options for emerging charge-based device technologies and interconnect options for spin-based technologies.

Our hope is that these essays will help to bridge the gap between research on emerging devices and their practical implementation in terascale integrated circuits by the mainstream semiconductor community.

Section I

CMOS circuits and technology limits

1 Energy efficiency limits of digital circuits based on CMOS transistors

Elad Alon

1.1 Overview

Over the past several decades, CMOS (complementary metal–oxide–semiconductor) scaling has come to be associated with dramatic and simultaneous improvements in functionality, performance, and energy efficiency. In particular, although the actual historical trends did not uniformly follow a single type of scaling, there was a relatively long period of "Dennard scaling" [1] during which the quadratic (with scale factor) improvements in transistor density were accompanied by a quadratic reduction in power per gate despite a linear increase in switching frequency. All of this was achieved by scaling the operating (i.e., supply) voltage of the circuitry linearly along with the lithographic dimensions of the transistor. Ideally, this would result in constant power consumption per unit chip area, making it relatively easy for chip architects and designers to exploit the increased transistor density with a fixed chip area (and hence power) to cram more functionality into a single die.

Unfortunately, however, as Dennard himself predicted, because of the fact that some intrinsic parameters associated with transistor operation – in particular, the thermal voltage kT/q – do not scale along with the lithographic dimensions, this type of scaling came to an end in the early 2000s. Up until that point, because leakage currents (and hence leakage energy) were essentially negligible, the transistor's threshold voltage had been treated as a scaling parameter that could be reduced with no significant consequence. However, since leakage current depends exponentially on the threshold voltage, this type of scaling indeed eventually came to a halt.

As will be described in detail in Section 1.2, for today's designs (and ever since roughly the 90 nm process technology node), both the threshold and supply voltages must be chosen to balance out the leakage and dynamic energy components at a given desired performance. The implication of this is that simple scaling no longer provides obvious benefits in all three dimensions (density, power, and performance); instead, one is forced to make direct trade-offs between energy and performance – even if given a more lithographically advanced process technology. This section will highlight that at the device level, transistors must achieve an on/off current ratio of ~10^4–10^6 in order to achieve optimal energy efficiency. Section 1.3 next discusses selected techniques – in particular, power gating and parallelism – utilized by architects and circuit designers to achieve the energy-efficiency potential of scaled CMOS technologies. Finally, in Section 1.4 we will highlight the fact that CMOS transistors have a well-defined minimum energy per operation, and thus even parallelism will eventually cease to be an effective means of keeping chip power consumption in check.

1.2   Energy–performance trade-offs in digital circuits

In order to explain why both the supply and threshold voltages must be balanced to achieve energy-efficient digital circuits, we must first briefly examine the composition of typical digital chips. As highlighted in Fig. 1.1, the largest contributor to the power consumed by a processor (which is a good representative for digital chip designs as a whole) is typically the control/datapath, and in fact, the overall performance and power of the chip generally track with those of the control/datapath as well. As also highlighted by the figure, the clock frequency (performance) of the design is set by the delay of the combinational logic between the clocked registers.

Although there are obviously extremely wide variations in the actual composition of the combinational logic within a digital chip, the behavior (in terms of energy and performance) of all such logic tracks very closely with the behavior of a cascade of inverters. To begin analyzing the underlying trade-offs, we can therefore utilize the simplified model shown in Fig. 1.2 as a proxy for the energy and performance of a generic digital circuit. As highlighted in the figure, the most relevant circuit-level parameters are the activity factor α – which is defined as the average probability of a given node in the circuit transitioning (i.e., changing its state) on any given clock cycle, the capacitive fanout[1] f, the capacitance per inverter (gate) C, and the logic depth (i.e., the number of stages of combinational logic between flip-flops) L_d.

With this model in hand, it is easy to show that the delay t_{delay} of the circuit is simply set by:

$$t_{delay} = \frac{1}{2} \frac{L_d \cdot f \cdot C \cdot V_{dd}}{I_{on}(V_{dd} - V_{th})}, \qquad (1.1)$$

where V_{dd} is the power supply voltage of the circuit, and $I_{on}(V_{dd} - V_{th})$ is the effective[2] drain current of the transistors within the inverter when they are in the on-state, driven by a given supply voltage V_{dd}, and with a given threshold voltage V_{th}. One can use a variety of different models to expand the functional relationship between I_{on}, V_{dd}, and V_{th} (e.g., alpha-power law [2], velocity saturation [3], etc.), but as we will see shortly, it is not necessary to do so to understand the underlying causes for the key trade-offs at hand; one must simply realize that the on-current increases if $(V_{dd} - V_{th})$ is increased.

Let us next consider the energy consumed by the chain of inverters during the completion of a single operation. For well-designed digital circuits, the energy will consist essentially of only two components: dynamic energy due to charge/discharging

---

[1] In this model the fanout may appear logical in that every inverter is driving f copies of itself, but for general digital circuits the fanout should be treated as capacitive – i.e., the ratio of the input capacitance of a given gate to the input capacitance of the succeeding gates in the chain.

[2] The drain current of the devices isn't actually constant during the output transition, but can be well approximated by a single number in most cases of interest.

(a)   (b)

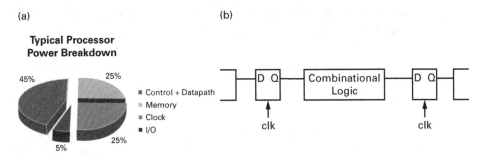

Fig. 1.1 (a) Power breakdown for a typical embedded processor. (b) Conceptual model for synchronous digital circuits.

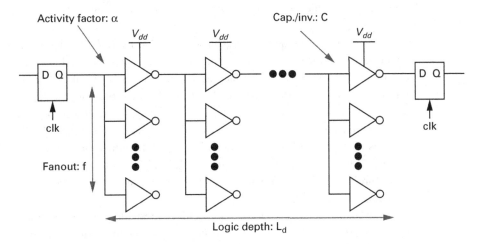

Fig. 1.2 Inverter-based model for combinational logic energy and performance.

the parasitic capacitance within the circuit, and leakage energy due to the fact that even the off switches within the logic gates still conduct current during the entire duration of the operation. Once again referring to the model in Fig. 1.2, the dynamic (E_{dyn}) and leakage (E_{leak}) energy components are:

$$E_{dyn} = \alpha \cdot L_d \cdot f \cdot C \cdot V_{dd}^2, \tag{1.2a}$$

$$E_{leak} = L_d \cdot f \cdot I_{off}(V_{th}) \cdot V_{dd} \cdot t_{delay}, \tag{1.2b}$$

where $I_{off}(V_{th})$ is the effective off-state leakage of the transistors within the inverter for a given device threshold voltage V_{th}.[3]

To highlight why one must now choose V_{dd} and V_{th} such that they balance out these two components of energy consumption at a given performance, it is instructive

---

[3] The supply voltage V_{dd} also affects the leakage current I_{off}, but for the purposes of this discussion this effect does not alter the underlying trade-offs/conclusions.

to combine Eqs. (1.1) and (1.2) as follows into a single expression for the total energy per operation:

$$E_{total} = \alpha \cdot L_d \cdot f \cdot C \cdot V_{dd}^2 + L_d \cdot f \cdot I_{off}(V_{th}) \cdot V_{dd} \cdot \frac{1}{2} \frac{L_d \cdot f \cdot C \cdot V_{dd}}{I_{on}(V_{dd} - V_{th})}$$

$$= \alpha \cdot L_d \cdot f \cdot C \cdot V_{dd}^2 \cdot \left[1 + \frac{L_d \cdot f}{2\alpha} \cdot \frac{I_{off}(V_{th})}{I_{on}(V_{dd} - V_{th})} \right]. \tag{1.3}$$

The most important point to notice about the expression in Eq. (1.3) is that, although one would like to use a low V_{dd} to reduce energy, one cannot do so without also lowering V_{th} if the same performance (i.e., $t_{delay} \propto CV_{dd}/I_{on}$) is to be maintained, thus increasing the leakage energy. The critical implication of this is that there are optimal V_{dd} and V_{th} values which balance out the two energy components such that the lowest total energy is achieved for a given delay target (or equivalently, the lowest delay for a given energy).

Notice also that the quantity I_{on}/I_{off} when scaled by $L_d f/\alpha$ (which is set purely by circuit-level parameters) is directly indicative of the ratio between dynamic and leakage energy for the whole circuit. In fact, as shown by Nose and Sakurai in [4] for super-threshold CMOS circuits, the optimal I_{on}/I_{off} (and therefore both the resulting optimal V_{dd} and V_{th} as well as the ratio of dynamic to leakage energy) is directly set by $L_d f/\alpha$, and remains relatively fixed regardless of the exact delay target. Furthermore, an analysis by Kam and his co-authors in [5] shows that this result essentially holds true for any CMOS-like device technology in essentially any operating region (i.e., sub- vs. super-threshold), even those with significantly steeper drain current vs. gate voltage than CMOS transistors.

Given the above observations, and in order to provide a numerical guideline for the optimal I_{on}/I_{off}, it is worthwhile to examine representative values for the circuit-level parameters L_d, f, and α, as well as the reasons underlying the selection of those values. Let's begin with the logic depth L_d, which is typically set to ~15–40. Much like the optimal V_{dd} and V_{th}, this selection is driven by balancing out the improved timing slack gained by further pipelining (i.e., reducing L_d) with the increased overhead from additional timing elements (i.e., flip-flops/registers) [6]. Similarly, the fanout f is typically set to greater than 2 to reduce the delay overhead associated with each gate stage and up to ~8 to ensure robust operation (gates with large fanout tend to be much more susceptible to noise/crosstalk). Finally, the overall activity factor α for most practical designs is ~10% down to 0.1%; these relatively low percentages can be understood by the fact that in most complex logic chains (and even more so in memory structures), the large majority of the states of the gates are not changing on any one clock cycle.

Taken together and with the appropriate scale factors, the optimal I_{on}/I_{off} for a wide variety of designs lies within the range 10^4–10^6. Since for reasonable performance levels CMOS transistors achieve ~100 mV/dec effective inverse slope (i.e., $V_{dd}/\log_{10}(I_{on}/I_{off})$, as defined in [5]), the supply voltage necessary to achieve this on/off current ratio is typically 500–600 mV. Note that the farther into the high-performance regime one wants to operate, the worse the effective overall slope will be, and hence many designs operate at closer to 1 V to achieve the desired (peak) performance.

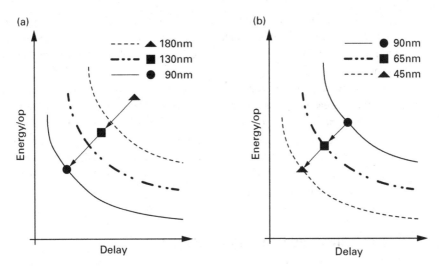

Fig. 1.3 Scaling of designs in the energy per operation vs. delay space using the nominal supply and threshold voltages under (a) traditional (Dennard) scaling and (b) modern (~sub-90 nm) scaling.

Before moving on to the next section, it is worth examining the implications of the above analysis on historical as well as future CMOS scaling. During the traditional (Dennard) scaling regime, simultaneously lowering V_{dd} and V_{th} caused a substantial and dramatic decrease in the I_{on}/I_{off} ratio from one process technology to the next. It turns out, however, that reducing the I_{on}/I_{off} ratio in this way was actually very desirable, because at that point the thresholds had been set so high that the leakage energy component was negligible. It was therefore beneficial to reduce the supply voltage and save on dynamic energy. In other words, the reason that scaling was able to proceed in this manner was that at that point, typical designs were actually not operating on optimal points in the energy vs. delay trade-off space.

To make this perspective clear, Fig. 1.3 uses markers to show where designs operating under the nominal supply and threshold voltages for a given process technology would lie relative to the optimal energy vs. delay curves. As shown in Fig. 1.3(a), typical designs were operating substantially above and to the right of the optimal curves, but as V_{dd} and V_{th} were reduced, scaling brought these designs closer to the actual optimal curves. In other words, a significant portion of the energy-efficiency benefits that came to be associated with scaling were not actually inherently due to the dimensional scaling itself – rather, they were the result of reducing the degree of sub-optimality.

This is of course not to say that dimensional scaling brings no benefits at all in energy and delay – it is simply that once designs were essentially operating on the optimal part of the curve, as highlighted in Fig. 1.3(b), purely dimensional scaling (with V_{dd} and V_{th} fixed) brings at best linear reductions in energy/operation and delay, both due to decreased capacitance/gate [7]. In practice, the poor scaling of interconnect parasitics and variation issues tend to make the capacitance/gate scale relatively poorly (i.e., the minimum total capacitance per gate does not reduce substantially from one process to the next).

Even in the best case, however, simple dimensional scaling does not provide sufficient benefit to enable scaled designs to achieve increased performance and functionality within a given power budget. Specifically, if one leaves the supply and threshold voltages fixed, the power per gate (which is proportional to E_{total}/t_{delay}) is also fixed. Nevertheless, if one actually exploited the increased density to integrate twice as many gates in each process generation, the power of the chip would double as well. In the vast majority of applications chip power must be kept constant from one generation to the next (due either to thermal or battery-life limitations), and thus designers have been forced to utilize other approaches to translate dimensional scaling into usable advances. The most prominent of these approaches – namely, parallelism – will be discussed further in the next section.

1.3   Design techniques for energy efficiency

Since many of the trade-offs between energy and performance discussed in the previous section can be traced backed to the fact that CMOS transistors leak when they are supposed to be off, it is natural to wonder whether a circuit- or system-level technique can be used to eliminate or at least mitigate the leakage energy. The most natural candidate for this is referred to as "power gating" or "sleep transistors" [8]. Figure 1.4 depicts the concept as applied to a chain of inverters, where the key idea is to disconnect an entire block from its power supply during periods of time where one knows that the block is not performing any useful work. The power switch itself must of course also be implemented by some kind of transistor (or more generally, whatever switch is available in the process technology), but if this switch is implemented with a higher I_{on}/I_{off} device (i.e., a device with higher V_{th} and/or larger gate voltage swing), turning this switch off can indeed reduce the leakage of the overall circuit vs. the original circuit in the off-state.

Continuing down the original line of thinking, one may then wonder if power gating could be utilized even more aggressively to cut off the power supply of each gate as soon as it has finished doing useful work, and hence break or at least improve upon the trade-offs described previously. In particular, if the gate was only "awake" whenever its output needs to transition, the activity factor α would effectively be much larger than the numbers quoted earlier. The issue with this idea, however, is that one must know when to turn the power gating switch on or off, and in the limit of power gating every single logic gate separately, one would need to replicate the functionality of the entire gate to

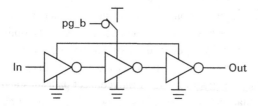

Fig. 1.4 Power gating applied to a chain of inverters.

compute this power gating signal. However, this replicated gate would then suffer from the exact same energy–performance trade-offs described earlier.

Clearly, attempting to power gate every single logic gate does not provide any benefit, but even for more moderate approaches (i.e., power gating individual sub-blocks), the key issue to keep in mind is that not only will the power gate itself introduce energy/performance overheads (due to voltage drops across the power gating device when it is active, and due to the energy consumed by driving the parasitic capacitance of the power gating device), the circuits to compute whether or not the power gate should be active will themselves introduce both static and dynamic energy overheads. Thus, power gating is usually only applied at relatively coarse levels of granularity where it is very straightforward to know (or be told by, e.g., the operating system) whether or not the underlying blocks are performing active work.

Even though power gating does not improve upon the fundamental energy–performance trade-offs described earlier, it is effective in dealing with the practical reality that in most applications, the required computations are bursty. For example, when a mobile phone is in standby mode, the applications processor is typically idle and/or only activated on regular intervals to perform some maintenance tasks. Only once the phone is turned on/being actively used would it be likely for the applications processor to have significant computational tasks to complete.

Continuing with the above example, let's assume that the applications processor as a whole is active only 10% of the time. Without power gating and in comparison to the case where the processor is being used continuously, the activity factor α is now effectively $10\times$ lower, forcing a nearly identical $10\times$ increase in the I_{on}/I_{off} ratio. With CMOS transistors and an 80 mV/dec sub-threshold slope, this would force one to increase the threshold voltage by approximately 80 mV, and hence the supply voltage by a similar percentage (to maintain the same performance). As shown in Fig. 1.5, the achievable energy/operation of this bursty processor would therefore be degraded relative to the case where the processor was used continuously. With an ideal (i.e., zero on-resistance, zero parasitic capacitance, and zero leakage) power gating device and "free" system-level cues to indicate when the processor is active or not, one could

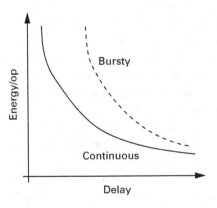

Fig. 1.5 Energy vs. delay implications of bursty vs. continuous usage of a digital circuit.

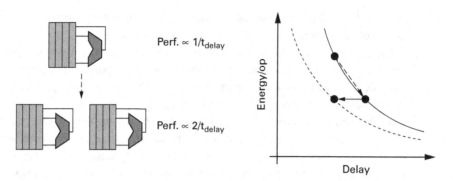

Fig. 1.6 Illustration of parallelism and how it improves the energy vs. performance trade-off on an example with two functional units compared to a single functional unit.

return the processor to the continuous-use energy-delay trade-off curve. In other words, the main benefit of power gating is that it reduces the penalty of the system-level variability in usage patterns.

Having examined the difficulties associated with eliminating or mitigating leakage within the logic gates themselves, we are still left with the fact that designers would like to utilize the dimensional scaling of transistors to simultaneously improve energy, performance, and functionality, but that scaling alone in the most straightforward manner (while leaving chip size fixed) would cause power consumption to increase substantially. Fortunately, there is a technique that designers can and have applied to exploit the availability of additional transistors to improve energy efficiency: parallelism [9].

The basic idea behind parallelism is quite straightforward, and is depicted in Fig. 1.6. In essence, if at the application level one has multiple pieces of data that can be operated on in parallel, replicating the digital hardware units and feeding them with the independent data inputs allows you to complete proportionally more operations within the same time period. Since our goal, however, is to improve energy efficiency, rather than simply increasing the throughput in this manner (but spending proportionally more power), we can instead run each unit more slowly – and therefore at lower energy/ operation. As also highlighted in Fig. 1.6, in comparison to a design where we tried to achieve the same performance by running a single unit at a higher frequency (i.e., lower delay), because each of its functional units can operate at a lower energy point of the curve, the parallel implementation can be significantly more energy efficient.

In practice, parallelism does not work quite as ideally as depicted in Fig. 1.6 – there are always some overheads involved in distributing/collecting the data to/from the various units, and not all applications (or even sections of code within a given application) naturally offer parallelism. These overheads can fortunately be made relatively minimal, and so for approximately the last decade, parallelism has indeed been the primary workhorse of the semiconductor industry to convert the availability of additional transistors in a scaled process technology into improved performance without breaking the power budget. In fact, it is very difficult to purchase a laptop PC without at least four cores integrated onto the central processing unit, and even within smartphones the vast majority of the applications processors utilize at least two cores. However, as we will describe next, even parallelism will soon cease (or perhaps even already has ceased) to be an effective tool for improving energy efficiency.

1.4   Energy limits and conclusions

As first described by Calhoun and Chandrakasan in [10], once a CMOS circuit is operated in the sub-threshold regime, for essentially any combination of L_d, f, and α of practical interest, there is a well-defined minimum energy/operation that the circuit must dissipate. To understand the reasons behind this, we can simply re-examine Eq. (1.3) from Section 1.2, and recall that in the sub-threshold region of operation, I_{off} is exponentially dependent upon $-V_{th}$, while I_{on} is exponentially dependent upon $(V_{dd} - V_{th})$. In this case, the I_{on}/I_{off} ratio depends purely on $V_{dd} - V_{th}$ specifically:

$$E_{total} = \alpha \cdot L_d \cdot f \cdot C \cdot V_{dd}^2 \cdot \left[1 + \frac{L_d \cdot f}{2\alpha} \cdot \frac{e^{\left(\frac{-V_{th}}{nkT/q}\right)}}{e^{\left(\frac{V_{dd}-V_{th}}{nkT/q}\right)}} \right] = \alpha \cdot L_d \cdot f \cdot C \cdot V_{dd}^2 \cdot \left[1 + \frac{L_d \cdot f}{2\alpha} \cdot e^{\left(\frac{-V_{dd}}{nkT/q}\right)} \right].$$

$$(1.4)$$

By plotting Eq. (1.4) above, it is easy to see that there is a specific value of V_{dd} that optimally balances the leakage and dynamic energy contributions. Reducing V_{dd} any further below this point actually increases the total energy because the exponential increase in the delay of the circuit causes the leakage energy actually to increase (despite the reduced supply voltage). The threshold voltage has no effect on the total energy because, even though increasing the threshold exponentially decreases the leakage current, it also exponentially increases the delay. So, increasing V_{th} simply allows the circuit to operate slower, but at no lower energy than before.

Parallelism relies on the principle that running a circuit slower will allow it to achieve lower energy/operation. As depicted in Fig. 1.7 and pointed out in [10], once each sub-unit operates at its minimum energy, slowing down each sub-unit further brings no further improvement in energy/operation – the supply voltage V_{dd} for each unit should remain fixed regardless of the degree of parallelization.

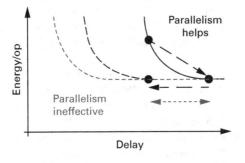

Fig. 1.7 Illustration of the limits of parallelism due to minimum energy/operation.

Although the large majority of commercial chips do not yet operate in sub-threshold, practical challenges such an ensuring functionality/yield in these conditions may hinder the practicality of even approaching this regime. Perhaps even more importantly, the performance penalties for truly operating in sub-threshold (as opposed to slightly above but near the threshold voltage) are substantial (i.e., the energy/op curve becomes very flat as you approach minimum energy), and so for cost considerations it may not make sense to utilize so much silicon area for such a small improvement in energy [7].

With the end of parallelism clearly in sight, chip designers and architects today are once again facing a crisis of how to reap the benefits they had once been accustomed to from scaling. While some design techniques to improve energy efficiency – in particular, the integration of specialized functional units – remain to potentially be exploited, it is increasingly difficult to find and apply such techniques in a general manner that would apply to the industry as a whole. In contrast, there is no question that if one were able to practically integrate a new device technology with steeper effective sub-threshold slope – such that the switch is able to achieve the required 10^4–10^6 on/off ratio as well as similar on-conductance with substantially lower voltage than the ~500–600 mV required of CMOS – it would bring immediate and widespread improvements in energy efficiency. The challenges facing the developers of such new switch technologies are immense – to compete with 14 nm CMOS, for example, one would need to integrate well over 10 billion such switches into a ~14 mm × 14 mm die with sufficiently high yield and low variability – but the potential benefits are dramatic enough that researchers in both industry and academia have and will continue to invest significant effort into making such beyond CMOS technologies a reality.

References

[1] R. H. Dennard, F. H. Gaensslen, V.L. Rideout, E. Bassous, and A. R. LeBlanc, "Design of ion-implanted MOSFET's with very small physical dimensions." *IEEE Journal of Solid-State Circuits*, **9**(5), 256–268 (1974).

[2] T. Sakurai and A. R. Newton, "Alpha-power law MOSFET model and its applications to CMOS inverter delay and other formulas." *IEEE Journal of Solid-State Circuits*, **25**(2), 584–594 (1990).

[3] K.-Y. Toh, P.-K. Ko, and R. Meyer, "An engineering model for short channel MOS devices." *IEEE Journal of Solid-State Circuits*, **23**(4), 950–958 (1988).

[4] K. Nose and T. Sakurai, "Optimization of V_{dd} and V_{th} for low-power and high-speed applications." In *Asia South Pacific Design Automation Conference, Proceedings of*, pp. 469–474 (2000).

[5] H. Kam, T.-J. King Liu, and E. Alon, "Design requirements for steeply switching logic devices." *IEEE Transactions on Electron Devices*, **59**(2), 326–334 (2012).

[6] V. Zyuban *et al.*, "Integrated analysis of power and performance for pipelined microprocessors." *IEEE Transactions on Computers*, **53**(8), 1004–1016 (2004).

[7] M. Horowitz, E. Alon, D. Patil, S. Naffziger, R. Kumar, and K. Bernstein, "Scaling, power, and the future of CMOS." In *IEEE International Electron Devices Meeting, Technical Digest*, pp. 7–15 (2005).

[8] J. W. Tschanz, S. G. Narendra, Y. Ye, B. A. Bloechel, S. Borkar, and V. De, "Dynamic sleep transistor and body bias for active leakage power control of microprocessors." *IEEE Journal of Solid-State Circuits*, **38**(110), 1838–1845 (2003).

[9] A. P. Chandrakasan, S. Sheng, and R. W. Brodersen, "Low-power CMOS digital design." *IEEE Journal of Solid-State Circuits*, **27**(4), 473–484 (1992).

[10] B. H. Calhoun and A. Chandrakasan, "Characterizing and modeling minimum energy operation for subthreshold circuits." In *Low-Power Electronics and Design, IEEE International Symposium on, Proceedings of*, pp. 90–95 (2004).

2 Beyond transistor scaling: alternative device structures for the terascale regime

Zachery A. Jacobson and Kelin J. Kuhn

2.1 Introduction

For more than 40 years, integrated-circuit device density has experienced exponential growth (a phenomenon known as Moore's law [1]). As traditional CMOS transistor scaling limits are being approached, there are many technologies that are being considered to supplant or integrate with CMOS to continue scaling into the terascale (10^{12} devices/cm^2) regime. This chapter reviews some of these future device technologies.

The scope of this chapter is confined to devices that could be direct replacements for (or complements to) to existing CMOS transistors and which are not presently mature enough for volume manufacturing (e.g., high electron mobility transistors and GaN were included, but not fully depleted silicon-on-insulator, or FinFET, devices). The use of other materials in conventional transistor structures is covered only for devices in which the basic operation of the device is vastly different than that of standard silicon-based MOS transistors (e.g., GaN-channel devices were included, but not III-V-channel MOS or germanium-channel MOS devices). Furthermore, the scope is restricted to devices based on charge transport. Although spin transport devices are of increasing interest, they would require a radical shift from the existing circuit architecture used today for CMOS technology.

Additionally, some devices were not included in this review due to other well-recognized limitations. For example, junction gate field effect transistors (JFETs) were not included, since the primary motivation of this work is extreme scalability of devices. Similarly, although organic semiconductor devices have excellent cost scaling per unit area, their potential for miniaturization and high-performance operation is poor. Carbon-based nanoelectronic structures, such as nanotubes and graphene–nanoribbon devices, also were not included due to current concerns about their manufacturability at the terascale level of integration.

2.2 Alternative device structures

2.2.1 HEMT

The high electron mobility transistor, or HEMT (see Fig. 2.1), achieves superior carrier mobility by separating mobile charge carriers from ionized dopant atoms, thus reducing ionized impurity scattering. This is accomplished by confining carriers in an undoped quantum well.

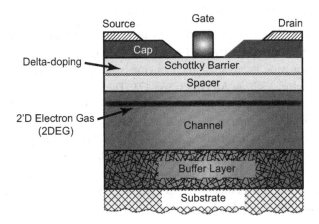

Fig. 2.1 A schematic cross-section of a typical high electron mobility transistor (HEMT) shows the two-dimensional/degree electron gas (2DEG) formed within an undoped channel region, which results in a high mobility for carriers.

2.2.1.1   History

Early work on HEMT devices occurred at Fujitsu in 1979 under the direction of Dr. Takashi Mimura [2, 3]. While working on creating a GaAs n-MOSFET, Dr. Mimura realized that electron inversion or accumulation was difficult due to the presence of a high concentration of surface states at the gate dielectric interface. Simultaneously, Bell Labs had developed a modulation-doped heterojunction superlattice where potential wells of undoped GaAs captured electrons from donors in AlGaAs layers [4]. These electrons move with high mobility in the undoped GaAs wells due to a lack of ionized impurity scattering. By combining these two concepts, Dr. Mimura realized that with a stack comprising a Schottky metal gate, doped n-AlGaAs region, undoped thin AlGaAs region, and GaAs, a structure similar to a MOS gate is formed and results in a device with reduced scattering and higher mobility. Additionally, by altering the thickness of the doped AlGaAs layer, a depletion mode device is formed. (A thicker AlGaAs layer results in an electron accumulation layer at the dielectric interface.)

In roughly the same time frame as the Mimura work, Delagebeaudeuf and Linh at Thomson-CSF demonstrated a two-dimensional electron gas effect in a metal–semiconductor field effect transistor (MESFET), similar to a HEMT [5, 6]. This device was the first "inverted" HEMT, where the Schottky gate is deposited on an undoped GaAs channel layer grown over a doped AlGaAs layer.

Further work resulted in new designs, such as AlGaAs/InGaAs pseudomorphic HEMTs (pHEMTs, not to be confused with p-type HEMTs) [3]. Traditional HEMTs are constrained to using materials with matching lattice constants; pHEMTs use very thin layers of materials with mismatched lattice constants, improving performance.

Most of the initial uses of HEMTs were for military and aerospace applications, but demand for HEMTs increased in the 1990s when Direct Broadcast Satellite television receivers began using HEMT amplifiers. More recent uses of HEMTs include radar systems, radio astronomy, and cell phone communications.

2.2.1.2   Device principles

HEMTs use the properties of a heterojunction to form a conductive channel with greater mobility than a traditional MOSFET. In a HEMT, as shown in Fig. 2.1, a heterojunction with a wide bandgap semiconductor is fabricated on top of a narrow bandgap semiconductor, such as AlGaAs/GaAs [7]. The electrons from the n-doped wide bandgap region (AlGaAs in this example) diffuse into the GaAs, which has a lower conduction band edge energy than AlGaAs. The GaAs is undoped, so carriers experience reduced scattering, resulting in increased mobility. This layer of carriers is called a two-dimensional electron gas, or 2DEG. Due to difficulties in forming a gate dielectric on these materials, HEMTs use a Schottky gate contact over the wide bandgap semiconductor. This Schottky contact results in higher gate leakage current in HEMTs than traditional MOSFETs.

2.2.1.3   Recent work

Several challenges exist for HEMTs. First, although drive currents are high, operating voltages for most HEMTs are much higher than traditional CMOS, which poses an issue for low-power operation. Gate leakage is a key concern, as Schottky gates have very high gate leakage due to the lack of a dielectric barrier. Band-to-band tunneling due to the narrow bandgap is an issue, as is high source and drain parasitic resistance [8]. There is also the issue of integration of p-type devices. Finally, the use of III-V wafers also adds fabrication cost and manufacturing complexity.

Recent work has focused on the use of HEMTs at lower operating voltages. Dewey *et al.* show drive currents that are able to match 40 nm MOSFETs at $V_{dd} = 1$ V as well as 0.5 V [9]. Figure 2.2 shows the device structure of [9], as well as relevant characteristics compared to a strained silicon device.

Gate leakage can be improved with new gate dielectric materials. Work from Radosavljevic *et al.* has shown improvements in gate leakage by using TaSiO$_x$ rather than a Schottky gate [10]. Kim *et al.* showed that by using a delta doping profile that is

(a)

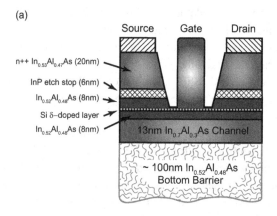

(b)

	V_{DS}	L_G	C_{Gi}@ V_{DS}=0.05V	PEAK G_M
	V	nm	µF/cm²	µS/µm
InGaAs QWFET	0.5	80	1.02	2013
Strained Si	0.5	40	2.45	1859
InGaAs QWFET	1	80	1.04	2082
Strained Si	1	40	2.45	2007

Fig. 2.2 (a) A schematic cross-section of an InGaAs/InAlAs quantum well structure. (b) Device characteristics at both $V_{DS} = 0.5$ V and $V_{DS} = 1$ V show reduced capacitance and excellent transconductance compared with Si CMOS [9].

located further away from the gate and removing a portion of it during etch a large reduction in gate leakage can be achieved while reducing drive current by only a small amount [11].

To reduce band-to-band tunneling, the bandgap of the channel material can be modified. Kim and del Alamo showed that using an InAs sub-channel sandwiched between two InGaAs layers reduces band-to-band tunneling due to energy level quantization within the InAs forming a larger effective bandgap [12]. In addition, they also showed that the source resistance is improved with good drain induced barrier lowering (DIBL) and sub-threshold swing (S). High source and drain resistance can also be addressed using thermal annealing or diffusion techniques [8].

To form p-type devices on the same wafer, wafer bonding has been used by Chung *et al.* to attach GaN to silicon wafers, where a p-type Si device can be used [13]. Since alignment issues are less of an issue with large power devices, this strategy may not be practical for logic applications for which high device density is critical.

2.2.1.4 Conclusions

HEMTs' main advantage when compared to the Si MOSFET is that the increased electron concentration allows for higher drive current. However, to be competitive with CMOS technology, certain significant challenges need to be resolved, such as gate leakage, device pitch, and the lack of equivalent p-type devices. Progress has been made in scaling L_G down to 30 nm, but strained silicon technology will continue to provide for improvements in MOSFET performance, possibly faster than HEMT technology can catch up. III-V MOSFETs that combine the high electron velocity of III-V materials with low gate leakage will also be a competing option [14]. Cost scaling is also expected to be a factor, since the substrates and/or specialized processing, such as metal-organic chemical vapor deposition (MOCVD), would likely be a significant additional manufacturing cost. HEMTs are expected to dominate in specialized applications where speed and frequency are more critical than power consumption and manufacturing costs, such as communications, military, and aerospace.

2.2.2 Gallium nitride

Gallium nitride is a III-V material with many properties that make it appealing as a channel material. It has a high breakdown voltage, high electron mobility, and high saturation velocity. Perhaps more importantly, a two-dimensional electron gas (2DEG) is induced by polarization at the AlGaN/GaN interface (creating a HEMT spontaneously, see Fig. 2.3), unlike an AlGaAs/GaAs HEMT that requires intentional doping to form charge.

2.2.2.1 History

Gallium nitride (GaN) crystals were synthesized in 1932 by W. D. Johnson by passing ammonia over heated gallium [15]. However, large crystals of GaN were not synthesized until 1969, when Maruska and Tietjen grew GaN on sapphire with hydride vapor phase epitaxy [16].

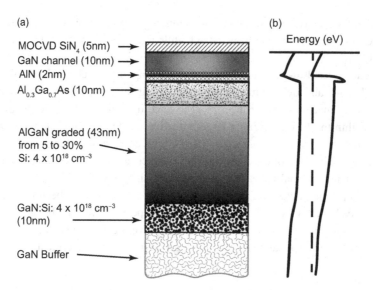

(a)

MOCVD SiN$_4$ (5nm) →
GaN channel (10nm) →
AlN (2nm) →
Al$_{0.3}$Ga$_{0.7}$As (10nm) →

AlGaN graded (43nm)
from 5 to 30%
Si: 4 x 10^{18} cm^{-3}

GaN:Si: 4 x 10^{18} cm^{-3}
(10nm)

GaN Buffer

(b)
Energy (eV)

Fig. 2.3 (a) GaN device layer cross-section schematic. The line of dots drawn in the GaN channel represents the high-mobility channel formed due to polarization at the GaN/AlN interface. (b) GaN band diagram beneath the gate showing why the channel forms at the interface in GaN [22].

A variety of devices can be constructed using the properties of GaN. An early switching device made using GaN was a MESFET created by Khan at APA Optics in 1993 [17]. Soon afterward, Khan demonstrated a GaN/AlGaN-based HEMT [18]. GaN nanotubes (similar to carbon nanotubes) have also been formed, with some of the earliest examples including Goldberger at UC Berkeley in 2003 [19] and Hu at NIMS [20].

2.2.2.2   Device principles

GaN devices can be considered as spontaneously formed HEMT devices. GaN and AlGaN are polar materials due to the large size difference between gallium and nitrogen. When AlGaN is deposited on GaN, the tensile stress which AlGaN induces in GaN causes piezoelectric polarization to occur. This polarization leads to the formation of electrons and holes, whose charges normally cancel each other out. However, due to the heterojunction at the interface as shown in Fig. 2.3, the AlGaN/GaN interface collects the electrons as an electron gas that can be used as a conduction channel, very similar to a traditional HEMT [21, 22].

2.2.2.3   Recent work

Challenges for GaN-based devices are similar to those of HEMTs. These include generating high drive current at low voltages, reducing gate leakage, and integrating p-type devices. Additional challenges include finding the best way to create enhancement mode devices, and reliability.

Significant work has been undertaken to tackle the challenges of GaN HEMT devices. Using N-face surfaces of GaN rather than Ga-face allowed Nidhi *et al.* to

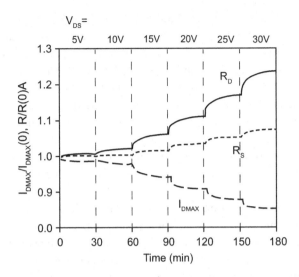

Fig. 2.4 GaN device characteristics, such as I_{DMAX}, R_D, and R_S change over time, with more change at high drain biases [30].

demonstrate depletion mode GaN devices producing about 1 mA/μm at $V_{DS} = 1$ V [22]. Xin and Chang have shown that high-k dielectrics such as atomic layer deposition (ALD) HfO_2 or Al_2O_3 can reduce the gate leakage [23, 24]. Chung *et al.* demonstrated a two-layer transfer process to integrate p-type silicon MOSFETs with n-type GaN [13].

To create enhancement-mode devices, researchers have used several methods to change the threshold voltage. Cai *et al.* used fluorination through the use of a CF_4 etch to passivate surface states, changing the threshold voltage by 5 V and allowing the creation of both enhancement and depletion mode devices, which was demonstrated by creating ring oscillators [25]. Ota *et al.* used piezoneutralization (a layer inserted beneath the gate to neutralize polarization charges underneath the gate) to adjust V_T [26]. Silicon nitride was used by Derluyn *et al.* to passivate the surface charge in AlGaN as another method of adjusting V_T [27]. Kanamura *et al.* used the piezoelectric effect of i-AlN on n-GaN to create an enhancement mode device while increasing the 2DEG density [28]. Finally, Im *et al.* demonstrated that a superlattice of AlN/GaN changes the biaxial stress to make enhancement mode devices with better on-state resistance [29].

Joh and del Alamo studied reliability concerns for GaN devices, and at $V_{DS} = 5$ V, hot carriers caused reductions in on-state current and changes in V_T, with greater voltages causing faster degradation (as shown in Fig. 2.4) [30]. In addition, lattice defects form due to excessive stress from the inverse piezoelectric effect.

2.2.2.4   Conclusions

GaN-based HEMTs' major advantages are high electron mobility (although not as high as GaAs), higher critical breakdown voltage, and a higher thermal conductivity than GaAs [21]. GaN-based devices have been suggested to be useful in radio-frequency (RF) or high voltage applications. Some groups have also thought that these devices could also be useful as traditional MOSFET replacements [24].

There are several challenges that would need to be overcome to replace traditional MOSFETs. First, GaN devices are typically depletion-mode rather than enhancement-mode. Next, to avoid HEMT gate leakage, gate dielectrics need to be developed that are compatible with GaN. Note that GaN only holds an advantage for n-type devices so p-type devices, such as silicon or germanium, would need to be fabricated on the same wafer. Additionally, reliability issues due to hot carriers need to be better studied. Scaling also needs further study, as devices have mostly been long channel up to this point. The inability to create GaN ingots as cost-effective substrates (or SiC ingots coupled with GaN deposition) means that sophisticated (and potentially expensive) techniques would need to be developed to integrate GaN on more conventional substrates.

GaN-based devices seem to perform best as power or RF solutions where voltages are too high for logic applications. Thus, GaN seems best suited for telecommunications and radar applications. For solutions like WiMax base stations and power electronics, GaN could also be useful. However, the limitations of GaN combined with integration challenges (particularly very thick buffer layers for growing defect-free GaN), GaN does not appear to be appealing for future development in logic.

2.2.3   Ferroelectric-dielectric gate stacks

Ferroelectric gate stacks (see Fig. 2.5) use a ferroelectric in series with a traditional gate oxide dielectric to achieve an operating region of negative small-signal gate capacitance C_{gate}, resulting in sub-threshold swing below 60 mV per decade.

2.2.3.1   History

In 2008, Sayeef Salahuddin at Purdue theorized that using a ferroelectric capacitor in series with a normal capacitor should stabilize the ferroelectric material to achieve a negative small-signal gate capacitance [31]. He predicted that with such a gate stack it would be possible to overcome the limitation that normally prevents operation with a sub-threshold swing less than 60 mV per decade at room temperatures [31, 32]. In 2010,

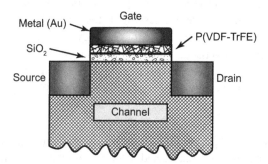

Fig. 2.5 Ferroelectric gate stack structure design with a ferroelectric material placed between traditional oxide and gate electrode [33]. This material creates negative small-signal capacitance over a portion of the gate voltage sweep, improving sub-threshold swing.

Rusu at EPFL reported experimental demonstration of sub-threshold swing smaller than 60 mV/dec at room temperature using internal voltage amplification in FETs with a metal–ferroelectric–metal–oxide gate stack [33]. Extending this concept in 2013, Lee demonstrated a ferroelectric negative capacitance hetero-tunnel FET using internal voltage amplification [34].

2.2.3.2   Device principles

For a standard dielectric capacitor, an energy versus charge curve shows a $Q^2/2C$ relationship. For a ferroelectric capacitor, the energy versus charge curve has two minima, resulting in a region of negative capacitance between the minima. By stacking a ferroelectric capacitor in series with a dielectric capacitor, the energy vs. charge curve is flattened in this region, i.e., the energy (hence voltage) corresponding to a large change in charge is reduced.

The equation for subthreshold swing (S) is given by

$$S = \ln(10)\frac{kT}{q}\left(1 + \frac{C_{\text{dep}}}{C_{\text{gate}}}\right), \tag{2.1}$$

where C_{dep} is the (small-signal) channel depletion capacitance. By making C_{gate} negative, Salahuddin predicted that it would be possible to overcome the limitation that normally prevents operation with a sub-threshold swing of less than 60 mV per decade at room temperature [31, 32]. The device would be fabricated by placing the ferroelectric layer between the gate electrode and conventional dielectric, as shown in Fig. 2.5.

2.2.3.3   Recent work

Research on ferroelectric-dielectric devices is very recent, with significant challenges today in simply observing the negative capacitance effect and achieving sub-60 mV behavior. Only a select few groups so far (primarily [33, 34]) have been able to achieve sub-60 mV/dec sub-threshold swing. However, Tanakamaru et al. looked at ferroelectrics for static-random access memory (SRAM) and were unable to achieve sub-60 mV/dec operation [35]. Khan et al. demonstrated a proof of concept device, which also does not show sub-60 mV/dec behavior, but was focused on demonstrating clear proof of the negative capacitance effect [36]. Complicating the research landscape, it is not possible to directly measure negative capacitance – only an enhancement in total capacitance [37, 38]. For example, Krowne et al. incorrectly interpreted a lack of measureable negative capacitance as an indication that ferroelectrics do not cause negative capacitance, but instead cause highly non-linear biasing behavior when in a series capacitor stack [37].

2.2.3.4   Conclusions

The potential for sub-60 mV/dec operation, if expanded over multiple decades of current, is a large potential advantage for ferroelectric-dielectric gate stack devices. However, the basic understanding of how to fabricate and design these devices still needs further study. Hysteresis is also a concern, as a large hysteresis would prevent

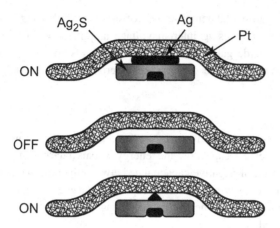

Fig. 2.6 Diagram of operation of an electrochemical device (ECD) using an Ag_2S wire with a Pt wire placed on top. These devices can be as small as molecules in size [41].

voltage scaling, further negating the underlying value of pursuing sub-60 mV/dec operation. While scaling studies have suggested that the technology scales poorly, more research is necessary, and the costs for incorporating a ferroelectric layer are relatively minimal. Even if there is a gate length limitation, if hysteresis can be reduced, these devices could find a use in a large gate length, ultra-low power application.

2.2.4   Electrochemical devices

Electrochemical devices (ECDs) use a chemical reaction (see Fig. 2.6) to control the flow of current through a device.

2.2.4.1   History

Devices with a single molecule were theorized as early as 1974 [39]. Modern ECDs were experimentally demonstrated by Collier in 1999 based on the concept of chemically assembled electronic nanocomputers (CAENs) [40]. Using Ag_2S as a filament, Terebe was able to produce various logic functions, including AND, OR, and NOT [41].

2.2.4.2   Device principles

Electro-chemical devices, like all switching devices, control the flow of current through a device. However, they use a chemical mechanism (e.g., the reaction of Cu ions precipitating out of Cu_2S as a voltage is applied, to create a conducting bridge) to do this. Some are 2-terminal devices, and others are 3-terminal devices. They are often used as memory devices, which can either be written once (irreversibly), or written many times (reversibly).

2.2.4.3   Recent work

A wide variety of circuit concepts have been explored. Molecular FETs, using molecules such as Rotaxane, have been shown for over a decade to be able to achieve basic logic functionality, such as AND and OR, but at very low currents (less than 1 nA) [40, 41]. Using $CuSO_4$, basic look-up tables can be created to mimic field programmable gate array (FPGA) technology [42–45]. (Unfortunately, endurance cycles of the FPGA in [44] are still low, at about 100 cycles, and retention and switching times need to be improved.) With a nanowire in a porous alumina membrane, Liang *et al.* were able to connect several nanowires in parallel to create similar FPGA-like devices [46]. A solid electrolyte in the back-end between a via and metal line have also been explored, but mostly as a nonvolatile memory on top of logic [47]. Endurance would again need to be improved beyond 100 cycles.

2.2.4.4   Conclusions

ECDs hold an advantage in their size, which can be on the order of individual atoms. However, these devices are still at too early a stage of development for use as a CMOS replacement without major breakthroughs. Unfortunately, electrochemical devices currently demonstrate low current, low endurance cycles, or poor switching speeds, and in some cases, all three negative qualities [48]. There have only been limited studies on scaling, and costs may or may not be significant, depending on the materials and process technologies needed for fabrication. Electrochemical devices may be more suited for use as nonvolatile memory devices.

2.2.5   Impact ionization devices

Impact ionization transistors (see Fig. 2.7) are gated p-n diodes that rely on avalanche breakdown to create carriers in the channel. This mechanism creates positive feedback, allowing for sub-60 mV/dec sub-threshold swing.

2.2.5.1   History

Impact ionization FET devices, also known as IMOS, were simulated and fabricated in 2002 at Stanford University [49]. Kailash Gopalakrishnan, under the direction of Professor James Plummer, was searching for a gain mechanism that was internal to the device with sufficient gate control. Gopalakrishnan simulated devices showing sub-threshold swings down to 5 mV/dec and fabricated devices with about a 10 mV/dec swing.

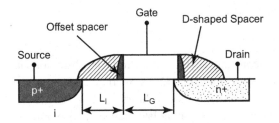

Fig. 2.7 Schematic diagram of an impact ionization device [50]. There is a large ungated region in the channel where impact ionization occurs after the gate inverts the gated portion of the channel. Impact ionization allows sub-threshold swings below 60 mV/dec to be seen.

2.2.5.2   Device principles

Impact ionization transistors use the avalanche mechanism of breakdown in reverse-biased diodes to achieve carrier transport. They are also gated p-i-n diodes, but they have a larger intrinsic region that is partially not gated, as Fig. 2.7 shows. The device works by modulating the intrinsic channel length with the gate voltage. At high gate voltages, the gate inverts a portion of the channel, reducing the length of the intrinsic region and increasing the electric field in this region. Avalanching occurs due to impact ionization, causing the current to abruptly increase.

2.2.5.3   Recent work

Challenges for IMOS devices include high drain voltage requirements, reliability, and circuit issues [50–54]. To cause avalanche breakdown, a high V_{DS} is required, which not only increases power dissipation but also reduces the efficacy of the IMOS device as a "pull-down" or "pull-up" switch for implementing complementary logic. Nematian *et al.* found that this drain voltage could be reduced if other materials with reduced bandgaps are used (such as SiGe) [51].

Reliability and variability can be an issue. Abelein *et al.* found that carriers with such high energy levels can cause large changes in V_T after multiple cycles [52]. IMOS has issues with increased C_{GD} due to high Miller capacitance (the drain couples to the entire intrinsic region of the device), as shown by Tura and Woo [53]. Also, the output current does not fully saturate at high drain voltages.

2.2.5.4   Conclusions

Impact ionization FETs have advantages of low sub-threshold swing with relatively high current compared to other sub-60 mV/dec devices, like TFETs. However, there are many challenges, such as reliability, avalanche onset delay, and high required drain voltage. Scalability is a significant concern, although cost would be similar to traditional MOSFETs. These devices could be useful for very specific circuit applications where a variable V_T can be used, such as a low-power write-once memory element.

2.2.6   Tunnel FET

Tunnel FETs (see Fig. 2.8) use quantum mechanical tunneling of electrons from the source to the channel as the primary carrier transport mechanism, allowing for sub-60 mV/dec sub-threshold swings.

2.2.6.1   History

The origins of 3-terminal tunneling devices come from the band-to-band tunneling component in a trench transistor cell (TTC) [55]. A 3-terminal tunnel device using this effect was proposed by Sanjay Banerjee at Texas Instruments in 1987 [56]. This device required gate overlap of the source, a situation later known as line tunneling. Later in 1992, Toshio Baba at NEC proposed a surface tunnel transistor using GaAs and AlGaAs that utilized point tunneling [57]. In 1995, William Reddick at Cambridge proposed a silicon device using point tunneling [58]. All of these devices showed low

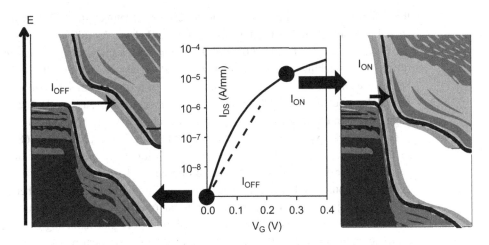

Fig. 2.8 Schematic diagram illustrating the operation of a generic n-channel TFET. Electrons tunnel from the valence band of the source to the conduction band of the channel when the device is on [59].

currents and no sub-threshold swing under 60 mV/dec. In 2004, Jorge Appenzeller at IBM showed experimental characteristics less than 60 mV/dec with carbon nanotube-based devices [60].

2.2.6.2 Device principles

Tunnel field effect transistors (TFETs) use the tunneling of electrons as the carrier transport method for device operation. They are generally designed as gated p-i-n diodes, where the gate is used to modulate an effective tunneling barrier height [61]. Ideally, these devices would have a very low off-state current (proportional to reverse-biased diode leakage), a very low sub-threshold swing, and acceptable on-current.

TFETs can be generally classified as point and/or line tunneling devices [62]. In a point tunneling device, the source does not appreciably deplete, but the gate causes the channel region to invert, resulting in tunneling from the source to the channel. In a line tunneling device, the source is inverted (generally by engineering an overlapped gate with an optimized source doping profile), resulting in tunneling into the inversion layer, similar to gate-induced drain leakage (GIDL).

2.2.6.3 Recent work

TFETs' major challenges are to achieve significantly better sub-threshold swing than 60 mV/dec and to provide drive current comparable to MOSFET devices. Miller capacitance is also a challenge due to the p-n diode nature of TFETs, similar to IMOS. Ambipolar operation (tunneling occurring at the drain when the gate is reverse biased) and circuit design challenges (due to asymmetric device operation) will also require further understanding.

Few devices are able to achieve sub-threshold swing of less than 60 mV/dec. Appenzeller *et al.* showed this with carbon nanotubes in 2004 [60]. Lu *et al.* also demonstrated this effect using DNA functionalization on carbon nanotubes [63]. Choi *et al.* demonstrated this with a purely silicon device in 2007 [64]. This was followed by a sub-60 mV/dec device by Mayer *et al.* in 2008 [65]. Jeon *et al.* used a silicided source to achieve sub-60 mV/dec switching in 2010 [66]. Leonelli *et al.* also demonstrated sub-60 mV/dec behavior with FinFET devices [67]. Kim *et al.* used a germanium source to achieve sub-60 mV/dec operation in 2009 [68]. Dewey *et al.* used thin gate oxide, heterojunction engineering, and high source doping to achieve sub-60 mV/dec operation in 2011 [69]

Even the best experimental devices (such as the device from Kim *et al.* [68]) show I_{on} in the μA/μm range, and do not meet the mA/μm requirements for future CMOS devices. Some strategies, such as that proposed by Kim *et al.* [68], use a recessed germanium source and have the potential for increased drive current. Another approach, from Mookerjea *et al.*, is to use one material (creating a homojunction rather than heterojunction) with a lower bandgap, allowing for a higher tunneling rate and hence, higher tunneling current [70]. An example of the improvement that potentially can be achieved with a heterojunction architecture is shown in Fig. 2.9.

In addition, drain capacitance may increase due to enhanced Miller capacitances. Since the entire channel is only electrically coupled to the drain (rather than the source and drain, as in a typical MOSFET), the drain experiences increased C_{GD} and the source experiences decreased C_{GS} [72, 73]. Increased voltage overshoot is therefore also possible using these devices.

A homojunction TFET (with a single source, channel, and drain material) has ambipolar characteristics (i.e., increasing current conduction with both increasingly positive and increasingly negative gate-to-source voltage). To reduce the ambipolar effect, the source and drain must be asymmetric, either by use of a heterostructure or offset drain [74, 75].

As an asymmetric device, a TFET can only conduct tunneling current in one direction, making circuit design more difficult. Groups have examined new SRAM

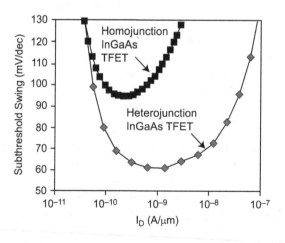

Fig. 2.9 Comparison of sub-threshold swing for homojunction and heterojunction devices showing the dramatic improvement in sub-threshold swing with heterojunction architectures [71].

and logic layouts and found that additional transistors (e.g., a 7T SRAM cell) may be necessary to have sufficient noise margins for operation [76].

2.2.6.4   Conclusions

Experimental TFET results show very low sub-threshold swing at very low currents, in sharp contrast to the simulation results which show low sub-threshold swing but with on-currents in the $0.1\,\text{mA/\mu m}$ to over $1\,\text{mA/\mu m}$ range. Unfortunately, in practice, devices have not been able to simultaneously show both sub-threshold swings below $60\,\text{mV/dec}$ and high on-state currents. Devices that do achieve reasonably high on-currents do not see sub-threshold swings below $60\,\text{mV/dec}$, making off-state currents very high and eliminating the advantage over conventional MOSFETs. It is important to note that TFETs are not symmetric, so additional challenging lithographic steps are necessary, complicating fabrication. Scaling seems robust for TFETs, with few cost increases compared to MOSFETs. However, unless TFETs improve to better match their simulation results, they have limited application for logic devices.

2.2.7   Metal source/drain devices

Metal source/drain (MSD) technology (see Fig. 2.10) uses Schottky barriers instead of doped sources and drains to reduce parasitic resistance.

2.2.7.1   History

The use of Schottky contacts for the source and drain was proposed and demonstrated by M. P. Lepselter and S. M. Sze at Bell Labs in 1968 [78]. They used platinum silicide source and drain regions and an n-type body region to demonstrate the first Schottky-based S/D devices.

2.2.7.2   Device principles

MSD devices, also known as Schottky barrier field effect transistors (SB-FETs), traditionally use Schottky barriers rather than diode junctions as the source and drain. Using metal rather than a doped semiconductor reduces parasitic resistances, but requires band edge metal work functions to be able to match the on-currents of traditional MOSFETs [79].

2.2.7.3   Recent work

MSD devices have Schottky barriers at the source and drain junctions. To achieve drive current comparable to standard MOSFETs, the Schottky barrier height (SBH) must be reduced (the MSD Fermi level must approach the Si band edge), with the amount of reduction dependent on the drive current requirements. Methods for reducing the barrier height include device structure optimization, Fermi level depinning, implantation, and dopant segregation.

Connelly *et al.* have shown that a SBH less than $0.1\,\text{eV}$ is needed to compete with traditional MOSFETs [80]. Underlap (effective channel length longer than the physical gate length) is preferred for these devices rather than conventional overlap of the source

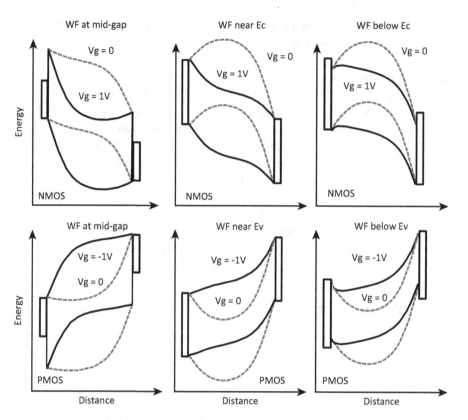

Fig. 2.10 Energy band diagram of NMOS and PMOS metal source/drain (MSD) devices illustrating the sensitivity of the band structure to the effective workfunction of the metal [77].

and drain, first due to parasitic capacitance, and second because the abrupt profile of a metal-semiconductor junction allows for a slight underlap while still maintaining gate control.

MSD devices utilizing a double-gated structure can meet ITRS benchmarks with larger SBH than single gate structures, as seen in Fig. 2.11, suggesting that gate-all-around (GAA) structures may be more attractive with MSD devices (neglecting the associated volume efficiency issues with GAA) [81].

Chen *et al.* created a planar structure where silicide was grown on top of the source and drain, and a small finger of silicide spread toward the gate to improve source/drain extension resistance. This method also allows strain to be induced with embedded SiGe [82].

Fermi level depinning can reduce the Schottky barrier height. A thin layer of nitride was used by Connelly *et al.* to de-pin the Fermi level and reduce SBH to 0.2 eV, at the cost of increased resistance [83, 84]. Another method by Tao *et al.* uses a selenium monolayer to reduce the SBH down to less than 0.08 eV [85].

Vega and Liu showed that a fluorine implant can reduce SBH close to 0 eV, but in FETs, the fluorine implant resulted in higher resistance, ultimately resulting in lower drive current [86, 87].

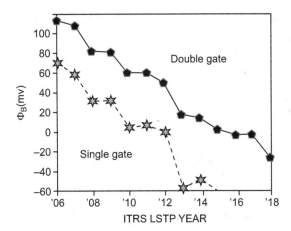

Fig. 2.11 Maximum Schottky barrier height needed to reach various ITRS LSTP specifications [81]. Double-gate devices extend the range at which metal source/drain (MSD) devices can meet performance specifications, but scaling is still limited.

Dopant-segregated Schottky (DSS) MOSFETs use dopants at the metal-semiconductor interface to reduce the SBH. Experimentally, this was demonstrated by Kinoshita *et al.* in 2004 and was shown by Qiu *et al.* to be achieved by either implantation into or implantation through silicide [88, 89]. An excellent summary of the different SBH values achieved for different annealing conditions is shown by Qiu *et al.* [89]. DSS FinFET and nanowires have also been demonstrated by Kaneko *et al.* and Chin *et al.* [90, 91].

2.2.7.4 Conclusions

MSD devices show some advantages if the SBH can be reduced to 0.1 eV or less. However, simulation-based studies of high-performance devices show that conventional RSD structures have better performance than DSS structures (even without accounting for the potential loss of strain effects with metal source/drain regions). Although some results suggest that MSD devices can outperform conventional MOSFETs (especially if the SBH is zero or negative [80]), lack of strain and increased parasitic capacitance effects are likely to result in lower performance overall. At very small gate lengths, while conventional raised-source/drain double-gate devices scale better, fabrication costs may make metal source/drain a more attractive option.

2.2.8 Relays

Mechanical relays (see Fig. 2.12) use physical movement from an off to an on position to regulate current.

2.2.8.1 History

Although mechanical relays pre-date solid-state devices by many decades, microelectromechanical systems (MEMS) technology that could compete with CMOS in

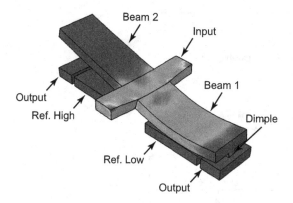

Fig. 2.12 A 4-terminal nanorelay inverter using a see-saw architecture [77].

scalability was first demonstrated in 1978 by Kurt E. Petersen [92]. Petersen demonstrated three devices: an optical display, a 4-terminal micromechanical switch, and a measurement method of Young's modulus.

2.2.8.2 Device principles

Relays use mechanical movement to physically short or open an electrical connection between two contacts. Relays can be placed either in the front-end or back-end of a traditional CMOS process [93]. Ideal MEMS relays show no off-state current, sharp sub-threshold swing, and low gate leakage. Low on-state resistance matters less in these devices than in CMOS because the relatively slow mechanical beam movement delay is the limiting factor, not the shorter RC delay time constant.

Relays have a pull-in voltage (V_{PI}) determined by the actuation force (usually electrostatic) overcoming a mechanical spring restoring force, and a pull-out voltage (V_{PO}) determined by mechanical force overcoming adhesive forces. To actuate relays, several different mechanisms can be used, including thermal, magnetic, piezo, and electrostatic [94–98]. While MEMS devices can be engineered for $V_{PI} = V_{PO}$, practical implementations frequently display voltage hysteresis (where $V_{PI} > V_{PO}$).

2.2.8.3 Recent work

Relays have the advantages of steep sub-threshold swing and negligible I_{off} (see Fig. 2.13). In addition, 4-terminal relays allow pass-gate logic, which potentially reduces the number of devices needed per function [99]. However, relay operating voltages are high and need to be reduced to achieve the benefits of low active power. In addition relays are currently very large (e.g., 7.5 μm × 7.5 μm [100]) and need to be reduced to sizes comparable with MOSFETs. Improved reliability needs to be demonstrated for use in high-activity-factor digital logic applications. Variability and hysteresis in switching voltages also need to be reduced to values comparable to MOSFETs.

In 2010, Kam *et al.* showed that relays have the potential to be 10× more energy efficient than CMOS (albeit at lower clock frequencies) [101]. Relay circuits were demonstrated by Spencer *et al.*, including a full adder comprising 12 relays with a

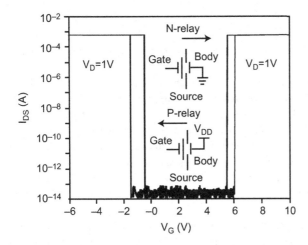

Fig. 2.13 4-terminal relay I_D-V_G characteristics [99]. The sub-threshold swing is extremely low, giving a very sharp transition from off-state to on-state.

single mechanical delay [102]. Fariborzi *et al.* designed a 16-bit relay multiplier, which promises to achieve lower energy per operation than CMOS, as well as experimentally demonstrating a 7:3 compressor composed of 98 relays [103].

High pull-in voltages remain an issue with relays. Lee *et al.* used an insulating liquid (such as oil) to reduce V_{PI} with the liquid's higher dielectric constant, but relay reliability was still worse than that achieved with an atomic layer deposition process [96]. Carbon nanotubes may also be used to reduce voltage, but can be difficult to fabricate, as discussed by Dadgour *et al.* [104, 105]. A suspended-gate MOS transistor fabricated by Abele *et al.* incorporates a mechanical gate electrode for enhanced switching steepness, but the drive current is low, and both mechanical and RC delay are issues for this device [106].

2.2.8.4 Conclusions

Relays offer extremely high I_{on}/I_{off} ratios with excellent sub-threshold swings. However, scalability and reliability need to be proven before these devices can find acceptance. Fabrication costs may be improved, as many steps (such as ion implantation and epitaxy) would not be necessary, although release etch processing may add cost and complexity. With current state-of-the-art relay technology, these devices could still find usage for non-volatile memory, or FPGA applications.

2.3 Summary

There are many alternative logic switch designs, each with its own trade-offs. Many are still in the research stage, but hold promise if solutions can be found to overcome some of their drawbacks. Further fundamental and engineering research (mostly in advanced materials) must continue to sustain inexorable increases in switching device density.

References

[1] G. E. Moore, "Cramming more components onto integrated circuits." *Electronics*, 114–117 (1965).

[2] T. Mimura, "The early history of the high electron mobility transistor (HEMT)." *IEEE Transactions on Microwave Theory and Techniques*, **50**, 780–782 (2002).

[3] "History of HEMTs." Available online at: www.iue.tuwien.ac.at/phd/vitanov/node11.html.

[4] R. Dingle, H. L. Störmer, A. C. Gossard, & W. Wiegmann, "Electron mobilities in modulation-doped semiconductor heterojunction superlattices." *Applied Physics Letters*, **33**, 665–667 (1978). http://link.aip.org/link/?APL/33/665/1.

[5] D. Delagebeaudeuf, P. Delescluse, P. Etienne, M. Laviron, J. Chaplart, & N. T. Linh, "Two-dimensional electron gas M.E.S.F.E.T. structure." *Electronics Letters*, **16**, 667–668 (1980).

[6] D. Delagebeaudeuf & N. T. Linh, "Charge control of the heterojunction two-dimensional electron gas for MESFET application." *IEEE Transactions on Electron Devices*, **28**, 790–795 (1981).

[7] "High electron mobility transistors (HEMTs)." Available online: www.mwe.ee.ethz.ch/en/about-mwe-group/research/vision-and-aim/high-electron-mobility-transistors-hemt.html.

[8] I. Ok, D. Veksler, P. Y. Hung *et al.*, "Reducing Rext in laser annealed enhancement-mode $In_{0.53}Ga_{0.47}As$ surface channel n-MOSFET." In *VLSI Technology Systems and Applications (VLSI-TSA), 2010 International Symposium on*, pp. 38–39 (2010).

[9] G. Dewey, R. Kotlyar, R. Pillarisetty *et al.*, "Logic performance evaluation and transport physics of Schottky-gate III-V compound semiconductor quantum well field effect transistors for power supply voltages (VCC) ranging from 0.5v to 1.0v." In *Electron Devices Meeting (IEDM), 2009 IEEE International*, pp. 1–4 (2009).

[10] M. Radosavljevic, B. Chu-Kung, S. Corcoran *et al.*, "Advanced high-K gate dielectric for high-performance short-channel $In_{0.7}Ga_{0.3}As$ quantum well field effect transistors on silicon substrate for low power logic applications." In *Electron Devices Meeting (IEDM), 2009 IEEE International*, pp. 1–4 (2009).

[11] T.-W. Kim, D.-H. Kim, & J. A. del Alamo, "30 nm $In_{0.7}Ga_{0.3}As$ inverted-type HEMTs with reduced gate leakage current for logic applications." In *Electron Devices Meeting (IEDM), 2009 IEEE International*, pp. 1–4 (2009).

[12] D.-H. Kim & J. A. del Alamo, "Scalability of sub-100 nm InAs HEMTs on InP substrate for future logic applications." *IEEE Transactions on Electron Devices*, **57**, 1504–1511 (2010).

[13] J. W. Chung, J.-K. Lee, E. L. Piner, & T. Palacios, "Seamless on-wafer integration of Si (100) MOSFETs and GaN HEMTs." *IEEE Electron Device Letters*, **30**, 1015–1017 (2009).

[14] J. A. del Alamo, "Nanometre-scale electronics with III-V compound semiconductors." *Nature*, **479**, 317–323 (2011).

[15] M. S. Shur, "GaN-based devices." In *Electron Devices, 2005 Spanish Conference on*, 15–18 (2005).

[16] H. P. Maruska & J. J. Tietjen, "The preparation and properties of vapor-deposited single crystalline GaN." *Applied Physics Letters*, **15**, 327–329 (1969).

[17] M. Asif Khan, J. N. Kuznia, A. R. Bhattarai, & D. T. Olson, "Metal semiconductor field effect transistor based on single crystal GaN." *Applied Physics Letters*, **62**, 1786–1787 (1993).

[18] M. Asif Khan, A. Bhattarai, J. N. Kuznia, & D. T. Olson, "High electron mobility transistor based on a GaN /$Al_xGa_{1-x}N$ heterojunction." *Applied Physics Letters*, **63**, 1214–1215 (1993).

[19] J. Goldberger, R. He, Y. Zhang, S. Lee, H. Yan, H.-J. Choi, & P. Yang, "Single-crystal gallium nitride nanotubes." *Nature*, **422**, 599–602 (2003).

[20] J. Hu, Y. Bando, D. Golberg, & Q. Liu, "Gallium nitride nanotubes by the conversion of gallium oxide nanotubes." *Angewandte Chemie International Edition*, **42**, 3493–3497 (2003).

[21] L. F. Eastman & U. K. Mishra, "The toughest transistor yet [GaN transistors]." *Spectrum, IEEE*, **39**, 28–33 (2002).

[22] L. Nidhi, S. Dasgupta, D. F. Brown, S. Keller, J. S. Speck, & U. K. Mishra, "N-polar GaN-based highly scaled self-aligned MIS-HEMTs with state-of-the-art fT.LG product of 16.8 GHz-μm." In *Electron Devices Meeting (IEDM), 2009 IEEE International*, pp. 1–3 (2009).

[23] X. Xin, J. Shi, L. Liu *et al.*, "Demonstration of low-leakage-current low-on-resistance 600-V 5.5-A GaN/AlGaN HEMT." *IEEE Electron Device Letters*, **30**, 1027–1029 (2009).

[24] Y. C. Chang, W. H. Chang, H. C. Chiu *et al.*, "Inversion-channel GaN MOSFET using atomic-layer-deposited Al_2O_3 as gate dielectric." In *VLSI Technology, Systems, and Applications, 2009. VLSI-TSA '09. International Symposium on*, pp. 131–132 (2009).

[25] Y. Cai, Z. Cheng, W. C. W. Tang, K. J. Chen, & K. M. Lau, "Monolithic integration of enhancement-and depletion-mode AlGaN/GaN HEMTs for GaN digital integrated circuits." In *Electron Devices Meeting, 2005. IEDM Technical Digest. IEEE International*, vol. **4**, p. 774 (2005).

[26] K. Ota, K. Endo, Y. Okamoto, Y. Ando, H. Miyamoto, & H. Shimawaki, "A normally-off GaN FET with high threshold voltage uniformity using a novel piezo neutralization technique." In *Electron Devices Meeting (IEDM), 2009 IEEE International*, pp. 1–4 (2009).

[27] J. Derluyn, M. Van Hove, D. Visalli *et al.*, "Low leakage high breakdown e-mode GaN DHFET on Si by selective removal of in-situ grown Si_3N_4." In *Electron Devices Meeting (IEDM), 2009 IEEE International*, pp. 1–4 (2009).

[28] M. Kanamura, T. Ohki, T. Kikkawa *et al.*, "Enhancement-mode GaN MIS-HEMTs with n-GaN/i-AlN/n-GaN triple cap layer and high-gate dielectrics." *IEEE Electron Device Letters*, **31**, 189–191 (2010).

[29] K.-S. Im, J.-B. Ha, K.-W. Kim *et al.*, "Normally off GaN MOSFET based on AlGaN/GaN heterostructure with extremely high 2DEG density grown on silicon substrate." *IEEE Electron Device Letters*, **31**, 192–194 (2010).

[30] J. Joh & J. A. del Alamo, "Mechanisms for electrical degradation of GaN high-electron mobility transistors." In *Electron Devices Meeting, 2006. IEDM '06. International*, pp. 1–4 (2006).

[31] S. Salahuddin & S. Datta, "Use of negative capacitance to provide voltage amplification for low power nanoscale devices." *Nano Letters*, **8**, 405–410 (2008).

[32] S. Salahuddin & Datta, S. "Can the subthreshold swing in a classical FET be lowered below 60 mV/decade?" In *Electron Devices Meeting, 2008. IEDM 2008. IEEE International*, pp. 1–4 (2008).

[33] A. Rusu, G. A. Salvatore, D. Jimenez, & A. M. Ionescu, "Metal-ferroelectric-metal-oxide-semiconductor field effect transistor with sub-60mV/decade subthreshold swing and internal voltage amplification." In *Electron Devices Meeting (IEDM), 2010 IEEE International*, pp. 16.3.1–16.3.4 (2010).

[34] M. H. Lee, J.-C. Lin, Y.-T. Wei, C.-W. Chen, H.-K. Zhuang, & M. Tang, "Ferroelectric negative capacitance hetero-tunnel field-effect-transistors with internal voltage amplification." In *Electron Devices Meeting, 2013, IEEE International*, pp. 104–107 (2013).

[35] S. Tanakamaru, T. Hatanaka, R. Yajima, M. Takahashi, S. Sakai, & K. Takeuchi, "A 0.5V operation, 32% lower active power, 42% lower leakage current, ferroelectric 6T-SRAM with VTH self-adjusting function for 60% larger static noise margin." In *Electron Devices Meeting (IEDM), 2009 IEEE International*, pp. 1–4 (2009).

[36] A. I. Khan, D. Bhowmik, P. Yu *et al.*, "Experimental evidence of ferroelectric negative capacitance in nanoscale heterostructures." *Applied Physics Letters*, **99**, 113501 (2011).

[37] C. M. Krowne, S. W. Kirchoefer, W. Chang, J. M. Pond, & L. M. B. Alldredge, "Examination of the possibility of negative capacitance using ferroelectric materials in solid state electronic devices." *Nano Letters*, **11**, 988–992 (2011).

[38] R. Jin, Y. Song, M. Ji *et al.*, "Characteristics of sub-100nm ferroelectric field effect transistor with high-k buffer layer." In *Solid-State and Integrated-Circuit Technology, 2008. ICSICT 2008. 9th International Conference on*, pp. 888–891 (2008).

[39] A. Aviram & M. A. Ratner, "Molecular rectifiers." *Chemical Physics Letters*, **29**, 277–283 (1974).

[40] C. P. Collier, E. W. Wong, M. Belohradský *et al.*, "Electronically configurable molecular-based logic gates." *Science*, **285**, 391–394 (1999).

[41] K. Terabe, T. Hasegawa, T. Nakayama, & M. Aono, "Quantized conductance atomic switch." *Nature*, **433**, 47–50 (2005).

[42] A. F. Thomson, D. O. S. Melville, & R. J. Blaikie, "Nanometre-scale electrochemical switches fabricated using a plasma-based sulphidation technique." In *Nanoscience and Nanotechnology, 2006. ICONN '06. International Conference on*, (2006).

[43] T. Sakamoto, N. Banno, N. Iguchi *et al.*, "A Ta$_2$O$_5$ solid-electrolyte switch with improved reliability." In *VLSI Technology, 2007 IEEE Symposium on*, pp. 38–39 (2007).

[44] T. Sakamoto, N. Banno, N. Iguchi *et al.*, "Three terminal solid-electrolyte nanometer switch." In *Electron Devices Meeting, 2005. IEDM Technical Digest. IEEE International*, pp. 475–478 (2005).

[45] S. Kaeriyama, T. Sakamoto, H. Sunamura, *et al.*, "A nonvolatile programmable solid electrolyte nanometer switch." *IEEE Journal of Solid-State Circuits*, **40**(1), 168–176 (2005).

[46] C. Liang, K. Terabe, T. Hasegawa, R. Negishi, T. Tamura, & M. Aono, "Ionic–electronic conductor nanostructures: template-confined growth and nonlinear electrical transport." *Small*, **1**, 971–975 (2005).

[47] M. N. Kozicki, C. Gopalan, M. Balakrishnan, & M. Mitkova, "A low-power nonvolatile switching element based on copper-tungsten oxide solid electrolyte." *IEEE Transactions on Nanotechnology*, **5**, 535–544 (2006).

[48] T. Hasegawa, K. Terabe, T. Sakamoto, & M. Aono, "Nanoionics switching devices: atomic switches." *MRS Bulletin*, **34**, 929–934 (2009).

[49] K. Gopalakrishnan, P. B. Griffin, & J. D. Plummer, "I-MOS: a novel semiconductor device with a subthreshold slope lower than kT/q." In *Electron Devices Meeting, 2002. IEDM '02. Digest. IEEE International*, pp. 289–292 (2002).

[50] A. Savio, S. Monfray, C. Charbuillet, & T. Skotnicki, "On the limitations of silicon for I-MOS integration." *IEEE Transactions on Electron Devices*, **56**, 1110–1117 (2009).

[51] H. Nematian, M. Fathipour, & M. Nayeri, "A novel impact ionization MOS (I-MOS) structure using a silicon-germanium/silicon heterostructure channel." In *Microelectronics, 2008. ICM 2008. International Conference on*, pp. 228–231 (2008).

[52] U. Abelein, M. Born, K. K. Bhuwalka *et al.*, "Improved reliability by reduction of hot-electron damage in the vertical impact-ionization MOSFET (I-MOS)." *IEEE Electron Device Letters*, **28**, 65–67 (2007).

[53] A. Tura & J. Woo, "Performance comparison of silicon steep subthreshold FETs." *IEEE Transactions on Electron Devices*, **57**, 1362–1368 (2010).

[54] C. Shen, J.-Q. Lin, E.-H. Toh, *et al.* "On the performance limit of impact-ionization transistors." In *Electron Devices Meeting, 2007. IEDM 2007. IEEE International*, pp. 117–120 (2007).

[55] S. Banerjee, J. Coleman, B. Richardson, & A. Shah, "A band-to-band tunneling effect in the trench transistor cell." In *VLSI Technology, 1987. Digest of Technical Papers. Symposium on*, pp. 97–98 (1987).

[56] S. Banerjee, W. Richardson, J. Coleman, & A. Chatterjee, "A new three-terminal tunnel device." *IEEE Electron Device Letters*, **8**, 347–349 (1987).

[57] T. Baba, "Proposal for surface tunnel transistors." *Japanese Journal of Applied Physics*, **31** (1992).

[58] W. M. Reddick & G. A. J. Amaratunga, "Silicon surface tunnel transistor." *Applied Physics Letters*, **67**, 494–496 (1995).

[59] U. E. Avci, R. Rios, K. Kuhn, & I. A. Young, "Comparison of performance, switching energy and process variations for the TFET and MOSFET in logic." In *VLSI Technology (VLSIT), 2011 Symposium on*, pp. 124–125 (2012).

[60] J. Appenzeller, Y.-M. Lin, J. Knoch, & P. Avouris, "Band-to-band tunneling in carbon nanotube field-effect transistors." *Physics Review Letters*, **93**, 196805 (2004).

[61] O. M. Nayfeh, C. N. Chleirigh, J. Hennessy, L. Gomez, J. L. Hoyt, & D. A. Antoniadis, "Design of tunneling field-effect transistors using strained-silicon/strained-germanium type-II staggered heterojunctions." *IEEE Electron Device Letters*, **29**, 1074–1077 (2008).

[62] W. Vandenberghe, A. S. Verhulst, G. Groeseneken, B. Soree, & W. Magnus, "Analytical model for point and line tunneling in a tunnel field-effect transistor." In *Simulation of Semiconductor Processes and Devices, 2008. SISPAD 2008. International Conference on*, pp. 137–140 (2008).

[63] Y. Lu, S. Bangsaruntip, X. Wang, L. Zhang, Y. Nishi, & H. Dai, "DNA functionalization of carbon nanotubes for ultrathin atomic layer deposition of high κ dielectrics for nanotube transistors with 60 mV/decade switching." *Journal of the American Chemical Society*, **128**, 3518–3519 (2006).

[64] W. Y. Choi, B.-G. Park, J. D. Lee, & T.-J. K. Liu, "Tunneling field-effect transistors (TFETs) with subthreshold swing (SS) less than 60 mV/dec." *IEEE Electron Device Letters*, **28**, 743–745 (2007).

[65] F. Mayer, C. Le Royer, J.-F. Damlencourt *et al.*, "Impact of SOI, Si1-xGexOI and GeOI substrates on CMOS compatible tunnel FET performance." In *Electron Devices Meeting, 2008. IEDM 2008. IEEE International*, pp. 1–5 (2008).

[66] K. Jeon, W.-Y. Loh, P. Patel *et al.*, "Si tunnel transistors with a novel silicided source and 46mV/dec swing." In *VLSI Technology (VLSIT), 2010 Symposium on*, pp. 121–122 (2010).

[67] D. Leonelli, A. Vandooren, Rooyackers, R. *et al.*, "Performance enhancement in multi gate tunneling field effect transistors by scaling the fin-width." *Japanese Journal of Applied Physics*, **49** (2010).

[68] S. H. Kim, H. Kam, C. Hu, & T.-J. K. Liu, "Germanium-source tunnel field effect transistors with record high I_{ON}/I_{OFF}." In *VLSI Technology, 2009 Symposium on*, pp. 178–179 (2009).

[69] G. Dewey, B. Chu-Kung, J. Boardman *et al.*, "Fabrication, characterization, and physics of III-V heterojunction tunneling field effect transistors (H-TFET) for steep sub-threshold swing." In *Electron Devices Meeting (IEDM), 2011 IEEE International*, pp. 785–788 (2011).

[70] S. Mookerjea, D. Mohata, T. Mayer, V. Narayanan, & S. Datta, "Temperature-dependent I-V characteristics of a vertical $In_{0.53}Ga_{0.47}As$ tunnel FET." *IEEE Electron Device Letters,* **31**, 564–566 (2010).

[71] U. E. Avci, S. Hasan, D. E. Nikonov, R. Rios, K. Kuhn, & I. A. Young, "Understanding the feasibility of scaled III-V TFET for logic by bridging atomistic simulations and experimental results." In *VLSI Technology (VLSIT), 2012 Symposium on,* pp. 183–184 (2012).

[72] S. Mookerjea, R. Krishnan, S. Datta, & V. Narayanan, "On enhanced Miller capacitance effect in interband tunnel transistors." *IEEE Electron Device Letters,* **30**, 1102–1104 (2009).

[73] S. Mookerjea, R. Krishnan, S. Datta, & V. Narayanan, "Effective capacitance and drive current for tunnel FET (TFET) CV/I estimation." *IEEE Transactions on Electron Devices,* **56**, 2092–2098 (2009).

[74] J. Wan, C. Le Royer, A. Zaslavsky, & S. Cristoloveanu, "SOI TFETs: suppression of ambipolar leakage and low-frequency noise behavior." In *Proceedings of the 2010 European Solid-State Device Research Conference (ESSDERC),* pp. 341–344 (2010).

[75] T. Krishnamohan, D. Kim, S. Raghunathan, & K. Saraswat, "Double-gate strained-Ge heterostructure tunneling FET (TFET) with record high drive currents and $<60\,mV/dec$ subthreshold slope." In *Electron Devices Meeting, 2008. IEDM 2008. IEEE International,* pp. 1–3 (2008).

[76] D. Kim, Y. Lee, J. Cai *et al.,* "Low power circuit design based on heterojunction tunneling transistors (HETTs)." In *Proceedings of the 14th ACM/IEEE International Symposium on Low Power Electronics and Design,* pp. 219–224 (2009).

[77] K. J. Kuhn, U. Avci, A. Cappellani *et al.,* "The ultimate CMOS device and beyond." In *Electron Devices Meeting (IEDM), 2012 IEEE International* pp. 171–174, (2012).

[78] M. P. Lepselter & S. M. Sze, "SB-IGFET: an insulated-gate field-effect transistor using Schottky barrier contacts for source and drain." *Proceedings of the IEEE,* **56**, 1400–1402 (1968).

[79] J. M. Larson & J. P. Snyder, "Overview and status of metal S/D Schottky-barrier MOSFET technology." *IEEE Transactions on Electron Devices,* **53**, 1048–1058 (2006).

[80] D. Connelly, C. Faulkner, & D. E. Grupp, "Optimizing Schottky S/D offset for 25-nm dual-gate CMOS performance." *IEEE Electron Device Letters,* **24**, 411–413 (2003).

[81] D. Connelly, P. Clifton, C. Faulkner, & D. E. Grupp, "Ultra-thin-body fully depleted SOI metal source/drain n-MOSFETs and ITRS low-standby-power targets through 2018." In *Electron Devices Meeting, 2005. IEDM Technical Digest. IEEE International,* pp. 972–975 (2005).

[82] H.-W. Chen, C.-H. Ko, T.-J. Wang, C.-H. Ge, K. Wu, & W.-C. Lee, "Enhanced performance of strained CMOSFETs using metallized source/drain extension (M-SDE)." In *VLSI Technology, 2007 IEEE Symposium on,* pp. 118–119 (2007).

[83] D. Connelly, C. Faulkner, P. A. Clifton, & D. E. Grupp, "Fermi-level depinning for low-barrier Schottky source/drain transistors." *Applied Physics Letters,* **88**, 012105–012105-3 (2006).

[84] D. Connelly, P. Clifton, C. Faulkner, J. Owens, & J. Wetzel, "Self-aligned low-Schottky barrier deposited metal S/D MOSFETs with Si_3N_4 M/Si passivation." *Device Research Conference,* pp. 83–84 (2008).

[85] M. Tao, S. Agarwal, D. Udeshi, N. Basit, E. Maldonado, & W. P. Kirk, "Low Schottky barriers on n-type silicon (001)." *Applied Physics Letters,* **83**, 2593–2595 (2003).

[86] R. A. Vega & T.-J. K. Liu, "DSS MOSFET with tunable SDE regions by fluorine pre-silicidation ion implant." *IEEE Electron Device Letters,* **31**, 785–787 (2010).

[87] R. A. Vega & T.-J. K. Liu, "Dopant-segregated Schottky junction tuning with fluorine pre-silicidation ion implant." *IEEE Transactions on Electron Devices*, **57**, 1084–1092 (2010).

[88] A. Kinoshita, Y. Tsuchiya, A. Yagishita, K. Uchida, & J. Koga, "Solution for high-performance Schottky-source/drain MOSFETs: Schottky barrier height engineering with dopant segregation technique." In *VLSI Technology, 2004. Digest of Technical Papers. 2004 Symposium on*, pp. 168–169 (2004).

[89] Z. Qiu, Z. Zhang, M. Ostling, & S.-L. Zhang, "A comparative study of two different schemes to dopant segregation at NiSi/Si and PtSi/Si interfaces for Schottky barrier height lowering." *IEEE Transactions on Electron Devices*, **55**, 396–403 (2008).

[90] A. Kaneko, A. Yagishita, K. Yahashi *et al.*, "High-performance FinFET with dopant-segregated Schottky source/drain." In *Electron Devices Meeting, 2006. IEDM '06. IEEE International*, pp. 1–4 (2006).

[91] Y. K. Chin, K.-L. Pey, N. Singh *et al.*, "Dopant-segregated Schottky silicon-nanowire MOSFETs with gate-all-around channels." *IEEE Electron Device Letters*, **30**, 843–845 (2009).

[92] K. E. Petersen, "Dynamic micromechanics on silicon: techniques and devices." *IEEE Transactions on Electron Devices*, **25**, 1241–1250 (1978).

[93] V. Joshi, C. Khieu, C. G. Smith *et al.*, "A CMOS compatible back end MEMS switch for logic functions." In *Interconnect Technology Conference (IITC), 2010 International*, pp. 1–3 (2010).

[94] V. Pott, H. Kam, R. Nathanael, J. Jeon, E. Alon, & T.-J. K. Liu, "Mechanical computing redux: relays for integrated circuit applications." *Proceedings of the IEEE*, **98**, 2076–2094 (2010).

[95] H. Kam, D. T. Lee, R. T. Howe, & T.-J. King, "A new nano-electro-mechanical field effect transistor (NEMFET) design for low-power electronics." In *Electron Devices Meeting, 2005. IEDM Technical Digest. IEEE International*, pp. 463–466 (2005).

[96] J.-O. Lee, M.-W. Kim, S.-D. Ko *et al.*, "3-terminal nanoelectromechanical switching device in insulating liquid media for low voltage operation and reliability improvement." In *Electron Devices Meeting (IEDM), 2009 IEEE International*, pp. 1–4 (2009).

[97] H. Kam, V. Pott, R. Nathanael, J. Jeon, E. Alon, & T.-J. K. Liu, "Design and reliability of a micro-relay technology for zero-standby-power digital logic applications." In *Electron Devices Meeting (IEDM), 2009 IEEE International*, pp. 1–4 (2009).

[98] F. Chen, M. Spencer, R. Nathanael *et al.*, "Demonstration of integrated micro-electro-mechanical switch circuits for VLSI applications." In *Solid-State Circuits Conference Digest of Technical Papers (ISSCC), 2010 IEEE International*, pp. 150–151 (2010).

[99] R. Nathanael, V. Pott, H. Kam, J. Jeon, & T.-J. K. Liu, "4-terminal relay technology for complementary logic." In *Electron Devices Meeting (IEDM), 2009 IEEE International*, pp. 1–4 (2009).

[100] I.-R. Chen, L. Hutin, C. Park *et al.*, "Scaled micro-relay structure with low strain gradient for reduced operating voltage." In *Electrochemical Society (ECS) Meeting, 221st*, p. 867 (2012).

[101] H. Kam, T.-J. K. Liu, V. Stojanović, & D. Marković , "Design, optimization, and scaling of MEM relays for ultra-low-power digital logic." *IEEE Transactions on Electron Devices*, **58**, 236–250 (2011).

[102] M. Spencer, F. Chen, C. C. Wang *et al.*, "Demonstration of integrated micro-electro-mechanical relay circuits for VLSI applications." *IEEE Journal of Solid-State Circuits*, **46**, 308–320 (2011).

[103] H. Fariborzi, F. Chen, V. Stojanovic,, R. Nathanael, J. Jeon, & T.-J. K. Liu, "Design and demonstration of micro-electro-mechanical relay multipliers." *In Solid State Circuits Conference (A-SSCC), 2011 IEEE Asian*, pp. 117–120 (2011).

[104] H. Dadgour, A. M. Cassell, & K. Banerjee, "Scaling and variability analysis of CNT-based NEMS devices and circuits with implications for process design." In *Electron Devices Meeting, 2008. IEDM 2008. IEEE International*, pp. 1–4 (2008).

[105] H. F. Dadgour & K. Banerjee, "Hybrid NEMS-CMOS integrated circuits: a novel strategy for energy-efficient designs." *Computers & Digital Techniques, IET*, **3**, 593–608 (2009).

[106] N. Abele, R. Fritschi, K. Boucart, F. Casset, P. Ancey, & A. M. Ionescu, "Suspended-gate MOSFET: bringing new MEMS functionality into solid-state MOS transistor." In *Electron Devices Meeting, 2005. IEDM Technical Digest. IEEE International*, pp. 479–481 (2005).

3 Benchmarking alternative device structures for the terascale regime

Zachery A. Jacobson and Kelin J. Kuhn

3.1 Introduction

In this chapter, the devices discussed in Chapter 2 are benchmarked against perform-ance targets set by the International Technology Roadmap for Semiconductors (ITRS), as well as against more conventional ultra-thin body (UTB), gate-all-around (GAA), junctionless accumulation mode (JAM) devices, and thin-film transistors.

The chapter begins with a short introduction to the scaling potential of the various devices used in the benchmarking discussion. The benchmarking metrics are then introduced, followed by the benchmarking results, discussion, and conclusions.

3.2 Scaling potential of alternative device structures

3.2.1 High electron mobility transistors (HEMT)

The high electron mobility transistor, or HEMT, increases device mobility by separating charge carriers from the ionized dopant atoms, thus reducing ionized impurity scatter-ing. This is accomplished by confining carriers in an undoped quantum well.

Several groups have addressed the dimensional scaling of HEMTs [1–4]. Using electron beam lithography and multiple etch steps, Waldron *et al.* showed that it is possible to reduce a HEMT down to 30 nm gate-to-contact spacing, but it is difficult to make the gate length small without improvements to the etch processes [1]. Kharche *et al.* found that InAs is projected to scale well, as quantum well width scaling brings improvements in I_{on}/I_{off} due to lower I_{off} [2]. The reduced well width brings the electron peak closer to the gate, allowing for better gate control. Oh and Wong showed that if issues with gate leakage and process integration at small gate lengths can be solved (along with finding a symmetric p-type device), HEMT devices can have lower delay or lower energy-delay product (EDP) [3]. However, others, including Skotnicki and Boeuf, have shown that when drain-induced barrier lowering (DIBL) and sub-threshold swing (S) are included in an effective current metric, strained silicon performs better than III-V HEMTs [4].

3.2.2 Gallium nitride (GaN) transistors

Gallium nitride (GaN) is a III-V material with a high breakdown voltage, high electron mobility, and high saturation velocity. Perhaps more importantly, a two-dimensional electron gas is induced by polarization at the AlGaN/GaN interface (creating a HEMT spontaneously), unlike an AlGaAs/GaAs HEMT that requires intentional doping to form charge.

Short channel devices with gate lengths down to 20 nm (with 40 nm source/drain offsets) have been developed by Shinohara *et al.* with record high on-current of 2.7 mA/μm [5]. Both enhancement and depletion mode devices were fabricated with high uniformity. However, voltages are still high to achieve these results (3–5 V). Uren *et al.* found punchthrough effects occurring in devices with a 0.17 μm gate length due to leakage through the GaN buffer layer. Uren proposed that buffer layers should be insulating to prevent this and confine the channel potential [6]. Park and Rajan found that N-polar GaN HEMTs suppressed DIBL better than Ga-polar HEMTs due to the N-polar device's superior electrostatics from its inverted structure [7].

3.2.3 Negative capacitance (Fe-gate) transistors

Ferroelectric gate stacks use a ferroelectric capacitor in series with a traditional dielectric capacitor to achieve an operating region of negative gate capacitance, resulting in sub-threshold swing below 60 mV per decade.

Jin *et al.* simulated the potential for scalability of these devices and found that the sub-threshold swing rises as the gate length is scaled down, with substantial increases below 50 nm negating their primary benefit (see Fig. 3.1) [8]. As the mechanisms for generating and operating negative capacitance are better understood, additional fundamental limitations may be uncovered for these devices.

3.2.4 Electrochemical transistors

Electrochemical devices (ECDs) use a chemical reaction to control the flow of current through a device.

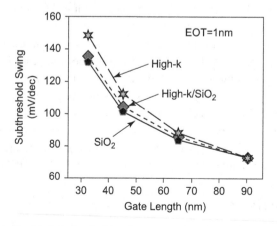

Fig. 3.1 Sub-threshold swing versus gate length for negative capacitance FETs suggests scaled devices may not have a sub-60 mV/dec swing, negating their primary benefit [8].

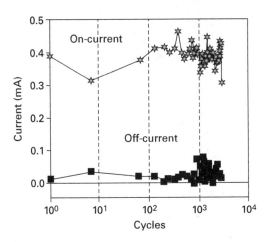

Fig. 3.2 Cycling of ECDs shows endurance to thousands of cycles, after which the switch is inoperable [13].

Scaling challenges for ECDs are focused on switching speeds, reliability (expressed in cycles, which is often called endurance), and circuit fabrication issues. Terabe *et al.* attained 1 MHz operation in 2005 using Ag_2S, although this is orders of magnitude from what would be necessary to compete with scaled CMOS logic [9]. Thomson *et al.* in 2006 showed switching speeds as fast as 0.1 microseconds [10]. Sakamoto *et al.* in 2007 switched to Ta_2O_5 and were able to increase $V_{program}$ to over 1 V while keeping switching times in the 10^{-5} to 10^{-4} seconds range [11].

Improvement of reliability, as expressed in mean cycles before failure, remains difficult. For 3-terminal devices, Sakamoto *et al.* showed that a gate isolated from a filament can be used to control the filament's conductivity, although the initial endurance was low at only 50 cycles [12]. Further work by Kaeriyama *et al.* [13] demonstrated significant reliability improvements using a Cu_2S electrolyte, as Fig. 3.2 demonstrates. In 2007, Sakamoto *et al.* were able to increase endurance to 10 000 cycles using a Cu_2S electrolyte [11].

3.2.5 Impact ionization metal-oxide semiconductor (IMOS) transistors

Impact ionization transistors are gated p-n diodes that rely on avalanche breakdown to create carriers in the channel. This mechanism creates positive feedback, allowing for sub-60 mV/dec sub-threshold swings.

Issues with scaling IMOS focus on the need of an ungated intrinsic region. As an example, Savio *et al.* showed that silicon IMOS will not scale well below 50 nm as there is no region where the device can exhibit transistor behavior (see Fig. 3.3) [14]. Additionally, as shown by Shen *et al.*, the need of time and space for the carriers to build enough energy for carrier multiplication to occur, limits both the fundamental scaling length of these devices, and the switching speed [15].

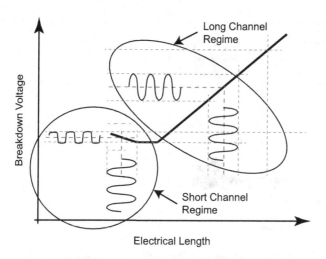

Fig. 3.3 Diagram for impact ionization devices showing how at (left) long L_G, there is a large region where the device can be tuned into and out of breakdown. For (right) short gate lengths, there is no region where the device can exhibit transistor behavior.

3.2.6   Tunnel field effect transistors (TFETs)

Tunnel FETs use quantum mechanical tunneling of electrons from the source to the channel as the primary carrier transport mechanism, allowing for sub-60 mV/dec sub-threshold swing.

Although simulations of scaled device structures show increased on-state current, many simulated cases require extremely abrupt junctions or doping profiles that have not been achieved in TFETs to date [16]. Some of these device structures require multiple junctions underneath the gate, which would also reduce future scalability. In general, the tunneling current is proportional to the barrier height (determined by bandgap/heterojunctions) and tunneling width (dependent on electrostatics and doping concentrations/gradients). TFETs with gate overlap have improved tunneling area (as well as electric field in the tunneling region), but reduced scalability. One solution is to decouple the overlap area from the gate length by using a raised source where tunneling is contained completely within the source [17].

3.2.7   Metal source/drain transistors

Metal source/drain (MSD) technology uses metal instead of doped semiconductor in the source and drain regions to reduce parasitic resistance.

Vega and Liu examined the scaling potential of dopant-segregated Schottky-junction (DSS) devices for double gate structures. For high-performance applications, at a gate length of 10 nm, a conventional double gate raised source/drain (RSD) structure will achieve higher performance than a DSS structure unless the epitaxial layer is doped at less than 10^{20} cm^{-3} [18]. For low operating power, Vega *et al.* showed the advantages of a dual high-*k*/low-*k* spacer technology for DSS structures, which allows fringing fields to be enhanced such that the source and drain can be underlapped, reducing parasitic capacitance and increasing effective gate length [19].

3.2.8 Relays

Mechanical relays use physical movement from an off to an on position to regulate current.

Spencer *et al.* produced a theoretical layout for a 90 nm technology node relay [20]. At the 90 nm node, a 32-bit adder was projected to require 7000 µm² of area versus 2000 µm² for a traditional CMOS Sklansky adder. However, for equivalent delay, parallelism is required that would make the area penalty approximately 100×. Spencer notes that a more optimized device layout could reduce this penalty.

Chen *et al.* used cantilever relays to achieve similar simulated throughputs with only a 6–25× area overhead [21]. Lee *et al.* calculated the scaling limits of these cantilever beams and found poly-Si would be difficult to scale to beam lengths below 80 nm [22]. TiN is more elastic and able to scale to ~30 nm. Lee notes that vertical structures may be advantageous for area efficiency.

Shen *et al.* used simulation tools to show that scaled relays with feature sizes down to 10 nm will have pull-in voltages of less than 0.25 V [23]. Pott *et al.* explained that relays will ultimately be limited by contact asperities (surface roughness), but scaling to the 65 nm node would result in an actuation area 67% larger than a comparable MOSFET [24].

Reliability up to 65 billion cycles has been demonstrated with appropriate contacts (10^9 as shown in Fig. 3.4). Joshi *et al.* demonstrated back-end-of-line process compatible relays to 10^{11} cycles (in comparison, for a device operating at 100 MHz with an activity factor of 1%, 10^{15} cycles would be needed for 10 years of operation) [25].

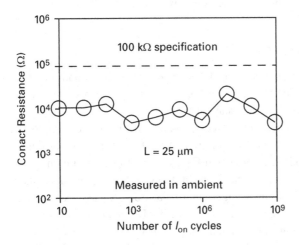

Fig. 3.4 Evolution of relay on-state resistance with the number of on/off switching cycles [26]. This device was tested to 10^9 cycles. 10^{15} cycles would be needed for a device operating at 100 MHz with an activity factor of 1% for 10 years.

Variation in contact resistance has been shown to have little impact on energy–performance characteristics, even when comparing best- and worst-in-class contact materials [21]. Dadgour *et al.* showed that a 10% variation in beam length and width (for carbon nanotubes) has a dramatic effect on pull-in voltage distribution [27].

3.3   Scaling potential of comparison devices

3.3.1   Ultra-thin body (UTB) transistors

Ultra-thin body (UTB) devices are formed with an underlying oxide layer. By making the body region very thin, gate control is enhanced, so that the body is fully depleted of mobile charge carriers in the off-state.

Dimensional scaling, especially of the body thickness, is a concern for UTB devices. As channel length scales, body thickness must decrease. Severe mobility degradation is seen below 3.5 nm thickness [28]. Also, quantization effects lead to an increase in threshold voltage and thus a decrease in on-state current. The effect of just a single unintentional impurity atom can degrade the drive current and increase threshold voltage, as shown by Vasileska and Ahmed [29]. Self-heating is an additional concern. In 1989, McDaid *et al.* showed that negative differential resistance in SOI MOSFET output characteristics was due to the reduced thermal conductance of the buried oxide [30]. More recently, Fiegna *et al.* showed that thermal resistance increases as the gate length and body thickness are scaled [31]. Reducing the buried oxide thickness reduces thermal resistance, potentially offering some room for improvement. Threshold voltage control is also a concern in UTB devices. Ren *et al.* demonstrated excellent variability and mismatch control with gate lengths down to 30 nm [32]. Liu *et al.* and Andrieu *et al.* demonstrated UTB with a thin buried oxide (BOX) layer (UTBB), which allows back biasing to tune the threshold voltage with low variability [33, 34].

3.3.2   Gate-all-around (GAA) transistors

Gate-all-around (GAA) (or nanowire) devices improve electrostatic integrity by wrapping the gate around the channel (see Fig. 3.5). This improves short channel control, reducing leakage and improving scalability.

Conventional GAA devices show comparable performance to partially depleted SOI devices for p-channel, but still lag in drive current for n-channel, as demonstrated by Bangsaruntip *et al.* [35]. Yeo *et al.* and Fang *et al.* demonstrated twin silicon nanowire FETs (TSNWFETs) that show excellent sub-threshold swing and drain-induced barrier lowering (DIBL) due to their high-quality gate oxides and excellent gate control [36, 37]. Three-dimensional stacked nanowires by Dupre *et al.* show excellent possibilities for scaling, although independent gate control of several layers is unlikely [38]. Singh *et al.* demonstrated that for p-type nanowires, the <010> crystalline orientation is superior to the <110> orientation, yielding 1.84× higher mean I_{on} [39]. Mobility is degraded in silicon nanowires due to phonon-scattering as nanowires get thinner, especially below 6 nm [40]. Mobility has been shown to decrease linearly with wire radius for InAs nanowires in the 7–18 nm range [41].

3.3.3 Junctionless accumulation mode (JAM) transistors

The junctionless accumulation mode (JAM) transistor is a fully depleted device where the source/drain regions are of the same doping type as the channel region (see Fig. 3.6). The devices have high doping (in the 10^{19} cm^{-3} range), ideally with uniform doping throughout the channel, source, and drain regions [43]. Generally the devices require

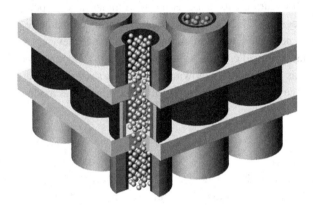

Fig. 3.5 Gate-all-around (GAA) devices improve electrostatic integrity by wrapping the gate around the channel. This improves short channel control, reducing leakage and improving scalability [42].

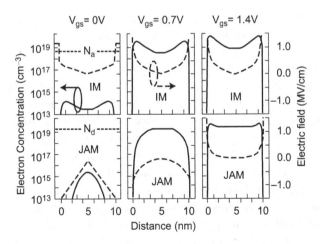

Fig. 3.6 Junctionless accumulation mode (JAM) devices versus conventional inversion mode (IM) devices [47]. Simulated (solid lines) electron concentration and (dashed lines) electric field profile differences between (top) IM and (bottom) JAM devices. Note the electric field is lower in the JAM devices and the carrier concentrations are more centered in the device. The gate work functions are set for identical off-state leakage current.

very small cross-sections (on the order of 5 nm × 5 nm) to achieve the desired threshold voltage. In addition to being relatively easy to fabricate, these devices also have a lower electric field than a conventional inversion-mode MOSFET [44]. Both traditional top-down and bottom-up fabricated nanowire devices have been studied for their advantages in fabrication complexity [45, 46].

The key trade-off in these devices is between the mobility gain due to reduced electric field and the mobility loss due to increased doping. Rios *et al.* showed experimentally that for a low-doped case, the low field effects win and the mobility is ~30% higher [47]. However, for a high-doped case, the mobility is reduced due to impurity scattering. Rios also points out that these devices have a mixed threshold behavior where a low value governs the sub-threshold turn-on and a higher one determines the extrapolated threshold of the accumulation regime [47]. In addition, when measuring temperature dependence, dV_T/dT is very poor compared to traditional inversion mode devices, which would reduce voltage scaling, a critical criteria in future devices [48]. Geometric variation will also be problematic, as junctionless nanowires have been shown to have significantly higher threshold voltage variation than a comparably sized inversion mode device as the width scales [49].

3.3.4   Thin-film transistors

There are alternative approaches to device miniaturization for improving transistor density. For example, recrystallization techniques can be used to form crystalline or polycrystalline semiconductor material for devices to be built over traditional MOSFETs. The value of such thin-film transistors (TFTs) is not only to increase device density, but also to lower parasitic interconnect resistance and capacitance for improved energy efficiency and performance

Varadarajan *et al.* used metal-induced crystallization to form crystalline semiconductor regions embedded within aluminum wires in the back-end-of-line process [50]. Unfortunately, aluminum heavily doped the silicon, resulting in a negative threshold voltage so that the device did not turn off properly. Other materials or structures could make this technique more competitive. However, care would need to be taken to make sure that the active region is thin enough to be fully depleted, while the contacts are wide enough to prevent parasitic resistance issues, similar to challenges for JAM devices.

Another method to induce crystallization is the use of poly-germanium seeds. Subramanian and Saraswat used this method in 1997 to create laterally crystallized TFTs [51]. In 1999, Subramanian *et al.* demonstrated TFTs scaled down to 100 nm, which were single grain and showed very low leakage (below 1 pA/μm) [52]. Mobility and on-state current of these devices remains lower than traditional MOSFET, but the ease of fabrication could allow for inexpensive additional layers of devices.

Metal-induced crystallization through a cap layer (MICC) is another method of forming polycrystalline silicon. Oh *et al.* demonstrated this technique using nickel mediated crystallization, although currents were below 100 μA/μm at high voltages [53].

Recrystallization techniques also have been investigated for high-density memory due to the challenges of scaling memory. As an example, Jung *et al.* have shown that laser-induced epitaxy can be used to form high-density three-dimensional (3D) crystalline-silicon SRAM cells [54]. Further evaluation of techniques used for novel SRAM and other memory devices may prove useful for logic applications.

3.4   Evaluation metrics

Benchmarking alternative logic devices can be quite complex. To begin with, it is difficult to select common metrics for devices that operate on different principles and at dramatically different voltages. Furthermore, alternative devices are in the early stages of their development and typically display large discrepancies between simulation results and early experimental results as models are refined and new physical effects are discovered. Additionally, parasitic effects may be poorly understood and difficult to include accurately. Finally, circuit-level performance metrics using standard CMOS circuits may not fully comprehend unique advantages (or disadvantages) of novel circuit architectures enabled by alternative devices [55–60].

To benchmark these devices for the terascale regime, a set of six metrics was developed to comprehensively benchmark their potential benefits. These are I_{on}, I_{off}, switching energy, fabrication complexity, fabrication cost, and density scaling. Four of the metrics are device level; the remaining two are circuit level. For device types with significant differences between simulation and experimental results, both sets of results were sometimes used. Since best-in-class performance can change over time, the published work with the most well-behaved totality of current devices is used. In some cases, the resulting scores were modified using guidance from industrial and academic experts, so all numbers besides I_{on} and I_{off} have a partially qualitative nature.

I_{on} and I_{off} are used as initial metrics for evaluation. These values are easily found using published data and form a concrete basis for the other metrics. Note that some works normalize current differently, especially for non-planar devices. I_{on} and I_{off} for all of the devices shown in Table 3.1 have been recalculated using gate perimeter normalization. For stacked devices such as nanowires, a single gate perimeter is used to allow for the advantageous drawn pitch of these devices. To determine the on-current value, I_{DSAT} is used. Supply voltage is determined by either the published work's typical voltage or the typical voltage value used for that class of devices. The off-current value is determined by the device's minimum quoted I_{off} if possible.

Energy in Joules per switching event is used as the next metric of evaluation. This was determined by inputting each device's current characteristics into a Verilog-A lookup table, which was then used in a set of SPICE simulations. For capacitance, experimental data were used when possible and calculated when not possible. Some device types are n-type only. In these cases, a standard p-type germanium MOSFET was used. The energy dissipated was measured and averaged for a high-to-low switching event and a low-to-high switching event. Intrinsic delay was not used as a metric because CV/I would not be appropriate for some of the devices; for example, relay delay is determined by mechanical delay, and ferroelectric-electric transistor delay is

Table 3.1 Normalized I_{off} and I_{on} with references for each class of device

	Polarity	I_{off} [nA/µm]	I_{on} [mA/µm]	V_{GS}/V_{DS} [V]	Ref
ITRS 2018 HP	N/P	100	1.805	0.73	[55]
ITRS 2018 LOP	N/P	5	0.794	0.57	[55]
ITRS 2018 LSTP	N/P	0.01	0.643	0.72	[55]
HEMT	N/Ge-P*	100	0.5	0.5	[62]
GaN	N/Ge-P*	100	0.25	0.6/5	[5]
Fe-gate	N/P	0.001	0.002	2/0.2	[63]
ECD	N/P	0.4	0.01	0.15/0.001	[64]
IMOS	N/P	100	1	1/5.3	[14]
TFET simulation	N/P**	0.001	0.2	0.5	[17]
TFET experiment	N/P**	0.1	0.005	0.5/0.3	[65]
UTB	N/P	100	1	1	[66]
Tri-gate	N/P	100	1.2	1	[67]
Stacked NW	N/P	0.42	2.546	1.2	[37]
Metal S/D (DSS)	N/P	100	1.2	1	[18]
JAM	N/P	0.0002	0.02	1	[68]
Thin film	N/P	0.01	0.055	2	[52]

* A p-type germanium FET was used in energy simulations.
** P-type TFETs have been fabricated, but require a different structure than n-type TFETs.

determined by polarization. (Designers will need to develop optimal circuit topologies for each of these devices.)

Fabrication complexity is the next metric, as manufacturability is also a concern for future devices. A detailed comparison of major processing steps was completed for each device. Steps needed by all devices, such as isolation, were not included.

Fabrication cost is a related metric. Moore's law can also be considered an economic motivation, as devices have not only increased performance at smaller sizes, but also cost less per device to manufacture. Each fabrication step was evaluated and assigned a value, depending on projected manufacturing cost. These values were totaled and used as the normalized fabrication cost per wafer. For devices that require a p-type germanium MOSFET for p-type operation, these costs were also included.

Finally, the last metric is area scaling. To determine area scaling, design rules scaled to 2018 ITRS projections were assumed (design rules were taken from [61]). These design rules were then used for each device to determine the size of a standard 6T SRAM cell. These areas were added to form the area metric, as both memory and logic elements will be necessary in a circuit design.

3.5   Benchmarking results

Benchmarking figures and tables (Figs. 3.7–3.9, Table 3.1) illustrate the various trade-offs associated with each technology using the metrics defined in Section 3.4.

Looking first at the I_{on}/I_{off} current ratios in Fig. 3.7, high I_{on} is necessary not only for intrinsic delay, but also to drive metal lines in integrated circuits. Low I_{off} and large

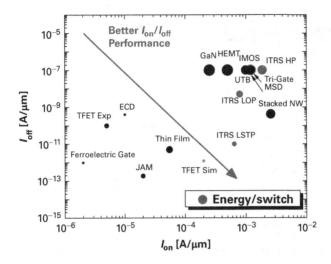

Fig. 3.7 Benchmarking normalized current and energy. Devices in the bottom right have excellent I_{on}/I_{off} ratios. Devices with small bubbles have lower energy per switching event. Grey dots labeled ITRS indicate ITRS targets.

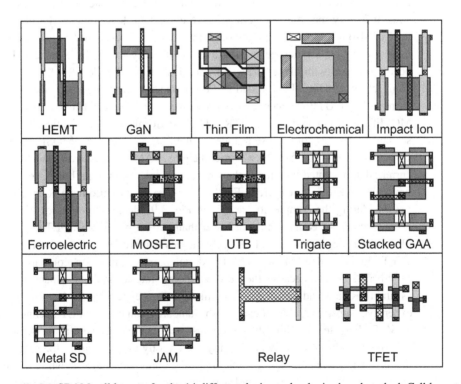

Fig. 3.8 SRAM cell layouts for the 14 different device technologies benchmarked. Cell layouts are not comparable to scale (i.e., individual width to length ratios are accurate, but each device is scaled to fit in the same space).

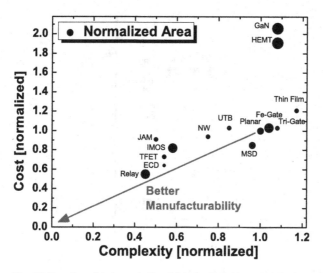

Fig. 3.9 Benchmarking normalized fabrication cost, complexity, and cell area. Devices that require less manufacturing complexity and cost are located in the bottom left corner, with the size of the bubble indicating normalized area.

I_{on}/I_{off} ratios are needed for low-power operation. HEMT and GaN devices both have some difficulty matching modern Si drive currents at low voltages. Ferroelectric gate devices also have low current, although they can have very low I_{off}. ECD devices with the highest I_{on}/I_{off} ratios generally have significantly lower drive current. IMOS has been fabricated with higher drive current, close to modern devices. Simulated TFETs come close to ITRS Low Standby Power specifications, but still need more drive current. Most UTB devices have I_{on} relatively close to the ITRS targets for planar MOSFETs. Compared to the ITRS High Performance specifications, tri-gate and stacked nanowires come close to meeting the I_{on}/I_{off} target. DSS MOSFETs also come close with optimistic Schottky barrier heights. JAM devices have had limited drive current demonstrated thus far. Relays have very low off-current, but currently on a per μm scale have low drive current. Finally, thin-film devices also have poor drive current.

Considering the complexity metrics in Fig. 3.9, ideally the complexity and cost of devices would be equivalent to or lower than current planar MOSFET technology. As illustrated in Fig. 3.9 many of the technologies that are less optimized such as IMOS, TFET, ECD, and JAM, show reduced complexity and costs compared to planar MOSFETs, largely because they do not incorporate technology enhancements (such as strain). Relays and nanowires could likely be less expensive, as they may not require expensive substrates. Modifications to standard planar MOSFET technology, such as MSD, Fe-gate, and thin-film, bring costs and complexity similar to planar devices. Tri-gate and ultra-thin body are slightly more costly than planar. In the case of tri-gate this is due to more complex fabrication for the non-planar process, while in the case of ultra-thin body there is less complexity but higher cost for the FDSOI wafers. HEMTs and GaN devices, which use expensive III-V wafers or III-V epitaxy, are significantly more expensive than planar devices.

Cell area is also illustrated in Fig. 3.9. Several devices are projected to scale more poorly due to larger spacers, such as in IMOS, HEMT, and GaN devices. A few devices simply are not projected to be able to continue scaling, such as IMOS, ferroelectric gate, UTB, and planar CMOS. Relays show respectable scaling down to the 65 nm node, but contact pitch between devices may still be larger, since the minimum feature size and material properties limit how small serpentine springs or cantilever beams can be produced. Devices using multiple gates, such as tri-gate, must account for fin pitch when designing cell layouts.

3.6   Conclusions

Alternative device structures were benchmarked against ITRS performance targets, as well as against more conventional UTB, GAA, JAM devices, and thin-film transistors. In the evaluation of I_{on}/I_{off} current ratios, simulated TFETs are closest to the ITRS Low Standby Power specification and tri-gate and stacked nanowires are closest to ITRS High Performance specification. In the evaluation of combined fabrication cost, complexity, and cell area, technologies such as IMOS, TFET, ECD, and JAM show reduced complexity and costs compared to planar MOSFETs, but this is largely because they do not incorporate the technology enhancements of more mature technologies (such as strain). Modifications to standard technologies, such as MSD, Fe-gate, and thin film, result in added costs compared to planar MOSFETs, but these costs are relatively small. HEMTs and GaN devices, which use expensive III-V wafers or III-V epitaxy, are anticipated to be significantly more expensive than Si-based devices.

References

[1] N. Waldron, D.-H. Kim, & J. A. del Alamo, "A self-aligned InGaAs HEMT architecture for logic applications." *IEEE Transactions on Electron Devices*, **57**, 297–304 (2010).

[2] N. Kharche, G. Klimeck, D.-H. Kim, J. A. del Alamo, & M. Luisier, "Performance analysis of ultra-scaled InAs HEMTs." In *Electron Devices Meeting (IEDM), 2009 IEEE International*, pp. 1–4 (2009).

[3] S. Oh & H.-S. P. Wong, "Effect of parasitic resistance and capacitance on performance of InGaAs HEMT digital logic circuits." *IEEE Transactions on Electron Devices*, **56**, 1161–1164 (2009).

[4] T. Skotnicki & F. Boeuf, "How can high mobility channel materials boost or degrade performance in advanced CMOS." In *VLSI Technology (VLSIT), 2010 Symposium on*, pp. 153–154 (2010).

[5] K. Shinohara, D. Regan, A. Corrion *et al.* "Deeply-scaled self-aligned-gate GaN DH-HEMTs with ultrahigh cutoff frequency." In *Electron Devices Meeting (IEDM), 2011 IEEE International*, pp. 19.1.1–19.1.4 (2011).

[6] M. J. Uren, K. J Nash, R. S. Balmer *et al.* "Punch-through in short-channel AlGaN/GaN HFETs." *IEEE Transactions on Electron Devices*, **53**, 395–398 (2006).

[7] P. S. Park & S. Rajan, "Simulation of short-channel effects in N- and Ga-polar AlGaN/GaN HEMTs." *IEEE Transactions on Electron Devices*, **58**, 704–708 (2011).

[8] R. Jin, Y. Song, M. Ji, H. Xu, J. Kang, R. Han, & X. Liu, "Characteristics of sub-100nm ferroelectric field effect transistor with high-k buffer layer." In *Solid-State and Integrated-Circuit Technology, 2008. ICSICT 2008. 9th International Conference on*, pp. 888–891 (2008).

[9] K. Terabe, T. Hasegawa, T. Nakayama, & M. Aono, "Quantized conductance atomic switch." *Nature*, **433**, 47–50 (2005).

[10] A. F. Thomson, D. O. S. Melville, & R. J. Blaikie, "Nanometre-scale electrochemical switches fabricated using a plasma-based sulphidation technique." In *Nanoscience and Nanotechnology, 2006. ICONN '06. International Conference on* (2006).

[11] T. Sakamoto, N. Banno, N. Iguchi *et al.* "A Ta_2O_5 solid-electrolyte switch with improved reliability." In *VLSI Technology, 2007 IEEE Symposium on*, pp. 38–39 (2007).

[12] T. Sakamoto, N. Banno, N. Iguchi *et al.* "Three terminal solid-electrolyte nanometer switch." In *Electron Devices Meeting, 2005. IEDM Technical Digest. IEEE International*, pp. 475–478 (2005).

[13] S. Kaeriyama, T. Sakamoto, H. Sunamura *et al.* "A nonvolatile programmable solid electrolyte nanometer switch." *IEEE Journal of Solid-State Circuits*, **40**(1), 168–176 (2005).

[14] A. Savio, S. Monfray, C. Charbuillet, & T. Skotnicki, "On the limitations of silicon for I-MOS integration." *IEEE Transactions on Electron Devices*, **56**, 1110–1117 (2009).

[15] C. Shen, J.-Q. Lin, E.-H. Toh *et al.* "On the performance limit of impact-ionization transistors." In *Electron Devices Meeting, 2007. IEDM 2007. IEEE International*, pp. 117–120 (2007).

[16] V. Nagavarapu, R. Jhaveri, & J. C. S. Woo, "The tunnel source (PNPN) n-MOSFET: a novel high performance transistor." *IEEE Transactions on Electron Devices*, **55**, 1013–1019 (2008).

[17] S. H. Kim, S. Agarwal, Z. A. Jacobson, P. Matheu, C. Hu, & T.-J. K. Liu, "Tunnel field effect transistor with raised germanium source." *IEEE Electron Device Letters*, **31**, 1107–1109 (2010).

[18] R. A. Vega & T.-J. K. Liu, "A comparative study of dopant-segregated Schottky and raised source/drain double-gate MOSFETs." *IEEE Transactions on Electron Devices*, **55**, 2665–2677 (2008).

[19] R. A. Vega, K. Liu, & T.-J. K. Liu, "Dopant-segregated Schottky source/drain double-gate MOSFET design in the direct source-to-drain tunneling regime." *IEEE Transactions on Electron Devices*, **56**, 2016–2026 (2009).

[20] M. Spencer, F. Chen, C. C. Wang *et al.* "Demonstration of integrated micro-electro-mechanical relay circuits for VLSI applications." *IEEE Journal of Solid-State Circuits* **46**, 308–320 (2011).

[21] F. Chen, H. Kam, D. Markovic, T.-J. K. Liu, V. Stojanovic, & E. Alon, "Integrated circuit design with NEM relays." In *Computer-Aided Design, 2008. ICCAD 2008. IEEE/ACM International Conference on*, pp. 750–757 (2008).

[22] D. Lee, T. Osabe, & T.-J. K. Liu, "Scaling limitations for flexural beams used in electro-mechanical devices." *IEEE Transactions on Electron Devices*, **56**, 688–691 (2009).

[23] X. Shen, S. Chong, D. Lee, R. Parsa, R. T. Howe, & H.-S. P. Wong, "2D analytical model for the study of NEM relay device scaling." In *Simulation of Semiconductor Processes and Devices (SISPAD), 2011 International Conference on*, pp. 243–246 (2011).

[24] V. Pott, H. Kam, R. Nathanael, J. Jeon, E. Alon, & T.-J. K. Liu, "Mechanical computing redux: relays for integrated circuit applications." *Proceedings of the IEEE*, **98**, 2076–2094 (2010).

[25] V. Joshi, C. Khieu, C. G. Smith *et al.* "A CMOS compatible back end MEMS switch for logic functions." In *Interconnect Technology Conference (IITC), 2010 International*, pp. 1–3 (2010).

[26] H. Kam, V. Pott,, R. Nathanael, J. Jeon, E. Alon, & T.-J. K. Liu, "Design and reliability of a micro-relay technology for zero-standby-power digital logic applications." In *Electron Devices Meeting (IEDM), 2009 IEEE International*, pp. 1–4 (2009).

[27] H. Dadgour, A. M. Cassell, & K. Banerjee, "Scaling and variability analysis of CNT-based NEMS devices and circuits with implications for process design." In *Electron Devices Meeting, 2008. IEDM 2008. IEEE International*, pp. 1–4 (2008).

[28] L. Gomez, I. Aberg, & J. L. Hoyt, "Electron transport in strained-silicon directly on insulator ultrathin-body n-MOSFETs with body thickness ranging from 2 to 25 nm." *IEEE Electron Device Letters*, **28**, 285–287 (2007).

[29] D. Vasileska & S. S. Ahmed, "Narrow-width SOI devices: the role of quantum-mechanical size quantization effect and unintentional doping on the device operation." *IEEE Transactions on Electron Devices*, **52**, 227–236 (2005).

[30] L. J. McDaid, S. Hall, P. H. Mellor, W. Eccleston, & J. C. Alderman, "Physical origin of negative differential resistance in SOI transistors." *Electronics Letters*, **25**, 827–828 (1989).

[31] C. Fiegna, Y. Yang, E. Sangiorgi, & A. G. O'Neill, "Analysis of self-heating effects in ultrathin-body SOI MOSFETs by device simulation." *IEEE Transactions on Electron Devices*, **55**, 233–244 (2008).

[32] Z. Ren, S. Mehta, J. Cai *et al.* "Assessment of fully-depleted planar CMOS for low power complex circuit operation." In *Electron Devices Meeting (IEDM), 2011 IEEE International*, pp. 15.5.1–15.5.4 (2011).

[33] Q. Liu, A. Yagishita, N. Loubet *et al.* "Ultra-thin-body and BOX (UTBB) fully depleted (FD) device integration for 22nm node and beyond." In *VLSI Technology (VLSIT), 2010 Symposium on*, pp. 61–62 (2010).

[34] F. Andrieu, O. Weber, J. Mazurier *et al.* "Low leakage and low variability ultra-thin body and buried oxide (UT2B) SOI technology for 20nm low power CMOS and beyond." In *VLSI Technology (VLSIT), 2010 Symposium on*, pp. 57–58 (2010).

[35] S. Bangsaruntip, G. M. Cohen, A. Majumdar *et al.* "High performance and highly uniform gate-all-around silicon nanowire MOSFETs with wire size dependent scaling." In *Electron Devices Meeting (IEDM), 2009 IEEE International*, pp. 1–4 (2009).

[36] K. H. Yeo, S. D. Suk, M. Li *et al.* "Gate-all-around (GAA) twin silicon nanowire MOSFET (TSNWFET) with 15 nm length gate and 4 nm radius nanowires." In *Election Devices Meeting, 2006. IEDM 2006. IEEE International*, pp. 1–4 (2006).

[37] W. W. Fang, N. Singh, L. K. Bera *et al.* "Vertically stacked SiGe nanowire array channel CMOS transistors." *IEEE Electron Device Letters*, **28** 211–213 (2007).

[38] C. Dupre, A. Hubert, S. Becu *et al.* "15nm-diameter 3D stacked nanowires with independent gates operation: ΦFET." In *Electron Devices Meeting, 2008. IEDM 2008. IEEE International*, pp. 1–4 (2008).

[39] N. Singh, F. Y. Lim, W. W Fang *et al.* "Ultra-narrow silicon nanowire gate-all-around CMOS devices: impact of diameter, channel-orientation and low temperature on device performance." In *Electron Devices Meeting, 2006. IEDM '06. International*, pp. 1–4 (2006).

[40] R. Kotlyar, B. Obradovic, P. Matagne, M. Stettler, & M. D. Giles, "Assessment of room-temperature phonon-limited mobility in gated silicon nanowires." *Applied Physics Letters*, **84**, 5270–5272 (2004).

[41] A. C. Ford, J. C. Ho, Y.-L. Chueh *et al.* "Diameter-dependent electron mobility of InAs nanowires." *Nano Letters*, **9**, 360–365 (2009).

[42] K. J. Kuhn,, U. Avci,, A. Cappellani *et al.* "The ultimate CMOS device and beyond." In *Electron Devices Meeting (IEDM), 2012 IEEE International*, pp. 171–174 (2012).

[43] J. P. Colinge,, C. W. Lee, A. Afzalian *et al.* "SOI gated resistor: CMOS without junctions." In *SOI Conference, 2009 IEEE International*, pp. 1–2 (2009).

[44] J.-P. Colinge, C.-W. Lee, I. Ferain *et al.* "Reduced electric field in junctionless transistors." *Applied Physics Letters*, **96** (2010), doi.org/10.1063/1.3299014.

[45] C.-W. Lee, A. Afzalian, N. D. Akhavan, R. Yan, I. Ferain, & J.-P. Colinge, "Junctionless multigate field-effect transistor." *Applied Physics Letters*, **94** (2009), doi.org/10.1063/1.3079411.

[46] Y. Cui, Z. Zhong, D. Wang, W. U. Wang, & C. M. Lieber, "High Performance silicon nanowire field effect transistors." *Nano Letters*, **3**, 149–152 (2003).

[47] R. Rios, A. Cappellani, M. Armstrong *et al.* "Comparison of junctionless and conventional trigate transistors with down to 26 nm." *IEEE Electron Device Letters*, **32**, 1170–1172 (2011).

[48] C.-W. Lee, A. Borne, I. Ferain *et al.* "High-temperature performance of silicon junctionless MOSFETs." *IEEE Transactions on Electron Devices*, **57**, 620–625 (2010).

[49] S.-J. Choi, D.-I. Moon, S. Kim, J. P. Duarte, & Y.-K. Choi, "Sensitivity of threshold voltage to nanowire width variation in junctionless transistors." *IEEE Electron Device Letters*, **32**, 125–127 (2011).

[50] V. Varadarajan, Y. Yasuda, S. Balasubramanian, & T.-J. K. Liu, "WireFET technology for 3-D integrated circuits." In *Electron Devices Meeting, 2006. IEDM '06. International*, pp. 1–4 (2006).

[51] V. Subramanian & K. C. Saraswat, "Laterally crystallized polysilicon TFTs using patterned light absorption masks." In *Device Research Conference Digest, 1997. 5th*, pp. 54–55 (1997).

[52] V. Subramanian, M. Toita, N. R. Ibrahim, S. J. Souri, & K. C. Saraswat, "Low-leakage germanium-seeded laterally-crystallized single-grain 100-nm TFTs for vertical integration applications." *IEEE Electron Device Letters*, **20**, 341–343 (1999).

[53] J. H. Oh, D. H. Kang, M. K. Park, & J. Jang, "Low off-state drain current poly-Si TFT with Ni-mediated crystallization of ultrathin a-Si." *Electrochemical and Solid-State Letters*, **12**, J29–J32 (2009).

[54] S.-M. Jung, J. Jang, W. Cho *et al.* "The revolutionary and truly 3-dimensional 25F2 SRAM technology with the smallest S3 (stacked single-crystal Si) cell, $0.16\mu m^2$, and SSTFT (atacked single-crystal thin film transistor) for ultra high density SRAM." In *VLSI Technology, 2004. Digest of Technical Papers. 2004 Symposium on*, pp. 228–229 (2004).

[55] International Technology Roadmap for Semiconductors (ITRS) (2011). Available at: http://public.itrs.net/.

[56] V. V. Zhirnov, R. K. Cavin, J. A. Hutchby, & G. I. Bourianoff, "Limits to binary logic switch scaling – a gedanken model." *Proceedings of the IEEE*, **91**(11), 1934–1939 (2003).

[57] K. Bernstein, R. K. Cavin III, W. Porod, A. Seabaugh, & J. Welser, "Device and architecture outlook for beyond-CMOS switches." *Proceedings of the IEEE*, **98**, 2169–2184 (2010).

[58] D. E. Nikonov & I. A. Young, "Overview of beyond-CMOS devices and a uniform methodology for their benchmarking." *Proceedings of the IEEE*, **101**(12), 2498–2533 (2013).

[59] A. C. Seabaugh, & Q. Zhang, "Low-voltage tunnel transistors for beyond CMOS logic." *Proceedings of the IEEE*, **98**, 2095–2110 (2010).

[60] W. G. Vandenberghe, B. Soree, W. Magnus, G. Groeseneken, & M. V. Fischetti, "Impact of field-induced quantum confinement in tunneling field-effect devices." *Applied Physics Letters*, **98** (2011), doi.org/10.1063/1.3573812.

[61] B. S. Haran, A. Kumar, L. Adam *et al.* "22 nm technology compatible fully functional 0.1 μm² 6T-SRAM cell." In *Electron Devices Meeting, 2008. IEDM 2008. IEEE International*, pp. 1–4 (2008).

[62] D.-H. Kim, & J. A. del Alamo, "30 nm E-mode InAs PHEMTs for THz and future logic applications." In *Electron Devices Meeting, 2008. IEDM 2008. IEEE International*, pp. 1–4 (2008).

[63] G. A. Salvatore, D. Bouvet, & A. M. Ionescu, "Demonstration of subthrehold swing smaller than 60 mV/decade in Fe-FET with P(VDF-TrFE)/SiO₂ gate stack." In *Electron Devices Meeting, 2008. IEDM 2008. IEEE International*, pp. 1–4 (2008).

[64] T. Sakamoto, N. Iguchi, & M. Aono, "Nonvolatile triode switch using electrochemical reaction in copper sulfide." *Applied Physics Letters*, **96**, (2010), doi.org/10.1063/1.3457861.

[65] G. Dewey, B. Chu-Kung, J. Boardman *et al.* "Fabrication, characterization, and physics of III-V heterojunction tunneling field effect transistors (H-TFET) for steep sub-threshold swing." In *Electron Devices Meeting (IEDM), 2011 IEEE International*, pp. 33.6.1–33.6.4 (2011).

[66] A. Majumdar, Z. Ren, S. J. Koester, & W. Haensch, "Undoped-body extremely thin SOI MOSFETs with back gates." *IEEE Transactions on Electron Devices*, **56**, 2270–2276 (2009).

[67] C. Auth, C. Allen, A. Blattner *et al.*, "A 22nm high performance and low-power CMOS technology featuring fully-depleted tri-gate transistors, self-aligned contacts and high density MIM capacitors." In *VLSI Technology (VLSIT), 2012 Symposium on*, pp. 131–132 (2012).

[68] J.-P. Colinge, C.-W. Lee, A. Afzalian *et al.* "Nanowire transistors without junctions." *Nature Nanotechnology*, **5**(15), 225–229 (2010).

4 Extending CMOS with negative capacitance

Asif Islam Khan and Sayeef Salahuddin

4.1 Introduction

It is now well recognized that energy dissipation in microchips may ultimately restrict device scaling – the downsizing of physical dimensions that has fueled the fantastic growth of the microchip industry so far [1–6]. But there is a fundamental limit to the dissipation that can be achieved in the transistors that are at the heart of almost all electronic devices. Conventional transistors are thermally activated. A barrier is created that blocks the current and then the barrier height is modulated to control the current flow. This modulation of the barrier changes the number of electrons following the exponential Boltzmann factor, $\exp(qV / kT)$. This, in turn, means that a voltage of at least $2.3kT / q$ (which translates to $60\,\text{mV}$ at room temperature) is necessary to change the current by an order of magnitude. In practice, a voltage many times this limit of $60\,\text{mV}$ has to be applied to obtain a good ratio of on- and off-currents. As a result, it is not possible to reduce the supply voltage in conventional transistors below a certain point, while still maintaining the healthy on/off ratio that is necessary for robust operation. On the other hand, continuous downscaling is putting an ever larger number of devices in the same area, thereby increasing the energy dissipation density beyond controllable and sustainable limits. This situation is often called Boltzmann's Tyranny [2], and it has been predicted that unless new principles can be found based on fundamentally new physics, then transistors will die a thermal death [4].

To overcome this problem in conventional transistors, a number of alternative approaches are currently being investigated. Examples include band-to-band tunneling field effect transistors (TFET) [7, 8], impact ionization metal–oxide–semiconductor transistors (IMOS) [9], and nanoelectromechanical (NEM) switches [10, 11]. In these approaches the mechanism of transport, i.e., the way electrons flow in a transistor, is altered such that the minimum limit of $2.3kT / q$ can be avoided. In contrast, Salahuddin and Datta showed that it may theoretically be possible to keep the mechanism of transport intact, but change the electrostatic gating in such a way that it steps up the surface potential of the transistor beyond what is possible conventionally [12, 13]. The basic principle of such "active" gating relies on the ability to drive the ferroelectric material away from its local energy minimum to a non-equilibirum state where its capacitance (dQ / dV) is negative and stabilizing it there by adding a series capacitance. In the following sections, we shall discuss this mechanism in detail.

4.2   Intuitive picture

4.2.1   Why negative capacitance?

To understand how negative capacitance may help reducing the supply voltage and hence energy dissipation in conventional transistors, let us imagine a series network of two capacitors: one an ordinary positive capacitor, C_s, and the other a negative capacitor, C_{ins}, as shown in Fig. 4.1(a). By simply using a capacitance divider formula, one would see that the total capacitance of this series network would be larger than C_s provided $|C_{ins}| > |C_s|$. This is surprising considering that in a series network of two ordinary capacitors the total capacitance must be smaller than either of the constituent capacitances. Figure 4.1(b) shows a schematic view of a transistor where the insulator capacitance (C_{ins}) and semiconductor capacitance (C_s) make an analogous series network to that shown in Fig. 4.1(a). Now the reduction in supply voltage can be simply understood in the following way: since the total capacitance is enhanced by having a negative C_{ins}, it requires a lower voltage to produce the same amount of charge Q across the capacitors, C_s and C_{ins}, both of which have the same Q due to being in series. The current in the channel is proportional to the charge across C_s. This means that the same amount of current can now be produced with smaller voltage. Perhaps a more intriguing aspect of the network in Fig. 4.1(a) is the fact that the internal node voltage, V_{in}, is larger than supply voltage, V, due to the presence of a negative C_{ins}. This makes the channel "see" a larger voltage than what was actually applied. Recognizing that the Boltzmann factor is given by $\exp(qV_{in} / kT)$, the minimum voltage required to increase current by one order of magnitude is $2.3kT / (rq)$. Conventionally, $r = V_{in} / V < 1$; but in this case, $r > 1$ since $C_{ins} < 0$. As a result the minimum voltage (to increase current by one order of magnitude) reduces below 60 mV at room temperature.

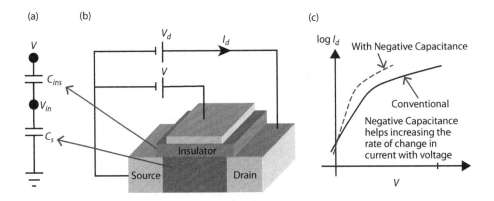

Fig. 4.1 (a) A series network of two capacitors. The proposed research is focused towards making $C_{ins} < 0$ so that $V_{in} > V$. (b) A schematic diagram showing how the transistor corresponds to the network shown in (a). (c) Schematic diagram showing how negative capacitance may reduce supply voltage requirement and thereby dissipation.

4.2.2   Reduction of sub-threshold swing

Mathematically, the rate of change in current as a function of voltage in a transistor is often quantified by the term "sub-threshold swing" (S), which is defined as:

$$S = \frac{\partial V}{\partial \log_{10}(I_D)} = \left(\frac{\partial V}{\partial V_{in}}\right)\left(\frac{V_{in}}{\partial \log_{10}(I_D)}\right), \tag{4.1}$$

where V is the gate voltage and I_D is the current flowing from source to drain. In Fig. 4.1(c), the region below where the current saturates is known as the sub-threshold region. S provides an estimate for how steeply the current is increasing with voltage. The lower the value of S, the steeper the curve and vice versa. Going back to Eq. (4.1), one would see that the expression can be written as a product of two terms. To understand these terms, look at Fig. 4.2, which shows a simplistic view of the potential profile in a nanoscale transistor. The capacitor network shown in Fig. 4.1(a) is redrawn in Fig. 4.2 to show its relation to the potential profile. Note that here we have not explicitly drawn the channel/drain or channel/source coupling capacitors. These capacitances are instead lumped into the semiconductor capacitance itself. (This treatment does not change the physical scenario that we explain here.) The internal node voltage, V_{in}, also called the surface potential, controls the current flow over the barrier. The second term determines the inverse of how much current flows as a function of V_{in}. This term is dictated by the Boltzmann factor, $\exp(qV_{in}/kT)$ and can only give an S of $2.3kT/q$ $(= 60\,mV/dec)$ at room temperature. Clearly, as long as the transport mechanism of electrons is not altered from a barrier modulated transport, the second term is a fundamental one and provides only $60\,mV/dec$ of sub-threshold swing. This is the motivation behind TFET [7, 8], IMOS [9], and NEMFET [10, 11], as mentioned above, where the mode of transport is changed.

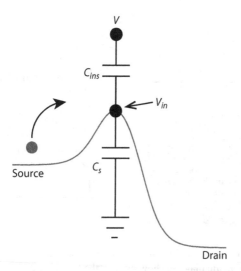

Fig. 4.2 Potential profile in a nanoscale transistor. The capacitor network shows how the applied gate voltage is divided between the oxide insulator and the semiconductor.

But what about the the first term? This term is simply the ratio of supply voltage, V, to the internal node voltage, V_{in}, which can be written as

$$m = \frac{\partial V_{gate}}{\partial \psi_s} = 1 + \frac{C_s}{C_{ins}}. \tag{4.2}$$

This ratio, often called the "body-factor" in MOSFET literature, will always be larger than 1 because of the voltage divider rule in conventional capacitors. Thus ordinarily S cannot be less than 60 mV/dec. However, if the conditions $C_{ins} < 0$ and $|\,C_{ins}\,| > |\,C_s\,|$, can be satisfied, m could be made to be less than 1, leading to an overall S of less than 60 mV/dec. Obtaining an effective negative C_{ins} is the main objective of this research.

4.2.3   How can a ferroelectric material give negative capacitance?

Now that we have explained how a negative capacitance could lead to $S < 60$ mV/dec, let us look at how a negative capacitance can be realized using a ferroelectric material. The energy landscape of a ferroelectric material is shown in Fig. 4.3(a). It has two degenerate energy minima. This means that the ferroelectric material could provide a non-zero polarization even without an applied electric field. In general, the total charge density in a given material can be written as $Q_A = \varepsilon E + P$, where ε is the linear permittivity of the ferroelectric, E is the external electric field and P is the polarization. In typical ferroelectric materials, $P \gg \varepsilon E$ leading to $Q_A \approx P$. For this reason we shall use P and Q_A interchangeably. Since charge density is what we are interested in, we shall also drop the subscript "A" and simply use Q for charge density.

Figure 4.3(a) shows that Q is an even polynomial function of E, which is a characteristic property of a ferroelectric material. If we compare this characteristic energy landscape with that of an ordinary capacitor shown in Fig. 4.3(b), we can see that the curvature around $Q = 0$ of a ferroelectric is the opposite of that of an ordinary capacitor.

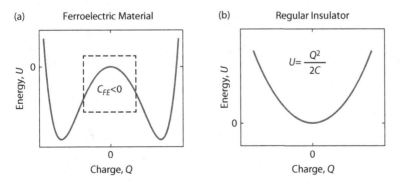

Fig. 4.3 Energy landscapes of (a) a ferroelectric and (b) an ordinary insulator. The negative curvature for the ferroelectric around $Q = 0$ gives rise to a negative capacitance around that charge range.

Remembering that the energy of an ordinary capacitor is given by $(Q^2/2C)$, this opposite curvature already hints at a negative capacitance for the ferroelectric material around $Q = 0$. Mathematically, the capacitance is defined as the inverse of the slope of the derivative of energy, U, with respect to Q:

$$C = \left[\frac{d^2U}{dQ^2}\right]^{-1}.$$

(4.3)

We see that this slope for the ordinary capacitor is always positive, while that for the ferroelectric capacitor is negative around $Q = 0$. Around this point, therefore, a ferroelectric material could provide a negative capacitance. However, the very fact that this negative capacitance occurs away from the local equilibrium (states of minimum energy) also means that the ferroelectric material will be unstable at this state. Even if it can be driven into negative capacitance, it will quickly come back to the equilibrium point where the capacitance is positive. So the next question would be: how can we stabilize the ferroelectric where its capacitance is negative?

4.2.4   How can we stabilize a ferroelectric material in a state of negative capacitance?

Intuitively, if we add the energies of a ferroelectric material and an ordinary capacitor together such that $Q = 0$ becomes the energy minimum for the total system, the ferroelectric will have to settle at $Q = 0$ where its capacitance is negative. This is shown in Fig. 4.4. In practice, this can be done by adding a series capacitor with a ferroelectric insulator as shown in Fig. 4.1(a). This means that if we can integrate a ferroelectric/dielectric bilayer on top of a MOSFET as a gate, the dielectric together with the semiconductor capacitance C_s may be utilized to stabilize the ferroelectric material at a state of negative capacitance. The resulting "step-up" of the surface potential V_{in} will then reduce the sub-threshold swing below 60 mV/dec.

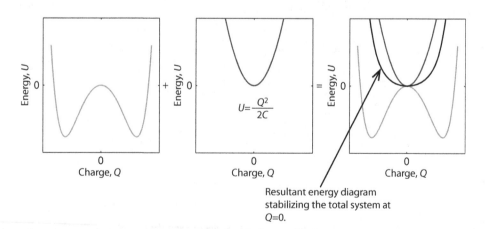

Fig. 4.4 An ordinary capacitor with the right amount of positive capacitance can stabilize a ferroelectric material where its capacitance is negative.

4.3   Theoretical framework

Before starting this section, note that throughout this discussion, whenever we write "capacitance," we mean dQ/dV and never Q/V. Thus the capacitance we calculate is in general a function of voltage. For some readers, it might be more familiar described as "differential capacitance." However, since capacitance is, by definition, a differential concept given by dQ/dV, we prefer using simply "capacitance." The theoretical model of ferroelectric negative capacitance is based on the Landau–Khalatnikov (LK) equation [14, 15], which describes the dynamics of a ferroelectric material. The equation is as follows:

$$\rho\frac{d\vec{P}}{dt} + \vec{\nabla}_{\mathrm{P}}U = 0, \tag{4.4}$$

where

$$U = \alpha P^2 + \beta P^4 + \gamma P^4 - \vec{P}\cdot\vec{E} \tag{4.5}$$

is the Gibb's free energy per unit volume given by the sum of the anisotropy energy and the energy due to the external field; and α, β, and γ are the material-dependent anisotropy constants. At a given lattice temperature, T, the parameter $\alpha = \alpha_0(T - T_C)$ is related to the Curie temperature, T_C, α_0 being a positive material-dependent constant. For a material to be ferroelectric at T, α is negative. β is either negative or positive depending on whether the material under consideration goes through a first-order or second-order phase transition respectively at the Curie temperature. γ is positive. It is evident that when the external field (E) is zero, the energy landscape given by Eq. (4.5) will be akin to the one shown in Fig. 4.3(a).

Next, let us consider a series combination of a ferroelectric (FE) and a dielectric (DE) capacitor. For this example, we shall consider PZT (lead zirconate titanate: PbZr$_{0.2}$Ti$_{0.8}$O$_3$) as the ferroelectric layer and STO (strontium titanate: SrTiO$_3$) as the dielectric layer. With l_f and l_d as the FE and DE layer thicknesses respectively, the total energy of the combination per unit surface area can be written as

$$U_{\mathrm{f+d}} = l_{\mathrm{f}}\left(\alpha_{\mathrm{f}}P_{\mathrm{f}}^2 + \beta_{\mathrm{f}}P_{\mathrm{f}}^4 + \gamma_{\mathrm{f}}P_{\mathrm{f}}^6 - E_{\mathrm{f}}P_{\mathrm{f}}\right) + l_{\mathrm{d}}\left(\alpha_{\mathrm{d}}P_{\mathrm{d}}^2 + \beta_{\mathrm{d}}P_{\mathrm{d}}^4 + \gamma_{\mathrm{d}}P_{\mathrm{d}}^4 - E_{\mathrm{d}}P_{\mathrm{d}}\right). \tag{4.6}$$

Here, the subscripts f and d refer to FE and DE respectively. The advantage of writing the energy function of a DE like that of an FE, as in Eq. (4.6), is the fact that this equation applies equally to a dielectric or a paraelectric material. Since STO is indeed paraelectric, Eq. (4.6) is particularly appropriate.

If the total voltage across the FE/DE series capacitor network is V, then Kirchhoff's law dictates that

$$V = V_{\mathrm{f}} + V_{\mathrm{d}} = E_{\mathrm{f}}l_{\mathrm{f}} + E_{\mathrm{d}}l_{\mathrm{d}}, \tag{4.7}$$

where $V_f = E_f l_f$ and $V_d = E_d l_d$ are the voltages across the FE and DE layers respectively. Applying Gauss's law at the interface between FE and DE, one obtains the following equation for the surface charge density, Q:

$$Q = \varepsilon_d E_d + P_d = \varepsilon_f E_f + P_f. \tag{4.8}$$

Combining Eqs. (4.6), (4.7), and (4.8), the equation for the total energy, U_{f+d} can be derived:

$$U_{f+d} = l_f\left(\alpha_f P_f^2 + \beta_f P_f^4 + \gamma_f P_f^6\right) + l_d\left(\alpha_d P_d^2 + \beta_d P_d^4 + \gamma_d P_d^6\right)$$
$$-V\frac{P_f l_f + P_d l_d}{l_f + l_d} + \frac{l_f l_d (P_f - P_d)^2}{\varepsilon_f l_f + \varepsilon_d l_d}. \tag{4.9}$$

For a given voltage V, the state of the combined FE/DE network can be found by minimizing the total energy given by Eq. (4.9) with respect to polarizations in FE and DE. This will provide the polarization and electric fields in individual layers. The total capacitance of the FE/DE series network can then be calculated from $C = dQ/dV$. The effective capacitance of the FE and DE layers at a given V can also be calculated using $C_i = dQ/dV_i$ ($i \equiv$ f,d).

For this particular case of a PZT/STO heterostructure, the values of the coefficients (α, β, γ) are taken from [16]. Figure 4.5(b) shows the simulated capacitance of PZT/STO heterostructure as a function of temperature and compares it to the capacitance of STO. Also shown in Fig. 4.5(b) is the FE capacitance in the heterostructure as well as the voltage amplification factor (r) at the FE/DE interface, $(1+C_d/C_f)^{-1}$. Note that an enhancement in capacitance is indeed observed, although it occurs only over a certain temperature range (between ~200°C and ~430°C). In the next section, we shall look into this temperature dependence in more detail.

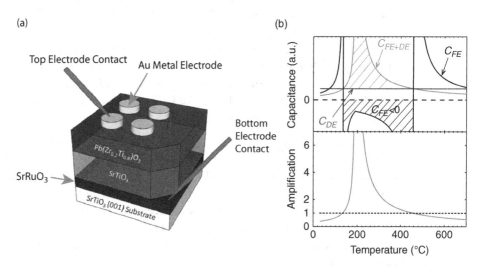

Fig. 4.5 (a) Schematic of the experimental stack. SrTiO$_3$ (STO) and Pb(Zr$_{0.2}$Ti$_{0.8}$)O$_3$ (PZT) form the bilayer. Au and SrRuO$_3$ (SRO) are used as top and bottom contacts respectively. (b) Capacitance of an FE (PZT)/DE $\varepsilon_r = 200$) bilayer capacitor (thickness ratio 4:1) as a function of temperature. Also shown in this figure is the capacitance of the constituent DE and FE in the heterostructure as well as the voltage amplification factor at the FE/DE interface. (Source: [17].)

4.3.1 Temperature dependence

A ferroelectric material is sensitive to the temperature due to the coefficient α being proportional to T (see the discussion above). This is the main reason why the negative capacitance is temperature dependent as seen in Fig. 4.5(b). For all practical purposes, in an FE/DE bilayer, the system would adopt a nearly uniform polarization throughout all the layers resulting in $Q \approx P_f \approx P_d$. This is due to the fact that the last term in Eq. (4.9), which represents the electrostatic energy arising due to the polarization mismatch at the interface, is costly in energy. Equation (4.9) can be simplified as

$$U_{f+d} = l_f\left(\alpha_f Q^2 + \beta_f Q^4 + \gamma_f Q^6\right) + l_d\left(\alpha_d Q^2 + \beta_d Q^4 + \gamma_d Q^6\right) - VQ. \qquad (4.10)$$

From Eq. (4.10), one can easily see that when a DE is added in series with an FE, a positive quantity ($\alpha_d l_d/l_f > 0$) is added to the α coefficient of the ferroelectric. One way to understand the impact of this addition of positive energy is to recognize that it reduces the "effective" Curie temperature of the ferroelectric material by reducing the total negative energy in the combined system. Indeed simulated capacitance vs. temperature shows a shift to the left of the effective Curie temperature for the FE/DE heterostructure in comparison to the isolated FE (see Fig. 4.6(a)).

In order to explain the temperature dependence in more detail, we resort to energy landscapes, as shown in Fig. 4.6(b) and (c) for two different temperatures, T_A and T_B. For a ferroelectric material, as the temperature increases, the double wells of the energy landscape get closer together and become shallower and flatter, making the ferroelectric capacitance increase. At the Curie temperature, the double wells develop into a single minimum, the capacitance peaks, and an FE-to-DE phase transition occurs. Beyond the Curie temperature, the energy landscape in the DE phase again becomes steeper around the single minima, making the capacitance decrease with the temperature. Interestingly, the exact same thing happens when, rather than changing the temperature, a dielectric is added in series. If temperature is increased, the α coefficient for the ferroelectric material increases by $\alpha_0 T$. On the other hand, if a dielectric material is added, the same coefficient increases by $\alpha_d l_d / l_f$ (see Eq. (4.9)). For a conventional dielectric $\alpha_d = 1/2\varepsilon$, $\beta_d = \gamma_d = 0$. Because of this duality, the negative capacitance effect in an FE/DE layer shows the temperature dependence seen in Fig. 4.6(a). On the other hand, this means that temperature can be used as an effective tuning parameter for characterization and probing of negative capacitance.

Note that the capacitance is not negative everywhere in the energy landscape of ferroelectric material. The portion where capacitance is indeed negative is shown by the dashed box in Fig. 4.6(b) and (c). The idea is to add enough positive energy in the system from the DE so that the total energy will be minimized at a (U, Q)-coordinate where the ferroelectric capacitance is negative. There are three distinct regions of operation that are of interest for negative capacitance. First at $T = T_A$, the added energy from the dielectric alone is not large enough to stabilize the ferroelectric material in a negative capacitance region (see Fig. 4.6(b)). If the temperature is increased from T_A to T_B, the combined energy moves to the point of minimum energy inside the dashed box and hence negative capacitance is expected (see Fig. 4.6(c)). But note that the total

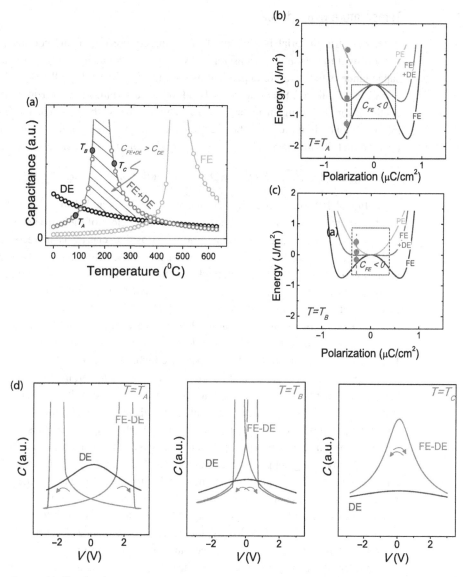

Fig. 4.6 (a) Simulated capacitance of a PZT/STO bilayer capacitor with PZT and STO thicknesses of t_{PZT} and t_{STO} respectively, an isolated STO with a thickness of t_{STO} and an isolated PZT with a thickness of t_{PZT} as a function of temperature. $t_{PZT}:t_{STO} = 4:1$. The temperature corresponding to the singularity in PZT capacitance corresponds to its Curie temperature. Note that capacitance–temperature characteristics of the PZT/STO bilayer capacitor has a shape similar to that of a PZT capacitor with a lower Curie temperature. (b) and (c) Energy landscapes of the series combination at two different temperatures, T_A (b) and T_B (c). (d) Calculated C–V characteristics of an STO capacitor and a PZT/STO bilayer capacitor (thickness ratio 4:1) at $T = T_A$, T_B, and T_C.

energy still shows small double wells. This means that, despite negative capacitance operation and capacitance enhancement, hysteresis should be observed. This is the second mode of operation. If the temperature is further increased to T_C, the positive energy will be large enough to completely eliminate the double well and make the combined energy look like a paraelectric or dielectric (shown by the tag "PE" in Fig. 4.6(b) and (c)). It is in this third region of operation that we find a capacitance enhancement and also the operation will be hysteresis free. Figure 4.6(d) shows capacitance–voltage simulations for a PZT/STO heterostructure at T_A, T_B, and T_C, as described above. At T_A, PZT/STO shows hysteresis in C–V characteristics and the combined PZT/STO capacitance is smaller than that of the STO capacitor around $V = 0$. The peaks in C–V characteristics in PZT/STO correspond to polarization switching and are not related to the negative capacitance effect. At T_B, the PZT/STO hysteresis has decreased and equivalent capacitance is larger than that of the STO capacitor around zero bias condition. Therefore, the negative capacitance operation can now be observed but it is accompanied by hysteresis. Finally, at a further elevated temperature, T_C, the PZT/STO capacitance becomes much larger than that of PZT and the hysteresis completely goes away. This is the most desired mode of operation.

4.4   Experimental work

To test the model predictions as described above, we fabricated FE/DE bilayer capacitors using PZT with a Zr/Ti ratio of 20/80 (PbZr$_{0.2}$Ti$_{0.8}$O$_3$) as the ferroelectric and STO as the dielectric material [17]. PZT is a robust room temperature ferroelectric with a remnant polarization of 80 μC/cm^2 [16, 18]. PZT has a tetragonal crystal structure with unstrained c- and a-axis lattice parameters of 0.413 and 0.393 nm and $c/a \approx 1.05$ [19]. Unstrained STO has a cubic crystal structure with $c = a = 0.3905$ nm [20] and remains paraelectric down to 0 K. These layers are grown on metallic SrRuO$_3$ (SRO) that has a pesudocubic crystal structure with lattice parameter of 0.393 nm. Well-matched lattice structures for all these materials allow coherent and epitaxial growth of atomically smooth PZT/STO/SRO heterostructures on STO (001) substrates.

Three different types of heterostructures were grown using the pulsed laser deposition technique: STO/SRO, PZT/SRO, and PZT/STO/SRO. Stoichiometric PZT, STO, and SRO targets were ablated at a laser fluence of ~1 J cm^{-2} and a repetition rate of 15, 5, and 15 Hz for PZT, STO, and SRO respectively. During growth, the substrate was held at 720°C for SRO and STO and at 630°C for PZT. The lower growth temperature for PZT is adopted to prevent the evaporation of volatile Pb. PZT and SRO were grown in an oxygen environment at 100 mTorr and STO in 250 mTorr. After growth, the heterostructures were slowly cooled down to room temperature at 1 atm of oxygen at a rate of –5°C/min. SRO thickness in all the samples is ~30 nm. Surface topography, X-ray diffraction spectrum, and TEM cross-sectional images of representative PZT/STO samples are shown in Fig. 4.7.

We first focus on the capacitance of a 28 nm PZT/48 nm STO bilayer capacitor. Figure 4.8(a) shows the C–V characteristics at 100 kHz of a PZT (28 nm)/STO (48 nm) bilayer capacitor at 30, 300, 400, and 500°C and compares it to the capacitance of a 48 nm STO. It is clear that the capacitance of PZT/STO sample is larger than that of the STO capacitor at elevated temperatures. The evolution of the C–V curves of PZT/STO and STO capacitors with temperature are very similar to those obtained by simulation,

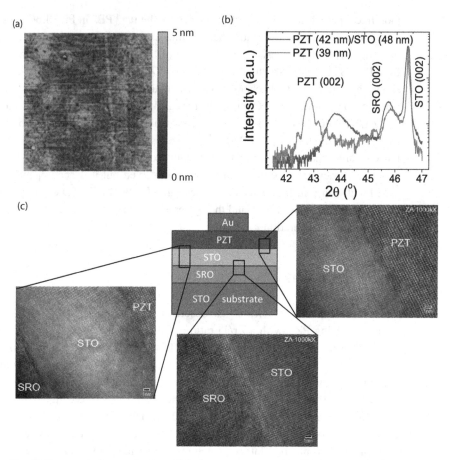

Fig. 4.7 Structural characterization of the heterostructures. (a) AFM topography image
of typical PZT/STO sample surfaces showing an RMS roughness less than 0.5 nm. (b) XRD
$\theta - 2\theta$ scans around (002) reflections of a PZT (42 nm)/STO (28 nm)/SRO (30 nm) and a PZT
(39 nm)/SRO (30 nm) sample. The (002) reflection from the STO thin film is buried within
the STO (002) substrate peak. (c) Cross-sectional HRTEM images of different interfaces of a
PZT/STO sample. PZT/STO film on SRO-buffered STO have in-plane epitaxy to the substrate
with atomically sharp interfaces. The structural characterization confirms the c-axis orientation
of the PZT films without any contribution from the a-axis oriented domains and impurity phases.
(Source: supplementary information in [17].)

shown in Fig. 4.6(d). Figure 4.8(b) shows the capacitance of the PZT/STO bilayer
and dielectric STO as a function of temperature. At ~225°C, the bilayer capacitance
exceeds the STO capacitance. This means that, beyond this temperature, the capacitance
of the 76 nm thick bilayer (STO: 48 nm + PZT: 28 nm) becomes larger than that of
48 nm STO itself. Figure 4.8(c) shows an effective capacitance for PZT that was
extracted by subtracting the STO capacitance from the measured total capacitance.
A corresponding voltage amplification factor is also shown. Figure 4.8(d) shows the
capacitance as a function of frequency at $T = 300$°C. We see that enhancement in
capacitance is retained even at 1 MHz, thereby indicating that the contribution of defect

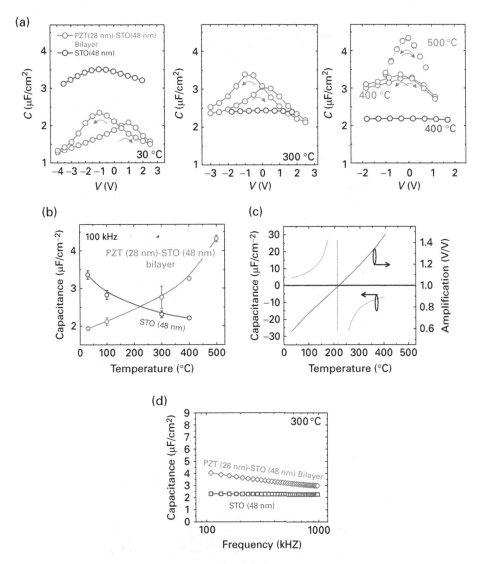

Fig. 4.8 (a) Comparison of the *C–V* characteristics of a PZT (28 nm)/STO (48 nm) and an STO (48 nm) sample at different temperatures. (b) Capacitances of the samples at the symmetry point as functions of temperature measured at 100 kHz. Symmetry point refers to the cross point of the *C–V* curves obtained during upward and downward voltage sweeps. (c) Extracted PZT capacitance in the bilayer and the calculated amplification factor at the FE/DE interface. (d) Capacitances of the samples as functions of frequency at 300°C. (Source: [17].)

mediated processes are minimal, if any, and therefore the enhanced capacitance cannot be attributed to such effects (see below).

Figure 4.9 shows the capacitance, permittivity, and impedance angle for three different samples (sample nos. 2–4) of STO/PZT bilayers of different thicknesses compared to that of an isolated STO (sample 1) and an isolated PZT sample (sample 5).

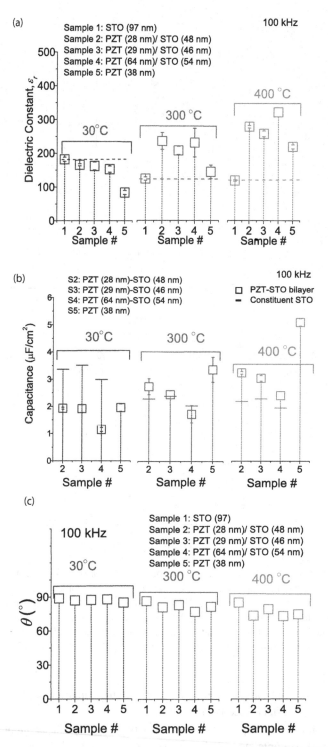

Fig. 4.9 Comparison of dielectric constant (a), capacitance (b) and the admittance angles, θ (c) of several PZT/STO samples with those of STO and PZT at 100 kHz at different temperatures. In (b), the capacitance of the constituent STO in each of the bilayers is shown by small horizontal line. (Source: [17].)

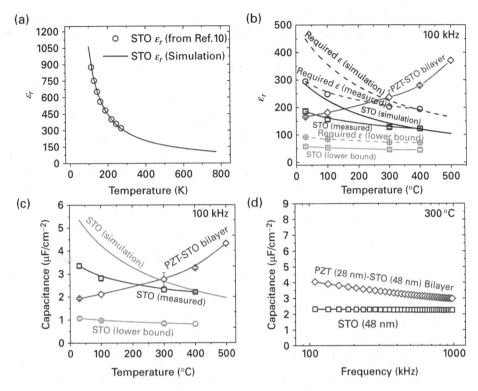

Fig. 4.10 (a) Comparison of STO dielectric constant simulated using Landau model with measured dielectric constant of 50 nm STO reported in [21]. (b) and (c) Comparison of dielectric constant (b) and capacitance (c) of a PZT (28 nm)/STO (48 nm) sample with those of STO samples at 100 kHz as functions of temperature. In (b), curves tagged "Required ε" refer to the required bilayer dielectric constant to achieve $C_{PZT-STO} = C_{STO}$ at a certain temperature. The lower bound of STO dielectric constant and capacitance correspond to those measured from the 48 nm STO sample. (d) Frequency dispersion of dielectric capacitance of PZT/STO and isolated STO samples at 300°C. (Source: supplementary information in [17].)

We see that all the bilayer samples show an enhancement in overall permittivity as the temperature is increased.

One point that needs to be considered is the thickness dependence of STO permittivity. Since the breakdown electric field of STO is smaller than that of PZT, it poses a problem when the bilayer capacitance is compared to that of an isolated STO layer over the same voltage range. To overcome this problem, we have used a 97 nm STO as our control sample. The permittivity for isolated STO was measured from this sample. Consequently the capacitance of 48 nm STO was calculated using the measured permittivity of the 97 nm sample. It is well known that STO permittivity goes down with decreasing thickness. Therefore, we have overestimated the capacitance of the isolated STO layer and *underestimated* the capacitance enhancement that we show in Fig. 4.8(a) and (b). If we directly use the measured capacitances from a 48 nm thick STO sample, the capacitance enhancement can be found at *room temperature*. In Fig. 4.10, this is what we illustrate by "lower bound."

Also note that a simple enhancement in relative permittivity ε_r is not good enough to provide a capacitance enhancement. Since, due to the formation of the bilayer, the total thickness of the heterostructure has increased, the enhancement in ε_r needs to offset the increase in thickness. By noting that $C_{PZT\text{-}STO} = \varepsilon_{PZT\text{-}STO} / (d_{PZT} + d_{STO})$ and $C_{STO} = \varepsilon_{STO} / d_{STO}$, $C_{PZT\text{-}STO} \geq C_{STO}$ requires $\varepsilon_{PZT\text{-}STO} > \varepsilon_{STO}(1 + d_{PZT} / d_{STO})$. In Fig. 4.10(b), $\varepsilon_{STO}(1 + d_{PZT} / d_{STO})$ curves are also plotted and tagged "Required ε."

To further test the robustness of enhancement, we have simulated the permittivity of STO as a function of temperature. This was necessary because independent experimental data on relative permittivity of STO up to 300°C were not available. The simulated permittivity matches excellently within the temperature range where data are available [21] (see Fig. 4.10(a)). In Fig. 4.10(b) and (c), the traces tagged by "STD simulation" indicate STO permittivity and capacitance obtained by the simulation in Fig. 4.10(a). The reason for calling this limit the "upper bound" is the fact that the simulated permittivity corresponds to the highest permittivity expected from STO at those temperatures. For example, the upper bound plot shows a $\varepsilon_r = 300$ at room temperature, the highest ever reported for STO at this temperature. We emphasize, again, that in most experiments with thin films, including ours, the measured permittivity is much lower than this value.

Finally, Fig. 4.10(d) shows that the capacitance enhancement is retained at frequencies up to 1 MHz. This is important because defects may play a role in determining the total capacitance of the structure. However, total defect dynamics usually die out beyond 10 kHz. Therefore, the fact that the enhancement is retained even at 1 MHz indicates that the negative capacitance effect observed in these bilayers is coming from the intrinsic properties of the materials. In the supplementary section of [17], we have presented a detailed discussion of the Maxwell–Wagner (MW) mechanism that is principally responsible for oxide dynamics in complex oxide systems. This discussion showed that the MW effect can be ignored for our samples.

4.5   Negative capacitance transistors

The challenge of using the negative capacitance effect for field effect transistors can be readily appreciated from Eq. (4.2). The goal is twofold: (i) to make m as close to 0 as possible and (ii) to make sure that m is not negative. The closer C_s / C_{ins} is to -1 and hence m is to 0, the steeper is S. But if $C_s / C_{ins} < -1$, m itself will become negative. This will lead to hysteresis, which is undesirable. Satisfying the above two conditions simultaneously over a large voltage range is very difficult for a conventional field effect transistor. This is due to the fact that the FET capacitance (C_s) is non-linear and a strong function of voltage. Starting from depletion to inversion the FET capacitance can change significantly. If, for example, C_{ins} is designed to match C_s at depletion, when close to threshold, $|C_{ins}| \ll C_s$. As a result, $C_s / C_{ins} \ll -1$, leading to large hysteresis. If C_{ins} is matched to the onset of threshold, then there will be no increase in the steepness of the swing in the sub-threshold region. For a detailed discussion of these issues refer to [22, 23] and references therein. For a conventional transistor, the onset of threshold seems to be the most appropriate region for matching $|C_{ins}|$ to C_s [22].

On the other hand, it may be possible to design the FET such that its capacitance becomes a weak function of voltage. If this were indeed possible, then it becomes much easier to match the FET capacitance with that of ferroelectric negative capacitance over a large voltage range. Such a design was recently proposed [24]. This work showed that by placing a highly doped body under an ultra-thin unspoiled channel, the dependence of capacitance on voltage can be sufficiently weakened and an excellent matching with negative capacitance over a large voltage range could be possible. TCAD simulations showed that for an intrinsic Si channel sitting on a p^+ doped body, a sub-threshold swing of less than 30 mV/dec can be achieved over more than five orders of magnitude of current modulation. Two-dimensional materials such as the transition metal-di-chalcogenides [25, 26] could be interesting in this aspect.

Experimental work on negative capacitance transistors was first reported by the group of Adrian Ionescu in 2008 [27]. This paper reported the measurement of less than 60 mV/dec sub-threshold swing in a Si transistor with a gate incorporating a polymer ferroelectric material (PVDF). However, the reduction of sub-threshold swing was only observed for very low current levels where noise was dominant. Improving on this work, the same group reported on another transistor the following year [28], where isolation techniques such as well formation were used to reduce noise. The measured $I_d - V_g$ clearly showed <60 mV/dec for multiple orders of magnitude up to 10 nA current amplitude while the noise floor was 10 pA. This measurement demonstrated the negative capacitance effect in a transistor for the first time and showed that it was indeed possible to reduce the voltage below $2.3kT/q$ using this mechanism. Nonetheless, a polymer ferroelectric is severely restricted in terms of speed (typical switching speed is at most a few kHz). Therefore, it is necessary to fabricate transistors with crystalline ferroelectric materials such as PZT whose switching speed could be in the tens of pico seconds [29].

4.6   Concluding remarks

Negative capacitance has now been demonstrated in both capacitor stacks and in a field effect transistor. Nonetheless, much remains to be explained in terms of its physical mechanism. Based on our own experimental work on single crystalline oxide heterostructures, one question that could be raised is why can the negative capacitance effect only be observed at temperatures beyond 200°C for the STO/PZT bilayer? Within the one-dimensional LK model presented in this article, if one uses the bulk Curie temperature of PZT (~430°C), one would expect the negative capacitance effect to show up at room temperature. There might be a number of issues that "conspire" to push the temperature above room temperature. Here we will mention two of those that we feel are the most dominant ones. First of all, the experimental structure is really not a one-dimensional problem. Two-dimensional effects such as domain nucleation and propagation significantly change the simple energy picture presented in this chapter. These two-dimensional effects need to be modeled accurately to quantify the temperature range where the negative capacitance effect would

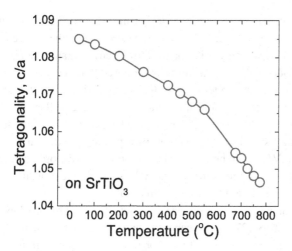

Fig. 4.11 X-ray diffraction measurement of lattice parameters as a function of temperature.

show up. One needs a full three-dimensional model to appropriately model such effects [30–32]. Another important aspect that significantly influences the behavior is substrate strain. Figure 4.11 shows X-ray diffraction measurements of the c/a ratio of PZT grown on an STO substrate. As the temperature increases, the tetragonal PZT becomes increasingly cubic. As a result the c/a ratio decreases. At some specific temperature, the thermal expansion of the lattice becomes more dominant and so the c/a ratio starts to increase. The inflection point provides an estimate for the transition temperature [33–36]. Note that for the PZT grown in STO substrate, we do not find this inflection point even at 600°C. This can be understood from realizing that STO exerts a compressive strain on PZT which makes it more tetragonal than bulk, thereby increasing its Curie temperature above the bulk temperature. Because the effective Curie temperature is now significantly increased, the temperature at which negative capacitance can be seen also increases beyond what is predicted from the model using the bulk Curie temperature. Thus strain effects are going to be very important when choosing appropriate materials systems for gate stacks. It may be worthwhile looking at ferroelectric materials such as $Ba_xSr_{1-x}TiO_3$, where by changing the stoichiometric ratio of Ba and Sr one could change the Curie temperature over a wide range. Such stoichiometric control could be used to counteract the effect of strain. Another common question about the STO/PZT bilayer has to do with the thickness of the dielectric STO layer. The one-dimensional model suggests that any material with arbitrary Curie temperature could still be made to give negative capacitance at room temperature simply by adjusting the thickness of the dielectric layer. The larger the Curie temperature, the larger the thickness of the dielectric layer should be. Experimentally, however, we have found that this is not the case. In fact, increasing the thickness beyond a certain point seems not to affect the overall behavior of the stack. We postulate that this is due to the screening of the electric field induced by ferroelectric polarization caused by the free charges in the STO. If this is indeed the case, making superlattices with a ferroelectric material sandwiched between thin dielectric layers may provide a pathway towards the room temperature negative capacitance effect. However, more studies need to be performed to properly understand the underlying physics.

It is important to realize that all ferroelectric materials are also piezoelectric. When one modulates the polarization in a ferroelectric material, one also modulates the piezoelectricity. Is it possible to use the piezoelectricity to aid the negative capacitance effect? This possibility is only starting to be investigated [37]. In fact, a recent experimental demonstration shows that the negative capacitance effect can be obtained in conventional nitride-based piezoelectric materials, leading to a 40 mV/dec sub-threshold swing [38]. Piezoelectric effects could provide an additional control mechanism of negative capacitance and positively influence its adoption for field effect transistors.

A number of other physical systems can be imagined that have negative terms in their energy profile and can behave as negative capacitance. One example is the exchange correlation between two closely spaced two-dimensional electron gases (2DEG). Experimentally, a negative compressibility was measured between two closely spaced 2DEG in a modulation doped GaAs/AlGaAs heterostructure at cryogenic temperature. Enhanced capacitance has recently been measured in an epitaxial LAO/STO heterostructure also at cryogenic temperature and was explained in terms of a negative capacitance that could arise due to similar exchange correlation [39, 40]. The advantage of ferroelectric-material based systems comes from the fact that the negative energy terms are reasonably large at room temperature, thereby removing the need for cryogenic operation.

To summarize, the negative capacitance effect is a potential way to overcome the limit of 60 mV/dec sub-threshold swing in conventional transistors without needing to change the conventional mechanism of electron transport. Experimental work in recent years, in both capacitor and transistor structures, has demonstrated the proof of the original concept. However, much remains to be understood about this effect, especially how practical material issues will influence the overall behavior. For transistor operation, integration with semiconductors is a major challenge. MBE grown crystalline oxides on semiconductors such as STO on Si [41, 42], barium titanate on GaAs [43], and barium titanate on Si [44] could be used as a starting template for fabrication. In addition, recently it was demonstrated that HfO_2 deposited by ALD (atomic layer deposition) and doped with various elements such as Si, Yttrium, Al, etc. could show robust ferroelectricity [45, 46]. This material system could prove to be the most suitable candidate for integration with silicon transistors.

Acknowledgments

This work was supported in part by Office of Naval Research (ONR), FCRP Center of Materials, Structures and Devices (MSD), the STARNET LEAST center, and Center for Energy Efficient Electronics Science (NSF Award 0939514).

References

[1] B. Nordman, "What the real world tells us about saving energy in electronics." In *Proceedings of 1st Berkeley Symposium on Energy Efficient Electronic Systems (E3S)* (2009).

[2] V. V. Zhirnov & R. K. Cavin, "Nanoelectronics: negative capacitance to the rescue?" *Nature Nanotechnology*, **3**(2), 77–78 (2008).

[3] S. Borkar, "Design challenges of technology scaling." *Micro, IEEE*, **19**(4), 23–29 (1999).

[4] L. B. Kish, "End of Moore's law: thermal (noise) death of integration in micro and nano electronics." *Physics Letters A*, **305**(3), 144–149 (2002).

[5] R. K. Cavin, V. V. Zhirnov, J. A. Hutchby, & G. I. Bourianoff, "Energy barriers, demons, and minimum energy operation of electronic devices." *Fluctuation and Noise Letters*, **5**(04), C29–C38 (2005).

[6] V. V. Zhirnov, R. K. Cavin III, J. A. Hutchby, & G. I. Bourianoff, "Limits to binary logic switch scaling-a gedanken model." *Proceedings of the IEEE*, **91**(11), 1934–1939 (2003).

[7] S. Banerjee, W. Richardson, J. Coleman, & A. Chatterjee, "A new three-terminal tunnel device." *IEEE Electron Device Letters*, **8**(8), 347–349 (1987).

[8] C. Hu, D. Chou, P. Patel, & A. Bowonder, "Green transistor – a VDD scaling path for future low power ICs." In *International Symposium on VLSI Technology, Systems and Applications, 2008. VLSI-TSA 2008*, pp. 14–15 (2008).

[9] K. Gopalakrishnan, P. B. Griffin, & J. D. Plummer, "Impact ionization MOS (i-MOS)-part I: device and circuit simulations." *IEEE Transactions on Electron Devices*, **52**(1), 69–76 (2005).

[10] H. Kam & T.-J. K. Liu, "Pull-in and release voltage design for nanoelectromechanical field-effect transistors." *IEEE Transactions on Electron Devices*, **56**(12), 3072–3082 (2009).

[11] M. Enachescu, M. Lefter, A. Bazigos, A. Ionescu, & S. Cotofana, "Ultra low power NEMFET based logic." In *Circuits and Systems (ISCAS), 2013 IEEE International Symposium on*, pp. 566–569 (2013).

[12] S. Salahuddin & S. Datta, "Use of negative capacitance to provide voltage amplification for low power nanoscale devices." *Nano Letters*, **8**(2), 405–410 (2008).

[13] S. Salahuddin & S. Datta, "Can the subthreshold swing in a classical fet be lowered below 60 mv/decade?" In *Electron Devices Meeting, 2008. IEDM 2008. IEEE International*, pp. 1–4 (2008).

[14] L. D. Landau & I. M. Khalatnikov, "On the anomalous absorption of sound near a second order phase transition point." *Doklady Akademii Nauk*, **96**), 469–472 (1954).

[15] V. C. Lo, "Simulation of thickness effect in thin ferroelectric films using Landau–Khalatnikov theory." *Journal of Applied Physics*, **94**(5), 3353–3359 (2003).

[16] K. M. Rabe, C. H. Ahn, & J.-M. Triscone, *Physics of Ferroelectrics: A Modern Perspective* (New York: Springer, 2007).

[17] A. Khan, D. Bhowmik, P. Yu, S. Joo Kim, X. Pan, R. Ramesh, & S. Salahuddin, "Experimental evidence of ferroelectric negative capacitance in nanoscale heterostructures." *Applied Physics Letters*, **99**(11), 113501–113503 (2011).

[18] H. N. Lee, S. M. Nakhmanson, M. F. Chisholm, H. M. Christen, K. M. Rabe, & D. Vanderbilt, "Suppressed dependence of polarization on epitaxial strain in highly polar ferroelectrics." *Physical Review Letters*, **98**(21), 217602 (2007).

[19] K. Hellwege & A. M. Hellwege, *Numerical Data and Functional Relationships in Science and Technology New Series*, vol. **3** (Berlin: Springer, 1969).

[20] K. Hellwege & A. M. Hellwege, *Numerical Data and Functional Relationships in Science and Technology New Series*, vol. **16a** (Berlin: Springer, 1981).

[21] H. W. Jang, A. Kumar, S. Denev et al., "Ferroelectricity in strain-free $SrTiO_3$ thin films." Physics Review Letters, **104**, 197601 (2010).

[22] A. I. Khan, C. W. Yeung, C. Hu, & S. Salahuddin, "Ferroelectric negative capacitance MOSFET: capacitance tuning & antiferroelectric operation." In Electron Devices Meeting (IEDM), 2011 IEEE International, pp. 11–13 (2011).

[23] G. A. Salvatore, A. Rusu, & A. M. Ionescu, "Experimental confirmation of temperature dependent negative capacitance in ferroelectric field effect transistor." Applied Physics Letters, **100**(16), 163504 (2012).

[24] C. W. Yeung, A. I. Khan, A. Sarker, S. Salahuddin, & C. Hu, "Low power negative capacitance FETs for future quantum-well body technology." In VLSI Technology, Systems, and Applications (VLSI-TSA), 2013 International Symposium on, pp. 1–2 (2013).

[25] B. Radisavljevic, A. Radenovic, J. Brivio, V. Giacometti, & A. Kis, "Single-layer MOS2 transistors." Nature Nanotechnology, **6**(3), 147–150 (2011).

[26] Y. Yoon, K. Ganapathi, & S. Salahuddin, "How good can monolayer MOS2 transistors be?" Nano Letters, **11**(9), 3768–3773 (2011).

[27] G. A. Salvatore, D. Bouvet, & A. M. Ionescu, "Demonstration of subthreshold swing smaller than 60mv/decade in Fe-FET with P(VDF-FrFE)/SiO_2 gate stack." In Electron Devices Meeting, 2008. IEDM 2008. IEEE International, pp. 1–4 (2008).

[28] A. Rusu, G. Salvatore, D. Jimenez, & A.-M. Ionescu, "Metal-ferroelectric-metal-oxide-semiconductor field effect transistor with sub-60mv/decade subthreshold swing and internal voltage amplification." In Electron Devices Meeting (IEDM), 2010 IEEE International, pp. 16.3.1–16.3.4, (2010).

[29] J. Li, B. Nagaraj, H. Liang, W. Cao, C. Lee, R. Ramesh et al., "Ultrafast polarization switching in thin-film ferroelectrics." Applied Physics Letters, **84**(7), 1174–1176 (2004).

[30] L.-Q. Chen, "Phase-field method of phase transitions/domain structures in ferroelectric thin films: a review." Journal of the American Ceramic Society, **91**(6), 1835–1844 (2008).

[31] Y. Li, L. Cross, & L. Chen, "A phenomenological thermodynamic potential for $BaTiO_3$ single crystals." Journal of Applied Physics, **98**(6), 064101–064101 (2005).

[32] K. Ashraf & S. Salahuddin, "Phase field model of domain dynamics in micron scale, ultrathin ferroelectric films: application for multiferroic bismuth ferrite." Journal of Applied Physics, **112**(7), 074102 (2012).

[33] D. D. Fong, G. B. Stephenson, S. K. Streiffer et al., "Ferroelectricity in ultrathin perovskite films." Science, **304**(5677), 1650–1653 (2004).

[34] M. Dawber, C. Lichtensteiger, M. Cantoni et al., "Unusual behavior of the ferroelectric polarization in $PbTiO_3$/$SrTiO_3$ superlattices." Physical Review Letters, **95**(17), 177601 (2005).

[35] D. Tenne, A. Bruchhausen, N. Lanzillotti-Kimura et al., "Probing nanoscale ferroelectricity by ultraviolet Raman spectroscopy." Science, **313**(5793), 1614–1616 (2006).

[36] K. Ishikawa, K. Yoshikawa, & N. Okada, "Size effect on the ferroelectric phase transition in $PbTiO_3$ ultrafine particles." Physical Review B, **37**(10), 5852 (1988).

[37] R. K. Jana, G. L. Snider, & D. Jena, "On the possibility of sub 60 mv/decade subthreshold switching in piezoelectric gate barrier transistors." Physica Status Solidi (C) (2013).

[38] H. Then, S. Dasgupta, H. Radosavljevic et al., "Experimental observation and physics of "negative" capacitance and steeper than 40mv/decade subthreshold swing in $Al_{0.83}In_{0.17}N$/ AlN/GaN MOS-HEMT on SiC substrate." In Electron Devices Meeting (IEDM), 2013 IEEE International, p. 4.5 (2013).

[39] L. Li, C. Richter, S. Paetel, T. Kopp, J. Mannhart, & R. Ashoori, "Very large capacitance enhancement in a two-dimensional electron system." Science, **332**(6031), 825–828 (2011).

[40] T. Kopp & J. Mannhart, "Calculation of the capacitances of conductors: Perspectives for the optimization of electronic devices." *Journal of Applied Physics*, **106**(6), 064504 (2009).

[41] R. McKee, F. Walker, & M. Chisholm, "Crystalline oxides on silicon: the first five monolayers." *Physical Review Letters*, **81**(14), 3014 (1998).

[42] J. Haeni, P. Irvin, W. Chang, R. Uecker, P. Reiche, Y. Li, S. Choudhury, W. Tian, M. Hawley, B. Craigo *et al.*, "Room-temperature ferroelectricity in strained $SrTiO_3$." *Nature*, **430**(7001), 758–761 (2004).

[43] R. Contreras-Guerrero, J. Veazey, J. Levy, & R. Droopad, "Properties of epitaxial $BaTiO_3$ deposited on GaAs." *Applied Physics Letters*, **102**(1), 012907 (2013).

[44] C. Dubourdieu, J. Bruley, T. M. Arruda *et al.*, "Switching of ferroelectric polarization in epitaxial $BaTiO_3$ films on silicon without a conducting bottom electrode." *Nature Nanotechnology*, **8**(10), 748–754 (2013).

[45] S. Mueller, J. Mueller, A. Singh, S. Riedel, J. Sundqvist, U. Schroeder, & T. Mikolajick, "Incipient ferroelectricity in Al-doped HfO_2 thin films." *Advanced Functional Materials*, **22**(11), 2412–2417 (2012).

[46] T. Boscke, J. Muller, D. Brauhaus, U. Schroder, & U. Bottger, "Ferroelectricity in hafnium oxide thin films." *Applied Physics Letters*, **99**(10), 102903 (2011).

Section II

Tunneling devices

5 Designing a low-voltage, high-current tunneling transistor

Sapan Agarwal and Eli Yablonovitch

5.1   Introduction

Tunneling field effect transistors (TFETs) have the potential to achieve a low operating voltage by overcoming the thermally limited sub-threshold swing voltage of 60 mV/dec [1], but results to date have been unsatisfying. The low-voltage operation is parameterized by the voltage required to obtain a $10\times$ change in output current, called the sub-threshold swing voltage, S. The best reported sub-threshold swing voltage has been measured at a low current density of ~1 nA/µm, but unfortunately becomes significantly larger as the current increases. When trying to design a new low-voltage switch to replace the transistor, there are three major requirements to be fulfilled:

- The sub-threshold swing voltage needs to be much steeper than 60 mV/dec and ideally only a few millivolts per decade to reduce the operating voltage.
- A large on/off ratio of around $10^6/1$ is needed to suppress leakage currents.
- A high conductance density around 1 mS/µm (or 1 mA/µm at 1 V) is needed so that the switch can be significantly smaller than the wire that it drives while maintaining a high speed.

While devices have been built that meet one or two of the three requirements, to date no logic switch meets all three requirements [1, 2]. No one has achieved a steep sub-threshold swing voltage at a high conductance.

To understand this, we first consider a simple tunneling diode in Sections 5.2–5.6 to understand the essential physics of tunneling and then in Sections 5.7–5.9 we consider the additional complexities of building a full transistor. In TFETs the challenge is complicated by the existence of two switching mechanisms. The gate voltage can be used to modulate the tunneling barrier thickness and thus the tunneling probability [3–6] as shown Fig. 5.1(a) and (b). The thickness of the tunneling barrier can be controlled by changing the electric field in the tunneling junction. Alternatively, it is also possible use energy filtering or density of states switching as illustrated in Fig. 5.1(c) and (d). If the conduction and valence band do not overlap, no current can flow. Once they do overlap, current can flow.

In Section 5.2 we describe the tunneling barrier thickness modulation mechanism. In Section 5.3 we look at the energy filtering switch or density of states switch. After presenting the two switching mechanisms, in Section 5.4 we analyze the existing device data which show that experimental performance is still far worse than the Boltzmann

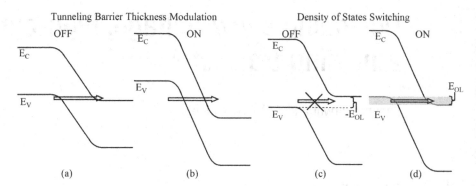

Fig. 5.1 The two different methods for achieving a steep tunneling transition are illustrated. First (a, b), the thickness of the tunneling barrier can be changed by changing the gate voltage and thus the electric field across the tunneling junction. Second (c, d), the alignment of the conduction and valence band can be used to cut off the available states for current to flow into.

limit 60 mV/dec. This is not even close to achieving a steep sub-threshold swing voltage at the current densities of interest. We propose some solutions to achieve better steepness in Section 5.5. To fulfill all three switch requirements, we introduce in Section 5.6 the benefits of quantum confinement or dimensionality. Up until this point in the chapter, the switch has been analyzed with respect to its 2-terminal properties. In Section 5.7, we consider the voltage, sub-threshold swing, and conductance of a full TFET. In Section 5.8 we analyze the relatively poor gate efficiency which leads to additional unfortunate trade-offs, and in Section 5.9 we consider what additional effects can limit the TFET performance.

5.2   Tunneling barrier thickness modulation steepness

First we consider the tunneling barrier thickness modulation mechanism. Applying a voltage bias across a tunneling junction can modulate the tunneling barrier thickness and thus the tunneling probability [3–6]. This is illustrated in Fig. 5.1(a) and (b). The thickness of the tunneling barrier can be controlled by changing the electric field in the tunneling junction. The difficulty in using this method is that at high conductivities there is already a large electric field across the tunneling junction and so the voltage bias cannot control the barrier width effectively. This results in a poor sub-threshold swing voltage at high conductivities. Consequently, we will now show that the tunnel barrier thickness modulation mechanism is incapable of achieving a steep sub-threshold swing voltage (S) at the required high current density.

To estimate how steep of a turn on this can give, we need to determine how many millivolts change in potential across the barrier, ϕ, it takes to change the tunneling probability, T, by a decade. Consequently, we define

$$\frac{1}{S_{\text{tunnel}}} \equiv \frac{d\log(\mathsf{T})}{d\varphi}, \tag{5.1}$$

where S_{tunnel} is the tunnel swing voltage in mV/dec resulting from tunneling barrier thickness modulation. The tunneling probability, T, is [7]

$$\mathsf{T}(\vec{F}) = \exp\left(\frac{-\pi \left(m^*_{tunnel}\right)^{1/2} E_G^{3/2}}{2\sqrt{2}\hbar qF}\right) \equiv \exp\left(\frac{-\alpha}{F}\right). \qquad (5.2)$$

For simplicity, we will assume that the electric field across the tunneling junction, F, is constant and equal to the peak electric field. The effective mass for tunneling [1, 8, 9] is m^*_{tunnel},[1] and E_G is the bandgap. All of the parameters can be collected into a single constant, α. Regardless of the exact shape of the barrier, there will be a constant α such that $\mathsf{T} = \exp(-\alpha/F)$. Combining Eqs. (5.1) and (5.2) gives

$$\frac{1}{S_{tunnel}} = \left(\log(e) \times \alpha \times \frac{1}{|\vec{F}|^2} \times \frac{d\vec{F}(\varphi)}{d\varphi}\right) = \left|\log(\mathsf{T}) \times \frac{dF(\varphi)/d\varphi}{F}\right|. \qquad (5.3)$$

To simplify, we solved Eq. (5.2) for F in terms of $\log(\mathsf{T})$.

Next we need to evaluate $F/(dF(\varphi)/d\varphi)$. For a doped p-n junction the potential will be parabolic, and so

$$\frac{F}{dF(\varphi)/d\varphi} = 2\varphi. \qquad (5.4)$$

In a MOSFET channel the voltage typically decays exponentially and is set by a screening length. This results in

$$\frac{F}{dF(\varphi)/d\varphi} = \varphi. \qquad (5.5)$$

In transistor structures such as a bilayer TFET [10, 11], the electric field is a constant and determined by the bias across the gates. Consequently,

$$\frac{F}{dF(\varphi)/d\varphi} = \varphi. \qquad (5.6)$$

---

[1] The tunneling mass can be computed from [8]:

$$m^*_{tunnel} = 2\left(\frac{1}{m^*_{e,z}} + \frac{1}{m^*_{h,z}}\right)^{-1}.$$

The WKB model and reduced mass work well in InAs where there are carriers in the conduction band tunneling to a single valence band. However, in silicon and germanium the band gap is indirect and there are many interacting bands and so the WKB model breaks down [9]. Consequently, we use an experimentally fitted tunneling effective mass derived in [1]. While in [1] a single-band tunneling model was used, we used a two-band tunneling model and consequently we need to adjust the mass accordingly:

$$m^*_{2\text{-band}} = \left(\frac{2\sqrt{2}}{\pi} \times \frac{4\sqrt{2}}{3}\right)^2 m^*_{1\text{-band}}.$$

This gives $m^*_{tunnel} = 0.043$ in InAs and 0.46 in Si.

In the best case, $F/(dF(\varphi)/d\varphi) = \varphi$. This gives

$$S_{\text{tunnel}} \approx \left| \frac{\varphi}{\log(\top)} \right|. \tag{5.7}$$

As we can see from Eq. (5.7), the lower the tunneling probability, the steeper the sub-threshold swing voltage. This simple equation is likely to be the explanation of all the experimental steep sub-threshold swing voltages that have been measured at extremely low current densities to date [3, 5, 12]! Since the steepness gets worse at high tunnel probability or higher currents, a steep sub-threshold swing voltage at low currents is insufficient for making a practical logic switch. For a reasonable on-state conductance, the tunneling probability should typically be >1%.

To align the conduction and valence bands, the on-state φ must be equal to at least the bandgap of the semiconductor. Consequently, the tunnel swing voltage S_{tunnel} is too large for different semiconductors. For $\top = 1\%$ in silicon, $S_{\text{tunnel}} = 560\,\text{mV/dec}$ at the on-state. In InAs, $S_{\text{tunnel}} = 177\,\text{mV/dec}$. These are worse than Boltzmann. Clearly, controlling the barrier thickness of a homojunction will not give a sub-threshold swing voltage steeper than 60 mV/dec at high current densities.

To get a sub-threshold swing voltage steeper than 60 mV/dec while maintaining a high on-state current, we need to reduce φ_s to less than 120 mV. This means that we need an effective tunneling barrier height less than 120 meV. This can be achieved using a type II heterostructure. Unfortunately, a small effective bandgap requires a steep band edge density of states otherwise the current will pass through the band tail states and never see the barrier. As we will see in the next section, there are states below the band edge that prevent the tunneling junction from fully turning off. At that point the switching becomes controlled by the energy filtering mechanism. Consequently, modulating the thickness of the tunneling barrier will never give a steep sub-threshold swing at high current densities.

5.3   Energy filtering switching mechanism

It is possible to use energy filtering as a switching mechanism. This is also called density of states switching. The energy filtering switch is illustrated in Fig. 5.1(c) and (d). If the conduction and valence band do not overlap, no current can flow. Once they do overlap, current can flow. An ideal density of states switch would be designed to switch abruptly from zero-conductance to the desired on conductance when the conduction and valence band overlap, thus displaying zero sub-threshold swing voltage [13]. Unfortunately, the band edges are not perfectly sharp and so there is a finite density of states extending into the bandgap.

In order to determine how steep the energy filter switching mechanism is, we need to determine how steep the electronic band edges are. While science has good knowledge on the magnitude of semiconductor bandgaps, there is not much information regarding the sharpness of the band edges. Although there are no good direct measurements of the band edge density of states, we can infer it from optical [14] and electronic measurements [15].

Typically the band edge density of states falls off exponentially below the band edge. We can parameterize this fall off with the term S_{DOS}, which represents how many millivolts you need to go below the band edge to reduce the density of states by a decade. Below the bandgap, the optical absorption coefficient also falls of exponentially and is called the Urbach tail [14]. In intrinsic GaAs, the absorption falls off at 17 meV/dec [16]. In intrinsic Si absorption falls off at 23 mV/dec [17]. For electrons, we can hope to see a similar limit on the band edge steepness, S_{DOS}. This may seem promising, but such a steep result has not been vindicated by electrical transport measurements.

Electrically measured joint density of states have generally indicated a steepness >90 mV/dec, unlike the intrinsic optical Urbach measurements, which are <60 mV/dec in good semiconductors. We attribute this broadening to the spatial inhomogeneity and on heavy doping that appears in real devices. Effectively, there are many distinct channel thresholds in a macroscopic device, leading to threshold broadening. Fortunately, this can be ameliorated. We can see this from the optical absorption in doped GaAs. When GaAs is doped with Si to 2×10^{18}/cm^3 the absorption falls off at a rate of 30 meV/dec [16]. If the doping is further increased to 10^{20}/cm^3, the absorption falls off at a rate worse than 60 meV/dec [18]. This means that if a tunnel switch is heavily doped, it will be unable to employ the density of states energy filtering mechanism to achieve a sub-threshold swing voltage smaller than 60 mV/dec! Furthermore, in the doped optical absorption measurements, the band edge density of states is reduced by the free carriers screening potential fluctuations [19, 20]. Unfortunately, in the depletion region of a TFET, there are no free carriers to screen the potential variations and so the band tails will be even worse in electronic devices.

5.3.1 Minimum effective bandgap

In addition to limiting the sub-threshold swing voltage, the band edge density of states can limit the on/off ratio if the effective bandgap (tunneling barrier height) is too small. If we want a particular on/off ratio, the barrier height in the off-state, $E_{g,eff}$, must be large enough to suppress the band edge density of states, S_{DOS}, by that on/off ratio. Consequently, we get the following limit:

$$E_{g,eff} \geq S_{DOS} \times \log(I_{on}/I_{off}). \tag{5.8}$$

For instance, if we want to use tunneling barrier width modulation we need a barrier height of less than 120 mV. For six decades of on/off ratio, S_{DOS} must be steeper than $120/6 = 20$ mV/dec, and this has not yet been achieved. Furthermore, the steepest turn-on will come from the band edge rather than the tunneling barrier thickness modulation if we had $S_{DOS} = 20$ mV/dec. Consequently, modulating the thickness of the tunneling barrier will never give a steep sub-threshold swing at high current densities.

5.4 Measuring the electronic transport band edge steepness

To interpret electrical transport measurements, we need to look at the absolute conductance, I/V, versus the bias voltage, V, in a tunneling diode. The absolute conductance is proportional to the tunneling joint density of states. This is discussed in detail in [15]. Investigating the electronic steepness in a 2-terminal p-n junction measurement allows the band alignment to be controlled directly, without concern for gate efficiency. Both barrier thickness modulation and density of states switching change the resistance of the tunneling junction. Consequently, we need to measure the change in resistance or conductance with bias rather than the change in current with bias. This can be seen from following model for the tunneling current:

$$I \propto \int (f_C - f_V) \times \top \times D_J(E) \times \partial E, \qquad (5.9)$$

where $f_C - f_V$ is the difference between the Fermi occupation probabilities on the p- and n-sides, $\top$ is the tunneling probability across the junction, and $D_J(E)$ is the joint density of states between the valence band on the p-side and the conduction band on the n-side. We are interested in measuring the voltage dependence of the tunneling joint density of states, $\top \times D_J(E)$, in the integrand of Eq. (5.9). Since $\int (f_C - f_V) \times \partial E = qV$, dividing the current by the voltage approximately eliminates the effect of the Fermi levels [15]. In a 3-terminal transistor measurement the source/drain bias would control the Fermi levels while the gate bias would control $\top \times D_J(E)$. This allows us to use a 2-terminal current voltage measurement to determine the joint density of states of a tunnel junction. Consequently, a 2-terminal measurement in a diode can be used to interpret the steepness of the tunnel joint density of states, without being limited by the gate efficiency in a TFET. We will now interpret some specific cases from the experimental literature by plotting I/V versus V.

First we consider the 2-terminal current–voltage characteristics of an InAs/AlSb/ $Al_{0.12}Ga_{0.88}As$ heterojunction backward diode [21]. In this diode the tunneling barrier thickness is fixed by the AlSb thickness and so the tunneling is entirely due to the density of states overlapping. The I–V curves are shown in Fig. 5.2(a) and the absolute conductance is in Fig. 5.2(b). As seen in Fig. 5.2(a), the current diverges on a semilog plot at $V = 0$, preventing direct interpretation. Likewise, the differential conductance diverges on a semilog plot at the Esaki peak. Thus a current or a differential conductance plot does not give us the information we want. By contrast, in Fig. 5.2(b), the absolute conductance, I/V, smoothly varies from reverse bias, through the origin, to forward bias.

The conductance is proportional to tunneling joint density of states, which can be parameterized by the inverse of the semilog slope of the conductance, called the semilog conductance swing voltage. This is equivalent to the steepness of the tunneling joint density of states in mV/dec shown by the inverse slope of the diagonal line in Fig. 5.2(b). In the figure, the semilog swing voltage of the absolute conductance is 98 mV/dec, and it measures the tunneling joint density of states. It has one of the steepest experimentally measured tunneling joint density of states. This is likely to be due to the type III band alignment permitting low doping levels ~$1.4 \times 10^{17}/cm^3$ near the junction region.

In Fig. 5.2(c), we consider a germanium backward diode [22]. This has the steepest semilog conductance swing voltage of 92 mV/dec that we could find in the literature. Next, in Fig. 5.2(d), we show the current and conductance for an InAs homojunction

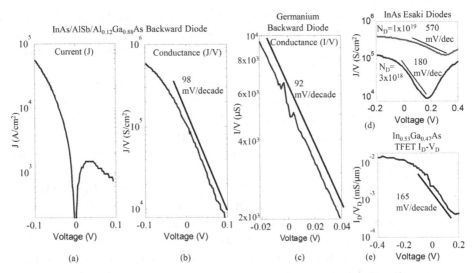

Fig. 5.2 (a)–(d) Graphs of current and conductance for (a, b) an InAs/AlSb/Al$_{0.12}$Ga$_{0.88}$As heterojunction diode [21], (c) a germanium diode [22], and (d) InAs diodes [23]. At $V = 0$, the current diverges on a log plot and so the logarithmic slope is meaningless. Fortunately, the conductance is proportional to the tunneling density of states. (e) Graph of $G = I_D/V_D$ versus V_D for an In$_{53}$Ga$_{0.47}$As TFET [24]. The measured sub-threshold swing voltage is 216 mV/dec while the semilog conductance swing voltage is 165 mV/dec. Since the I_D–V_D characteristic is not limited by the gate oxide, it reflects the junction's steeper intrinsic tunneling properties.

diode at two different doping levels ($N_A = 1.8 \times 10^{19}$; $N_D = 3 \times 10^{18}$ and 1×10^{19}) [23]. When the n-side doping is decreased from 1×10^{19}/cm^3 to 3×10^{18}/cm^3 the absolute conductance I/V swing improves from 570 mV to only 180 mV/dec, but still falls far short of our goal. This clearly illustrates that smearing the band edge by doping is very bad, but that even the lower doped samples perform poorly.

We can apply the same 2-terminal analysis to 3-terminal TFETs. In a TFET we can either make a 2-terminal source/drain measurement, or a 3-terminal I_D–V_G measurement. In a 3-terminal measurement, the sub-threshold swing voltage will not give the tunneling joint density of states, since some voltage is lost across the gate oxide.

We reduce gate issues by doing a 2-terminal measurement. If the critical tunneling junction is at the source/channel junction, we need to fix the $V_{gate} - V_{drain}$ voltage while measuring source/drain current versus source/drain voltage (I_D–V_S). Since we want to measure the band edge mechanism without being confounded by the gate modulation mechanism, we leave the $V_{gate} - V_{drain}$ voltage fixed. As the gate potential will have a strong influence on the channel potential, the drain will not be able to effectively control the source/channel junction and so it is best to vary V_{GS}. If on the other hand the critical tunneling is at the channel/drain junction we need to fix the $V_{gate} - V_{source}$ voltage while measuring source/drain current versus source/drain voltage (I_D–V_D).

In Fig. 5.2(e), we fix the $V_{gate} - V_{source}$ voltage while measuring source/drain current versus source/drain voltage (I_D–V_D) of an In$_{0.53}$Ga$_{0.47}$As TFET with a poor gate oxide [24]. Effectively, this is a 2-terminal measurement on a 3-terminal device. The semilog conductance swing voltage is 165 mV/dec. The corresponding 3-terminal measurement shows a worse sub-threshold swing voltage of ~216 mV/dec. This shows the value of a proper 2-terminal measurement to analyze a TFET's potential performance when the gate oxide is poor quality.

In TFETs, sub-threshold swing voltages of less than 60 mV/dec have been measured, but only at extremely low current densities of ~1 nA/μm. In backward diodes and Esaki diodes the low current densities have been obscured by trap-assisted tunneling and forward leakage current. Moreover, at current densities measured in tunneling diodes, tunneling barrier width modulation is weak, and I/V versus V reflects the band edge density of states.

Measuring the steepness of the conductance, I/V, in mV/dec of a tunneling diode, or of a TFET source/drain $I–V$, will give the tunneling joint density of states. This tells us the potential sub-threshold swing voltage that we can expect from a TFET based on that tunneling junction at reasonable current densities. Looking at the best tunneling diodes to date, we find that they all have a semilog conductance swing voltage worse than 60 mV/dec. This is because they are macroscopic devices with considerable threshold inhomogeneity leading to multiple channels, each with a different threshold, smearing the sub-threshold swing voltage. In the next section we suggest some remedies.

5.5   Correcting spatial inhomogeneity

So far the prospects of designing a steep tunneling junction at high current densities seems quite bleak. Modulating the thickness of the tunneling barrier will not work and a band edge density of states steeper than 60 mV/dec has not been measured in electrical transport. In order to get better performance, novel geometries that provide spatial homogeneity, eliminate doping, and promote atomic perfection are needed. Modulation doping that moves the dopants away from the tunneling junction may help, but the ideal geometry would use electrostatic doping through gates. This can be achieved through the bilayer [10, 11] structure in Fig. 5.3(a), or alternatively, one could try to make lateral double gate structure as shown in Fig. 5.3(b). Additional structures that remove doping from the tunneling junction and preserve the material quality are needed.

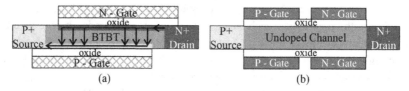

Fig. 5.3 (a) A bilayer TFET. By applying opposite biases to both the n- and p-gates, both an electron and a hole channel form in the undoped channel, allowing band to band tunneling as shown. (b) A lateral TFET with both p- and n-gates. The tunneling junction at the center of the channel is formed electrostatically without dopants. By eliminating doping in both stuctures, a steep density of states can be achieved.

In addition to eliminating the doping, we need to eliminate any other sources of spatial inhomogeneity. This can come from rough heterojunctions, atomic thickness fluctuations, or any other non-ideality. Small devices that encompass a single quantum wavefunction will still be inhomogeneous from device to device, but a single device is more likely to show the intrinsic energy sharpness which has not yet been measured by electrical transport.

Alternatively, the electrical transport measurements can be performed at low temperature, sharpening up the individual energies, and providing an opportunity to measure the discrete levels of the inhomogeneous distribution. Such a low-temperature device would not be practical for use as a switch, but would provide scientific information about the inhomogeneities.

Part of the inhomogeneity arises from thickness fluctuations in conventional quantum wells. There have now emerged monolayer semiconductors such as MoS_2 that can precisely define the layer thickness, hopefully eliminating the problems of spatial homogeneity.

5.6   p-n junction dimensionality

To further improve the performance of a tunneling junction, we need to maximize the on-state conductance and minimize the overdrive voltage. (The overdrive voltage is the extra voltage needed beyond the sub-threshold regime to get the desired conductance.) This is strongly dependent on the actual geometry of the tunneling junction. Fortunately, confining the carriers in the tunneling direction provides four benefits that help achieve this [25]:

1.   The carrier velocity is increased and set by the confinement energy.
2.   A higher electron energy can increase the tunneling probability.
3.   Shrinking the region in which the electron is allowed will cause a greater percentage of the electron density to be in the barrier and thus the tunneling wave-function overlap increases.
4.   Reducing the dimensionality results in a sharper density of states which reduces the overdrive voltage needed to get the full conductance.

Whenever specifying a p-n junction it is also necessary to specify the dimensionalities of the respective p- and n-regions. In Fig. 5.4 we show nine different possible p-n junction dimensional combinations. In the following sections we analyze each of these devices and investigate which are the most promising for adaptation into a TFET.

5.6.1   1D–1D$_{end}$ junction

A 1D–1D$_{end}$ p-n junction describes tunneling within a nanowire [26] or carbon nanotube [27] junction as schematically represented in Fig. 5.4(a). Tunneling is occurring from the valence band on the p-side to the conduction band on the n-side. For a

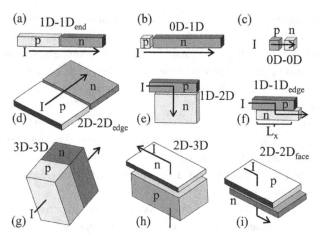

Fig. 5.4 We identify nine distinct dimensionality possibilities that can exist in p-n junctions. Each of the different tunneling p-n junction dimensionalities shown has a different effect on characteristics.

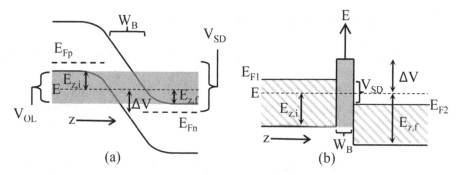

Fig. 5.5 (a) Energy band diagram for the tunnel p-n junction showing that the relevant voltage is the overlap voltage and not the source/drain voltage. (b) Energy band diagram for a typical 1D quantum of conductance showing that the relevant voltage is the source/drain voltage.

transistor, the gate is not shown as there are many possible gate geometries. The band diagram across this junction is given by Fig. 5.5(a).

In analyzing all of the devices, we consider a direct gap semiconductor with a small gate bias. In particular we consider the regime near the band overlap turn-on where a small change in voltage ($k_B T/q$ or less) will result in a large change in the density of states but only a small change in the tunneling barrier thickness. Consequently, we assume that the tunneling probability is roughly a constant, $\top$, and will not change significantly for small changes in the control voltage. Initially, we also assume the tunneling probability is independent of energy and can be given by an energy averaged tunneling probability. We will discuss the energy dependence $\top(E)$ in the next section. The tunneling probability, $\top$, is the probability that an electron in a given mode tunnels through the barrier and ends up on the other side. It is often given by a WKB approximation: $\top = \exp(\int k dx)$.

We also define $V_{OL} = qE_{OL}$ to be the overlap voltage between the conduction and valence bands as shown in Fig. 5.5(a). In order to keep the analysis as simple and general as possible we will use the band overlap voltage, V_{OL}, in all of the analyses instead of V_G or V_{SD}. Making these approximations allows us to focus on the effects of changing the dimensionality and discover some new insights into tunneling in reduced dimensionality systems.

The 1D–1D$_{end}$ current can be derived as an adaptation of the normal quantum of conductance, $2q^2/h$, approach. The band diagram for the typical quantum of conductance is shown in Fig. 5.5(b). The current flow is controlled by the difference in Fermi levels, which is V_{SD}, as shown. Current is given by charge × velocity × density of 1D states. Since the energy dependence of the velocity and 1D density of states exactly cancel, we get the quantum of conductance: $I = (2q^2/h) \times V_{SD} \times \mathsf{T}$.

Now to properly consider the transition from conduction band to valence band, we look at the band diagram given in Fig. 5.5(a). Initially, we consider the situation where the valence band on the p-side of the junction is completely full and the conduction band on the n-side is completely empty. This would correspond to non-degenerate doping, $V_{SD} > k_B T/q$ and $V_{SD} > V_{OL}$.

As shown in Fig. 5.5(a), the band edges cut off the number of states that can contribute to the current. Unlike a single-band 1D conductor, the overlap voltage V_{OL} determines the amount of current that can flow. Consequently, it is V_{OL} and not V_{SD} that controls the current:

$$I_{1D-1D} = \frac{2q^2}{h} \times V_{OL} \times \mathsf{T}. \tag{5.10}$$

5.6.1.1 Small source/drain bias limit

Instead of assuming that there is a large bias across the tunneling junction, we can also consider the opposite limit where $V_{SD} < 4k_B T/q$. To account for the small voltage we need to multiply by the Fermi occupation difference ($f_C - f_V$). In this small bias regime everything of interest occurs within a $k_B T$ or two of energy. Consequently, we can optionally Taylor expand $f_C - f_V$:

$$f_{C,V} = \frac{1}{e^{(E-E_{fC,V})/k_B T} + 1}, \tag{5.11}$$

$$f_C - f_V \approx \frac{(E_{fC} - E_{fV})}{4k_B T} \approx \frac{qV_{SD}}{4k_B T}. \tag{5.12}$$

Thus the ultimate effect of the small differential Fermi occupation factors is to multiply the low-temperature current by the factor $qV_{SD}/4k_B T$. We can therefore write a conductance for small source/drain biases:

$$G_{1D-1D} = I_{1D-1D} \times \frac{q}{4k_B T}, \tag{5.13a}$$

$$G_{1D-1D} = \frac{2q^2}{h} \times \mathsf{T} \times \frac{qV_{OL}}{4k_B T}. \tag{5.13b}$$

This is also true for all of the devices to be considered in the following sections. Thus we will continue to make the approximation that the valence band is full and the conduction band is empty when calculating the potential current flow. The exact integral over the Fermi functions is discussed in Chapter 6 for the 1D–1D$_{end}$ case.

Equation (5.13b) illustrates a fundamental trade-off between switching voltage and switch conductance. Even if the tunneling probability is 1, the conductance will be limited by $qV_{OL}/4k_BT$. For biases less than $4k_BT$, the conductance is reduced due to the thermal distribution of carriers. We can express this as a voltage–resistance product that will be limited to $>2hk_BT/q^3$. The voltage–resistance product says that low-voltage switches inherently have high resistance, while high-conductivity switches will also require high voltage.

5.6.1.2   Fermi's golden rule derivation

The current can be derived in a different manner using the transfer Hamiltonian method [25, 28–31]. We do this as an alternative to employing the more modern channel conductance approach. The transfer Hamiltonian method was first used by Oppenheimer to study the field emission of hydrogen [31]. It was then expanded by Bardeen [28] for tunneling in superconductors, and then the case of independent electrons was considered by Harrison [30]. The transfer Hamiltonian method is just an application of Fermi's golden rule with a clever choice of states and perturbing Hamiltonian. The current density is given by Fermi's golden rule,

$$J = 2q \times \frac{2\pi}{\hbar} \sum_{k_i, k_f} |M_{fi}|^2 \delta(E_i - E_f)(f_C - f_v). \tag{5.14}$$

The calculation of the matrix element M_{fi} is done in [25] and [30] and is given by

$$M_{fi} = \frac{\hbar^2}{2m} \sqrt{\frac{k_{z,f}k_{z,i}}{L_{z,f}L_{z,i}}} \times \sqrt{T} \times \delta_{k_x,i,k_x,f} \delta_{k_y,i,k_y,f}. \tag{5.15}$$

In this equation, $k_{\alpha,i}$ and $k_{\alpha,f}$ are the α-component of the wave-vector in the initial and final states respectively. $L_{z,i}$ and $L_{z,f}$ are the lengths along the tunneling direction of the initial and final sides of the junction. Using this method allows us to extend the transfer Hamiltonian approach to the trickier reduced dimensionality cases by simply summing over fewer states. When quantum confinement is used in the tunneling direction, two effects will result in a large matrix element and thus a higher conductance. First, k_z will be set to a large value corresponding to the increased velocity due to confinement. Second, L_z will also be shorter. By shrinking the region in which the electron is allowed, a greater percentage of the electron density is in the barrier and thus the tunneling wavefunction overlap increases.

5.6.2   Energy dependent tunneling probability

A significant energy dependence arises at low energies where the WKB approximation breaks down. The tunneling probability approaches zero as the energy approaches zero.

At small energies relative to the barrier height, the wave function begins to approach infinite barrier boundary conditions, where it is almost zero amplitude at the barrier. Therefore the tunneling probability has to approach zero at low energy.

The 1D band tunneling probability through a rectangular barrier, as shown in Fig. 5.5(b), can be found by matching boundary conditions using propagation matrices [32], and is given by

$$\mathsf{T} = \frac{1}{\frac{\left(\sqrt{E_{z,i}}+\sqrt{E_{z,f}}\right)^2}{4\sqrt{E_{z,i}E_{z,f}}} + \frac{(E_{z,i}+\Delta V)(E_{z,f}+\Delta V)}{4\Delta V\sqrt{E_{z,i}E_{z,f}}}\sin h^2(\kappa W_B)}. \tag{5.16}$$

We considered the situation where the initial and final energies, $E_{z,i}$ and $E_{z,f}$ respectively, are different as shown in Fig. 5.5(b). The barrier height relative to the tunneling energy, E, is given by ΔV. The barrier width is W_B. The wavevector in the tunneling barrier is given by: $\kappa = \sqrt{2m\Delta V}/\hbar$. For a typical barrier the sinh term will be large and so we get

$$\mathsf{T} \approx \frac{16\Delta V\sqrt{E_{z,i}E_{z,f}}}{(E_{z,i}+\Delta V)(E_{z,f}+\Delta V)}\exp(-2\kappa W_B). \tag{5.17}$$

At small energies, $E \ll \Delta V$, so we get

$$\mathsf{T} \approx \frac{16\sqrt{E_{z,i}E_{z,f}}}{\Delta V}\exp(-2\kappa W_B). \tag{5.18}$$

Thus we see that there is an energy-dependent pre-factor to the WKB exponential. As the energy goes to zero, the tunneling probability goes to zero.

Since the exact form of the tunneling probability will be dependent on the barrier shape, in the following sections we will continue to assume that an averaged tunneling probability, $\langle \mathsf{T}(E)\rangle$, can be used when calculating the current. This still captures the key voltage dependence. The main result to remember is that at small overlap voltages the initial turn-on will be limited by the tunneling turn-on.

In the band-to-band tunneling case, the probability will still be given by Eq. (5.17) if we assume a rectangular barrier. The only change is that $E_{z,i}$ is the hole energy and $E_{z,f}$ is the electron energy as shown in Fig. 5.5(a). This can be found by computing the tunneling matrix element used in Eq. (5.14). V_{OL} represents the available kinetic energy. When there is no confinement in the tunneling direction, such as the 1D–1D$_{end}$ case, the tunneling probability, Eq. (5.18), will be maximized when $E_{z,i} = E_{z,f} = qV_{OL}/2$:

$$\mathsf{T}_{max} \approx \frac{8V_{OL}}{\Delta V}\exp(-2\kappa W_B). \tag{5.19}$$

This means that the tunneling probability will be linearly proportional to V_{OL} at turn-on and a finite V_{OL} of $\Delta V/8$ is needed for the pre-factor to reach 1. Once the prefactor is 1, we assume the WKB approximation is valid.

There is also a secondary energy dependence that affects the tunneling probability. In the three-dimensional WKB approximation, a large transverse energy will reduce the tunneling probability. Since the transverse energy is limited by the available overlap voltage, V_{OL}, at threshold the tunneling problem becomes more one-dimensional, and the transverse energy can be neglected. The impact of accounting for the transverse energy on the 1D–1D$_{end}$, 2D–2D$_{edge}$, and 3D–3D cases is discussed in Chapter 7.

5.6.3   3D–3D bulk junction

A 3D–3D junction simply means a p-n junction or heterojunction where there is a bulk semiconductor on either side of the sample. A generalized schematic diagram of the tunneling junction is shown in Fig. 5.4(g).

The 3D bulk current can be derived from a few simple considerations. The junction is a large 2D surface and can be considered to be a 2D array of 1D channels. The 2D array is defined by the transverse k-states that can tunnel. Each 1D channel is equivalent to the 1D–1D case described in the previous section and will conduct with a quantum of conductance times the tunneling probability. The differential current density can therefore be written as

$$\partial I = N_{\perp \text{states}} \times \frac{2q}{h} \times \langle \mathsf{T} \rangle \times \partial E. \tag{5.20}$$

The number of transverse states is the number of k-states within the maximum transverse energy at a given energy and is given by the number of 2D states: $N_{\perp} = (AmE)/(2\pi\hbar^2)$, where A is the area of the tunneling junction. The transverse energy is limited by the closest band edge and peaks in the middle of the overlap. Integrating Eq. (5.20) gives

$$I_{3D-3D} = \frac{1}{2} \left(\frac{Am^*}{2\pi\hbar^2} \times \frac{qV_{OL}}{2} \right) \times \frac{2q^2}{h} V_{OL} \times \langle \mathsf{T} \rangle$$

$$= \text{no. of 2D channels} \times \text{1D conductance.} \tag{5.21}$$

This is the same as taking the appropriate limits of Kane's tunneling theory [7] (except for a factor of $\pi^2/9$).

5.6.4   2D–2D$_{edge}$ junction

A 2D–2D$_{edge}$ junction is shown in Fig. 5.4(d). The derivation of the current is almost identical to the 3D–3D case, except that instead of having a 2D array of 1D channels we now have a 1D array of 1D channels. Therefore the current is given by

$$I_{2D-2D,\,edge} = \frac{2}{3} \left(\frac{L_x \sqrt{m^*}}{\pi\hbar} \times \sqrt{qV_{OL}} \right) \times \left(\frac{2q^2}{h} \times V_{OL} \times \langle \mathsf{T} \rangle \right)$$

$$= \text{no. of 1D channels} \times \text{1D conductance,} \tag{5.22}$$

where L_x is the length of the junction.

5.6.5   0D–1D junction

A 0D to 1D junction represents tunneling from a quantum dot to a nanowire as shown in Fig. 5.4(b). We consider two different 0D–1D systems. First we will assume that there is an electron in the quantum dot and find the rate at which it escapes into the end of a 1D wire. We analyze this junction as building block for the 2D–3D and 1D–2D junctions. To build a real 0D–1D device, we would also need to electrically contact the quantum dot. Therefore we consider a more realistic situation that includes this. This becomes a single electron transistor (SET) as shown in Fig. 5.6.

The rate at which an electron escapes from the quantum dot into a nanowire is given by the field ionization of a single state such as an atom. In Gamow's model of alpha particle decay [33], the particle oscillates back and forth in its well and attempts to tunnel on each complete oscillation. If the dot has a length of L_z along the tunneling direction, the electron will travel a distance of $2L_z$ between tunneling attempts. Its momentum is given by $p_z = mv_z = \hbar k_z$, where $k_z = \pi/L_z$ in the ground state. Using $E_z = \hbar^2 k_z^2 / 2m$, the time between tunneling attempts is $\tau = 2L_z/v_z = h/2E_z$. The tunneling rate per second is $R = (1/\tau) \times \langle \mathsf{T} \rangle$. This can be converted to a current by multiplying by the electron charge and a factor 2 for spin to give

$$I = \frac{4q}{h} \times E_z \times \langle \mathsf{T} \rangle. \tag{5.23}$$

This is the same result that one obtains from the transfer Hamiltonian method.

To include coupling into the dot, we add a second nanowire to supply current, as shown in Fig. 5.6, and form a "single electron transistor" [34]. We assume that the second nanowire has the same tunneling probability/coupling strength to the quantum dot as the original one. Unlike a conventional SET, we want the current to be high enough that we do not see any Coulomb blockade effects. The tunneling event out of the dot follows sequentially after tunneling in. Consequently, the current is cut in half, as follows:

$$I_{0D-1D} = \frac{2q}{h} \times E_z \times \langle \mathsf{T} \rangle. \tag{5.24}$$

As seen in Fig. 5.6(e), the tunneling occurs at a single energy and will result in a sharp turn-on once the bands overlap. This is one of the key benefits of quantum confinement.

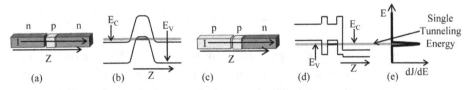

Fig. 5.6 (a) A 0D–1D junction converted into a more realistic 1D single electron transistor (SET) structure. (b) Band diagram corresponding to the SET. (c) An alternative SET structure with a p-type contact. (d) Band diagram corresponding to the alternative SET. (e) All the current is concentrated around a single energy, which allows for a small overlap voltage V_{OL}, and thus a small overdrive voltage V_{OV}.

The current density is concentrated in a narrow energy range which allows for a smaller V_{OL}. This can be contrasted with the 1D–1D$_{end}$ case, Eq. (5.10), where the current flows over the entire energy range corresponding to qV_{OL}. The width of the 0D–1D energy range will be given by the broadening of the energy level in the quantum dot. This broadening can be extrinsically caused by any inhomogeneities in the lattice such as defects, dopants, or phonons. Even without these effects, simply coupling the dot to the nanowires causes a significant amount of broadening. Each contact will broaden the level by γ_0 for a total broadening of $2\gamma_0$ [35]:

$$\gamma_0 = \frac{\hbar}{\tau} = \frac{1}{\pi} \times E_z \times \langle \mathsf{T} \rangle. \tag{5.25}$$

In the limit that $\mathsf{T} \to 1$, the 0D–1D case will become the 1D–1D$_{end}$ case with a perfect quantum of conductance: $I = 2q^2/h \times V_{OL}$. However, in a realistic situation $\mathsf{T} \ll 1$, and so we can use quantum confinement to concentrate the current at a single energy and significantly reduce V_{OL}.

The quantum confinement also has an added benefit of increasing the tunneling probability itself. This can be seen from Eq. (5.18). In the 1D–1D$_{end}$ case, $E_{z,i}$ and $E_{z,f}$ are both limited by V_{OL}. In the 0D–1D case, only $E_{z,f}$ will be limited by V_{OL}. $E_{z,i}$ can be set to a large value by the quantum confinement.

5.6.6   2D–3D junction

A 2D–3D tunneling junction is typical in vertical tunneling junctions where the tunneling occurs from the bulk to a thin confined layer [3]. The thin layer can either be a thin inversion layer or a physically separate material [3, 36, 37]. A generalized schematic diagram of this tunnel junction is shown in Fig. 5.4(h).

The derivation for this case is very similar to the 3D–3D case. As in that section, the junction is a large 2D surface and can be considered to be a 2D array of 1D tunneling problems. However, this case does not represent the typical 1D quantum of conductance. The 1D problem is better described by tunneling from a quantum dot to a nanowire as shown in Fig. 5.4(b).

To find the 2D–3D current, we simply multiply the 0D–1D result, Eq. (5.23), by the number of 2D channels to get a current of

$$I_{2D-3D} = \text{no. of 2D channels} \times \text{1D field ionization}$$

$$I_{2D-3D} = \left(\frac{Am}{2\pi\hbar^2} \times \frac{qV_{OL}}{2}\right) \times \left(\frac{4q}{h} \times E_z \times \langle \mathsf{T} \rangle\right), \tag{5.26}$$

were E_z is the confinement energy of the 2D layer. We have only included transverse states for a transverse energy up to $qV_{OL}/2$. For transverse energies larger than that, no state exists on both sides of the junction with the same total energy. This is the same result that comes from the transfer Hamiltonian method. Compared to the bulk 3D–3D case, confining one side of the junction resulted in the replacement of qV_{OL} with $4E_z$. The quantum confinement can also increase the tunneling probability by fixing the electron energy in the quantum well and thus $E_{z,i}$ in Eq. (5.18).

Current can flow in along the transverse direction as shown in Fig. 5.4(h). Other methods such as tunneling into the quantum well can also be considered for making electrical contact.

5.6.7   1D–2D junction

A 1D–2D junction describes tunneling between the edge of a nanowire and a 2D sheet as shown in Fig. 5.4(e). The derivation for this case is almost identical to the 2D–3D case. The only difference is that instead of a 2D array of 1D tunneling, we now have a 1D array of 1D tunneling. Thus the current is given by

$$I_{1D-2D} = \text{no. of 1D channels} \times \text{1D tunnel ionization}$$

$$I_{1D-2D} = \left(\frac{L_x}{\pi\hbar} \times \sqrt{qm^*V_{OL}}\right) \times \left(\frac{4q}{h} \times E_z \times \langle T \rangle\right). \tag{5.27}$$

Compared to the 2D–2D edge overlap formula, confining one side of the junction results in the replacement of qV_{OL} with $3E_z$ and an increased $\langle T \rangle$ by fixing $E_{z,i}$ in Eq. (5.18).

5.6.8   0D–0D junction

This case represents tunneling from a filled valence band quantum dot to an empty conduction band quantum dot and is schematically represented in Fig. 5.4(c). In order to create a meaningful device, the quantum dots need to be coupled to contacts to pass current into and out of the device. Consequently, we consider the structure in Fig. 5.7. Current will only flow when the confined energy levels in each dot are aligned. This can be seen from Fig. 5.7(d). The two dots will only couple if the density of states in each dot overlaps. This results in an *I–V* curve that resembles a delta-function as shown in Fig. 5.7(e). We can estimate the peak current by considering the coupling strength between each dot and its contact as well as the coupling between dots. For simplicity,

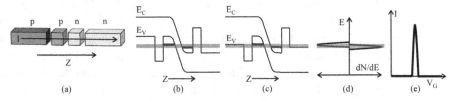

Fig. 5.7 The properties of a 0D–0D junction that is coupled to nanowire contacts are shown. (a) Schematic representation of the junction. (b) Band diagram of the junction. Tunneling only occurs at a fixed energy when the two confined levels overlap. The fixed tunneling energy results in a higher tunneling probability. The sub-threshold swing voltage will be determined by the sharpness of the levels. (c) Alternative band diagram that shows the nanowire band edges aligned with the confined levels. In this case, the sub-threshold swing voltage will be determined by the sharper of the level or the nanowire band edge. Conversely, the tunneling probability will be lower as the energy in the nanowires is low. (d) Tunneling only occurs when the density of states in each dot is aligned. (e) The *I–V* curve resembles a delta function when the levels align.

we will assume that the dots and contacts are symmetric. The coupling strength or broadening due to each contact is given by

$$\gamma_0 = \frac{\hbar}{\tau} = \frac{1}{\pi} \times E_z \times \langle \mathsf{T}_{\text{contact}} \rangle, \tag{5.28}$$

where $\mathsf{T}_{\text{contact}}$ represents the probability of tunneling between the contact and a dot.

The coupling strength between each dot is the matrix element between the dots and is given by Eq. (5.15). Since we have a single level in each dot, we can simplify the matrix element by using $k_z = \pi/L_z$ and $E_z = \hbar^2 k_z^2/2m^*$:

$$|M_{\text{fi,0D-0D}}| = \frac{1}{\pi} \sqrt{E_{z,\text{i}} \times E_{z,\text{f}} \times \langle \mathsf{T} \rangle}, \tag{5.29}$$

where T is the single barrier tunneling probability between the two dots. In order to maximize the current we want all the coupling strengths to be equal, i.e., $|M_{\text{fi}}| = \gamma_0$. Since $|M_{\text{fi}}| \propto \sqrt{\langle \mathsf{T} \rangle}$ it is possible to design the central barrier to have $|M_{\text{fi}}| > \gamma_0$. Unfortunately, doing this will cause the dots to strongly couple and will result in a level splitting that reduces the current. Consequently, we want to design γ_0 to be large and then design $|M_{\text{fi}}| = \gamma_0$. This means that the tunneling rate through each barrier will be the same and is given by $\gamma_0/\hbar$. Since we have a three-step tunneling process the peak current will be given by

$$I_{\text{peak}} \leq \frac{2q}{3\tau} = \frac{2\gamma_0}{3\hbar} = \frac{2}{3} \times \frac{2q}{h} E_z \langle \mathsf{T}_{\text{contact}} \rangle. \tag{5.30}$$

The width of the tunneling peak is given by the broadening of the confined level, $2\gamma_0$. Additional broadening mechanisms such as electron–phonon interactions can further broaden the turn-on and reduce the peak current by smearing out the levels and reducing the coupling strength between the dots. As with the 0D–1D case, in the limit that $\mathsf{T} \to 1$, the 0D–0D case will become the 1D–1D$_{\text{end}}$ case with a perfect quantum of conductance, i.e., $I = 2q^2/h \times V_{\text{OL}}$. However, in a realistic situation $\mathsf{T} \ll 1$, and so we can use quantum confinement to concentrate the current at a single energy and significantly reduce V_{OL}.

The 0D–0D peak current is almost identical to the 0D–1D case from Eq. (5.24) with the exception of a factor of 2/3 as this is now a three-step tunneling process instead of a two-step process. The key difference arises when evaluating $\mathsf{T}_{\text{contact}}$. If we design the 0D–0D system as shown in Fig. 5.7(b) both the initial and final tunneling energies are non-zero and set by the quantum confinement. This means that the prefactor in the tunneling probability in Eq. (5.18) can be close to 1 and so an additional overlap voltage is not needed to increase the tunneling probability. Nevertheless, this also means that the sub-threshold swing voltage will be determined only by the sharpness of the confined energy levels and not the band edge. Alternatively, we can design the 0D–0D system so that the nanowire band edges line up with the confined energy levels as shown in Fig. 5.7(c). In this case, the sub-threshold swing voltage will be determined by the sharper of the band edge or the confined level. However, we will lose the increased tunneling probability as the energy in the nanowire will be low. Overall we see that by using the 0D–0D structure in Fig. 5.7(b) we may be able to have an advantage over 0D–1D through the increased tunneling probability. On the other hand, the delta-function like shape of the *I–V* curve shown in Fig. 5.7(e) may make it difficult to design a conventional logic circuit.

5.6.9   2D–2D$_{face}$ junction

A 2D–2D$_{face}$ junction describes tunneling from one quantum well to another through the face of the quantum well. This can be seen in resonant interband tunnel diodes [38–40]. The junction is schematically represented in Fig. 5.4(i). This is one of the most interesting cases as it is the closest to a step function turn-on.

The step function turn-on can be seen by considering the conservation of transverse momentum and total energy. This is depicted in Fig. 5.8(a). The lower paraboloid represents all of the available states in *k*-space on the p-side of the junction and the upper paraboloid represents the available *k*-space states on the n-side of the junction. In order for current to flow, the initial and final energy and wave-vector *k* must be the same and so the paraboloids must overlap. However, as seen in the right-hand part of Fig. 5.8(a), they can only overlap at a single energy. Furthermore, the joint density of state pairs between valence and conduction band is a constant in energy. Thus the number of state pairs that tunnel is a constant regardless of the overlap energy as seen in Fig. 5.8(b).

The current can be computed by using Fermi's golden rule. Due to the conservation of transverse momentum, each initial state is coupled to only one final state. Current can flow into each quantum state along the quantum well, or through the face of the quantum well. We simply need to sum Eq. (5.14) over all initial or final states as follows:

$$I = 2q \times \frac{2\pi}{\hbar} \sum_{k_t} |M_{fi}|^2 \delta(E_C - E_V)(f_C - f_V). \tag{5.31}$$

Plugging in the 0D–0D matrix element, Eq. (5.29), converting the sum to an integral, and assuming a full valence band and empty conduction band gives

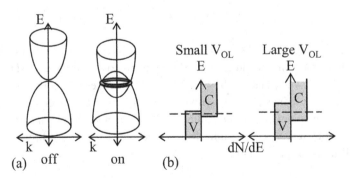

Fig. 5.8 Various characteristics of a 2D–2D$_{face}$ junction. (a) There is only a single tunneling energy because of the simultaneous conservation of energy and momentum. The energy versus wave vector paraboloids on each side of the junction only intersect at a single energy. (b) Even though the overlap of the density of states increases with increasing overlap voltage, there is only a single energy, indicated by the dotted line, at which the electrons tunnel.

$$I = 2q \times \frac{Am}{\pi^2 \hbar^3} \int E_{z,i} \times E_{z,f} \times \mathsf{T} \times \delta(E_i - E_f) dE_t. \tag{5.32}$$

Finally evaluating the integral over the delta function gives an additional factor of ½ as $E_i - E_f = 2E_t - qV_{OL}$:

$$I_{2D-2D,\,face} = \frac{qmA}{\pi^2 \hbar^3} \times E_{z,i} \times E_{z,f} \times \langle \mathsf{T} \rangle. \tag{5.33}$$

The main change in going from 3D–3D to 3D–2D is that the energy factor qV_{OL} became E_z. Likewise, in going from the 3D–2D to 2D–2D$_{face}$ the other energy factor qV_{OL} also became E_z. Thus for each confined side of the junction the relevant energy changes from the overlap energy to the confinement energy. Consequently the 2D–2D$_{face}$ case has the same current as a 3D–3D case if $qV_{OL} = 2\sqrt{2}E_z$. In practice, E_z can be much larger than qV_{OL}, providing the 2D–2D$_{face}$ case with a significant current boost. The quantum confinement also increases the tunneling probability itself as seen from Eq. (5.18). The pre-factor in the tunneling probability is no longer dependent on V_{OL} and is instead set to a large value by the quantum confinement.

Following the joint density of states, the current takes the form of a step function with respect to the gate voltage. This is because all of the tunneling current is concentrated near a single energy. This is similar to the step function case of quantum well optical transitions. As soon as the bands overlap, the current immediately turns on. However, various broadening mechanisms will smear out the step-like turn-on function and this will be discussed later.

5.6.10   1D–1D$_{edge}$ junction

A 1D–1D$_{edge}$ junction represents two nanowires overlapping each other along the edge as shown in Fig. 5.4(f). This junction is similar to the 2D–2D$_{face}$ junction. The current can be found by summing over 1D set of transverse states in Eq. (5.31). The resulting current is

$$I_{1D-1D,\,edge} = 2\frac{qL_X}{\pi^2 \hbar^2} E_{z,i} \times E_{z,f} \times \sqrt{\frac{m}{qV_{OL}}} \times \langle \mathsf{T} \rangle. \tag{5.34}$$

As in the 2D–2D$_{face}$ case the tunneling only occurs at a single energy due to the conservation of momentum and energy. Since we are now dealing with 1D nanowires, the number of transverse states follows a 1D density of states which follows a $1/\sqrt{V_{OL}}$ dependence. This predicts a step function turn-on followed by a reciprocal square root decrease. This seemingly implies that the initial conductance will be infinite. However, the contact series resistance will limit the conductance and various broadening mechanisms will limit the peak conductance.

5.6.11   Trade-off between current, device size, and level broadening

When a level on the p-side of a junction interacts with a level on the n-side of the junction it is possible for the two levels to interact strongly and repel each other. In most cases this is not a problem as the interaction between any two *particular* levels goes to zero when the devices get larger and any small amount of level broadening will wash out the level repulsion. In the case of very large contact regions leading to the tunnel junction, the large normalization volume of the wave functions guarantees that individual level repulsion matrix elements are negligible.

In contrast, the 0D–0D, 1D–1D$_{edge}$, and 2D–2D$_{face}$ cases have a finite extent along the tunneling direction, restricting the normalization volume. This means that the tunnel interaction matrix element, $|M_{fi}|$, can take on a large finite value. If this interaction is too large, the two interacting levels will be strongly coupled and all the perturbation results in this chapter will fail. The strong coupling will cause the interacting levels to repel each other and consequently limit the current. To prevent this and wash out the level repulsion, the level broadening, γ, needs to be greater than the level repulsion matrix element:

$$\gamma > |M_{fi}| = \frac{1}{\pi}\sqrt{E_{z,i} \times E_{z,f} \times \langle \mathsf{T} \rangle}. \tag{5.35}$$

The broadening γ is typically caused by coupling to the contacts or by various scattering mechanisms. Unfortunately, this level broadening also smears out the sharp turn on of the 1D–1D$_{edge}$ and 2D–2D$_{face}$ junctions. Since the tunneling current is proportional to the matrix element, $|M_{fi}|$, there is a fundamental trade-off between the level broadening and the tunneling current. The greater the on-state current, the more the level needs to be broadened to allow the electrons to escape into the contact. The 1D–1D$_{end}$, 2D–2D$_{edge}$, and 3D–3D junctions can be thought of as the limit where the levels are completely broadened into a continuous band. In the limit that $\mathsf{T} \rightarrow 1$, we cannot do better than a perfect quantum of conductance from the 1D–1D$_{end}$, 2D–2D$_{edge}$, and 3D–3D cases. However, any realistic device will have $\mathsf{T} \ll 1$, and so we can use the quantum confinement to increase the matrix element and to engineer the trade-off between the broadening and the on-current to get a sharper turn on.

Another major broadening limit occurs for the 1D–2D, 2D–3D, 1D–1D$_{edge}$, and 2D–2D$_{face}$ cases when the transverse dimensions are reduced. Consider the 1D–1D$_{edge}$ case shown in Fig. 5.4(f). The current is flowing in along extended nanowires in the x-direction. At low energies, the wavelength along the x-direction will be very long and so only the tail of the wavefunction will fit overlap region between the two quantum wires. In order to get a good transverse momentum matching at least half a wavelength needs to fit in the overlap region. Consequently, the turn-on will be broadened by the energy corresponding to $\lambda = 2L_x$:

$$\gamma \approx \frac{\hbar^2 \pi^2}{2mL_x^2} = E_x. \tag{5.36}$$

The same limit applies for the 1D–2D, 2D–3D, and 2D–2D$_{face}$ cases. This means that these cases will lose their abrupt turn-on if the transverse dimensions are too small. For extremely small dimensions 0D–1D or 0D–0D may be more favorable. Alternatively, tunneling contacts could be used, but a different broadening limit given by Eq. (5.28) will apply.

5.6.12   Comparing the different dimensionalities

Now that we have considered many different tunneling junction geometries, we've plotted a comparison of the different cases in Fig. 5.9. To plot the figures we used a reasonable tunneling probability of 1%. We assumed confinement energies of 130 meV, an effective mass of 0.1, and overlap lengths of 20 nm. There are four different broadening mechanisms that will limit the initial turn-on, as indicated by the dotted lines. Using the above constants, the broadening mechanisms and the affected dimensionalities are summarized below:

- Transverse momentum matching: Eq. (5.36).
    - 1D–2D, 2D–3D, 1D–1D$_{edge}$, 2D–2D$_{face}$, $\gamma = 9.4$ meV
- Matrix element broadening: Eq. (5.35).
    - 0D–0D, 1D–1D$_{edge}$, and 2D–2D$_{face}$, $\gamma = 4.1$ meV
- Contact broadening: Eq. (5.28)
    - 0D–0D and 0D–1D, $\gamma = 0.8$ meV
- Tunneling probability turn-on: Eq. (5.18)
    - 1D–1D$_{end}$, 2D–2D$_{edge}$, and 3D–3D, $\gamma = 12.5$ meV
    - 0D–1D, 1D–2D, and 2D–3D, $\gamma = 0.6$ meV

To estimate the tunneling probability turn-on, we assume a barrier height of $\Delta V = 100$ meV, which assures a good on/off ratio as discussed in Section 5.3.1. The 1D–1D$_{end}$, 2D–2D$_{edge}$, and 3D–3D junctions are unconfined in the tunneling direction and so we can set the prefactor in Eq. (5.19) equal to $1 : 8 V_{OL} / \Delta V = 1$. This gives $V_{OL} = \gamma = 12.5$ meV. For the 0D–1D, 1D–2D, and 2D–3D cases, one side of the junction is confined and so we need to use Eq. (5.18) with $E_{z,f} = q V_{OL}/2$ and set the prefactor in Eq. (5.18) equal to $1 : 16\sqrt{E_{z,i} q V_{OL}/2}/\Delta V = 1$. This gives $V_{OL} = \gamma = 0.6$ meV. The broadening due to contacts is twice the broadening from a single contact given by Eq. (5.28).

For each dimensionality, the largest form of broadening will dominate. For the 1D–2D, 2D–3D, 1D–1D$_{edge}$, and 2D–2D$_{face}$ junctions, the turn-on will be limited by the transverse momentum matching. The 1D–1D$_{end}$, 2D–2D$_{edge}$, and 3D–3D junctions are limited by the turn-on of the tunneling probability. The 0D–1D and 0D–0D junctions are limited by the contact broadening.

The turn-on conductance versus overlap control voltage V_{OL} can be seen in Fig. 5.9 for all of the cases. The initial broadened turn-on is represented by the dotted lines. For the 0D–0D case the entire line shape is due to the broadening and is thus unknown, but the calculated width and height are still represented in the figure.

The nanowire-based devices shown in Fig. 5.9(a) have the lowest conductance as they only tunnel at a single point. However, we see that introducing quantum confinement can still help increase the conductance when the tunneling probability is low. We also see that for the parameters chosen, the 0D–1D case captures all the benefits of the quantum confinement, while the 0D–0D case is a narrow pulse with a slightly lower peak conductance. In some situations, when the tunneling probability requires a larger voltage to turn on, the 0D–0D case can have a higher initial peak.

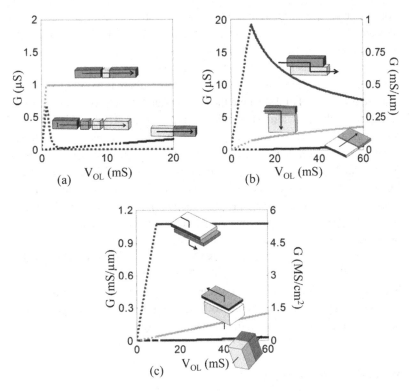

Fig. 5.9 The conductance curves for the different dimensionalities are plotted using the following parameters: $\langle \mathsf{T} \rangle = 1\%$, $E_z = 130$ meV, $L_X = 20$ nm, $m^* = 0.1 m_e$, and $\Delta V = 100$ meV. The dotted lines represent the initial broadened turn-on where the shape of the line is uncertain. Graphs show (a) the 1D–1D$_{\text{end}}$, 0D–1D, and 0D–0D cases; (b) 2D–2D$_{\text{edge}}$, 1D–2D, and 1D–1D$_{\text{edge}}$ cases; and (c) 3D–3D, 2D–3D, and 2D–2D$_{\text{face}}$ cases.

The edge tunneling devices shown in Fig. 5.9(b) have a higher conductance as they have a larger tunneling length. Consequently, we also normalize the current to the tunneling length. In these cases, maximizing the quantum confinement on both sides of the junction results in the highest conductance. The same applies for the area tunneling devices shown in Fig. 5.9(c).

Overall, we see that using quantum confinement in the tunneling direction can significantly increase the conductance and thus reduce the overdrive voltage when the tunneling probability is low.

5.7   Building a full tunneling field effect transistor

Now that we have analyzed how to make a good tunneling junction we can consider what happens when we try to build a TFET. In considering the performance of a full TFET, one has to consider the sub-threshold regime, gate efficiency, and any overdrive voltage needed to achieve the full on-state. While the voltage response is exponential over most of the drive range, it invariably saturates as the switch approaches the on-state. The overdrive voltage represents the extra voltage needed to achieve the full on-state after saturation has set in.

We can understand many of the design issues by considering three simple TFET structures that capture the essential physics behind most TFETs to date. We only analyze n-channel TFETs, where the gate modulates the n-type side of the p-n junction, since the analysis is almost identical for p-channel TFETs. Figure 5.10(a) shows a double gate TFET with identical gates such that the tunneling occurs laterally at a single point at the source/channel junction. The key design issues will be similar for other point tunneling devices such as nanowires or even single gate TFETs [12, 27]. Figure 5.10(b) shows a vertical tunneling TFET where the gate overlaps the source. The gate inverts the source and a tunneling junction is formed within the source. A variety of schemes to optimize the vertical tunneling such as creating a doped pocket or using a vertical heterojunction have been tested, but they all operate on similar principles and will face similar issues [3, 36, 37]. Figure 5.10(c) shows a newer device concept called the electron hole bilayer TFET [10, 11]. By applying opposite biases to both the n- and p-gates, both an electron and a hole channel form, allowing band-to-band tunneling as shown in Fig. 5.10(f). In all of the designs, the gate workfunction can be engineered to correctly set the threshold voltage.

5.7.1   Minimum voltage required

The operating voltage is given by

$$V_{DD} = V_{OV} + S \times \log(I_{on}/I_{off}). \tag{5.37}$$

Here we explicitly include the overdrive voltage, V_{OV}, and the sub-threshold swing voltage, S, as V_{OV} can be a significant part of the total voltage in a steep slope device. I_{on}/I_{off} is the ratio of the on- and off-state currents. V_{OV} is the extra voltage required after the conduction and valance band are aligned to get the desired on-state conductivity and is given by

$$V_{OV} = \frac{dV_{gate}}{dE_{OL}} \times \frac{1}{q} E_{OV} = \frac{1}{\eta_{gate}} \times \frac{1}{q} E_{OV}, \tag{5.38}$$

where E_{OL} is the overlap between the conduction and valence band edge as shown in Fig. 5.1(c) and (d), and E_{OV} is the minimum energy overlap, E_{OL}, required to get the desired on-state conductivity. Typically, the conductance increases as E_{OL} increases as there are more states that contribute to the tunneling current and so a minimum energy overlap E_{OV} is needed. Fortunately, E_{OV} can be significantly reduced by confining the carriers in the tunneling direction as discussed in Section 5.6.

The next factor in computing V_{OV} is the gate efficiency, η_{gate}. Only a fraction of the voltage applied on the gate, V_{gate}, contributes to changing E_{OL}. The gate efficiency can divided into two terms:

$$\eta_{gate} = \frac{1}{q} \frac{dE_{OL}}{dV_{gate}} = \frac{1}{q} \frac{d\varphi_s}{dV_{gate}} \frac{dE_{OL}}{d\varphi_s} = \eta_{el} \times \eta_{quant}, \tag{5.39}$$

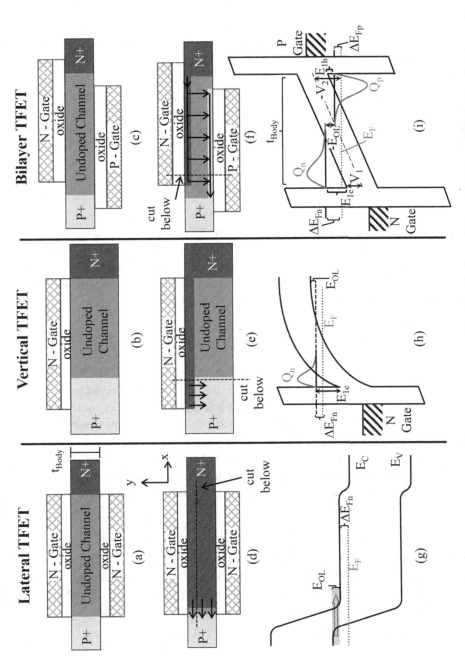

Fig. 5.10 Three representative tunneling FET designs: illustration of (a) lateral, (b) vertical, and (c) bilayer TFETs. The directions of tunneling for the lateral, vertical, and bilayer TFETs are shown in (d), (e), and (f) respectively. The band diagrams parallel to the directions of tunneling are given in (g), (h), and (i) respectively.

where φ_s is the surface potential under the gate. The standard electrostatic gate efficiency (η_{el}) is just the change in φ_s with respect to the gate bias (V_{gate}):

$$\eta_{el} = d\varphi_s/dV_{gate}. \tag{5.40}$$

There is also a quantum confinement efficiency (η_{quant}) for vertical and bilayer TFETs. In these structures a triangular quantum well is formed under the gate and the effective band edge is raised to the quantum level E_{1e} shown in Fig. 5.10(h) and Fig. 5.10(i). In the bilayer, both sides of the tunneling junction are confined as shown in Fig. 5.10(i). If the bias on the gate is increased, the triangular quantum well gets narrower and the energy level E_{1e} increases. This works against the gate bias and reduces the change in the confined eigenstate energy overlap (E_{OL}). Consequently, we need to multiply by an additional quantum confinement efficiency:

$$\eta_{quant} = \left| \frac{1}{q} \frac{dE_{OL}}{d\varphi_s} \right|. \tag{5.41}$$

So far we have assumed the tunneling process limits the current, but there is an additional requirement that the channel should have enough charge to conduct the current (i.e., the MOSFET in a TFET needs to be on as well), which is typically true in short channel TFETs. In a long channel device where a significant amount of charge is needed, the standard MOSFET electrostatics could dominate the overdrive voltage and needs to be accounted for. This would result in a very low η_{el} and increased E_{OV}.

5.7.2   Sub-threshold swing voltage

The next step is to find the sub-threshold swing voltage. The ideal TFET would rely upon a sharp band edge and would switch abruptly from zero-conductance to the desired on-conductance when the electron and hole eigenstate energies overlap. Unfortunately the band edges are not perfectly sharp and thus there is a finite density of states extending into the bandgap, smearing out the desired abrupt response. Conventional TFET modeling does not account for the smeared band edge density of states.

We consider the following simple model for the tunneling current in order to understand the various contributions to the sub-threshold swing voltage (S):

$$I \propto \int (f_C - f_V) \times \top \times D_J(E) \times \partial E, \tag{5.42}$$

where $f_C - f_V$ is the difference in the Fermi occupation probabilities of the conduction and valence bands, $\top$ is the transmission probability of a tunneling electron, and $D_J(E)$ is the joint density of states. Below the band edge, E'_C, the density of states in the conduction band, $D_C(E)$, is given by:

$$D_C(E) = D_{C0} \times e^{-(E'_C - E)/qV_0}, E < E'_C, \tag{5.43}$$

where D_{C0} is a constant prefactor for the electron density of states and E'_C is the electron eigenstate energy. We assume that the density of states falls off exponentially below the band edge with a semilog slope of V_0 and has a prefactor of D_{C0}. An exponential falloff is typical of band edges as seen in the optical absorption edge [18]. Above the band edge the density of states will simply be given by the 1D, 2D, or 3D density of states depending on the device geometry. Similarly, the valence band edge density of states above the valence band edge, E'_V, will be given by

$$D_V(E) = D_{V0} \times e^{-(E-E'_V)/qV_0}, E > E'_V, \tag{5.44}$$

where D_{V0} is a constant prefactor for the hole density of states and E'_V is the hole eigenstate energy. For simplicity, we take the exponential slope, V_0, to be the same for conduction and valence band edges.

Now we consider the situation where the electron and hole eigenstates are not aligned, as shown in Fig. 5.1(c). Ideally, no current would flow, but due to the band tails an overlapping density of states exists as shown in Fig. 5.11. Combining the conduction and valence band density of states to get a joint density of states gives

$$D_J(E) \propto e^{-|E_{OL}|/qV_0} \times \begin{cases} e^{-(E-E'_C)/qV_0}, & E \geq E'_C, \\ 1, & E'_C > E > E'_V, \\ e^{-(E'_V-E)/qV_0}, & E \leq E'_V, \end{cases} \tag{5.45}$$

where E_{OL} is the overlap energy between the electron and hole eigenstates and is given by $E'_V - E'_C$ as shown in Fig. 5.1(c). Since the joint density of states has a maximum plateau in the bandgap region between E'_C and E'_V, we can approximate the current integral as

$$I \propto \left(\int_{E'_V}^{E'_C} (f_C - f_V) \times \top \times \partial E \right) \times e^{-|E_{OL}|/qV_0}, \tag{5.46a}$$

$$I \propto I_0 \times e^{-|E_{OL}|/qV_0}, \tag{5.46b}$$

where the tunneling pre-factor is

$$I_0 \equiv \int_{E'_V}^{E'_C} (f_C - f_V) \times \top \times \partial E. \tag{5.47}$$

Thus we have arrived at a simplified model for the tunneling current when band tails are present. Now we can compute the sub-threshold swing voltage (S) using the definition

$$S \equiv dV_{gate}/d \log(I). \tag{5.48}$$

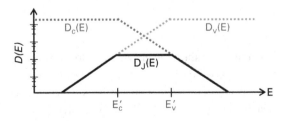

Fig. 5.11 The conduction and valence band density of states, $D_C(E)$ and $D_V(E)$, are shown. Below the band edges the density of states falls off exponentially. The joint density of states, $D_J(E)$, is also shown.

Plugging Eq. (5.46) into Eq. (5.48) gives:

$$S = \left(\frac{d\varphi_s}{dV_{gate}} \times \frac{d\log(I_0)}{d\varphi_s} + \frac{d\varphi_s}{dV_{gate}} \frac{dE_{OL}}{d\varphi_s} \times \frac{d\log(e^{-|E_{OL}|/qV_0})}{dE_{OL}} \right)^{-1}. \quad (5.49)$$

In the first term we took the derivative with respect to the surface potential, φ_s, since tunneling transmission probability, T, depends sensitively on this potential. In the second term we took the derivative with respect to E_{OL} as the band edge density of states depends on the band alignment. Finally, the sub-threshold swing voltage in Eq. (5.49) can be expressed in the following form by replacing each term with the appropriate symbol to highlight the four contributing factors:

$$S = \left(\eta_{el} \times \frac{1}{S_{tunnel}} + \eta_{el} \times \eta_{quant} \times \frac{1}{S_{DOS}} \right)^{-1}, \quad (5.50a)$$

$$S = \frac{1}{\eta_{el}} \times \left(\frac{1}{S_{tunnel}} + \frac{\eta_{quant}}{S_{DOS}} \right)^{-1}, \quad (5.50b)$$

where η_{el} and η_{quant} are the electrostatic and quantum confinement efficiencies given by Eq. (5.40) and Eq. (5.42) respectively, and S_{DOS} is the semilog slope of the joint band edge density of states in mV/dec that was discussed in Section 5.3:

$$S_{DOS} \equiv \frac{1}{q} \frac{dE_{OL}}{d\log(e^{-|E_{OL}|/qV_0})} = V_0/\log(e). \quad (5.51)$$

We redefine S_{tunnel} to be the semilog slope measuring how steeply the tunneling conductance pre-factor changes with respect to the surface potential, φ_s:

$$S_{tunnel} \equiv \frac{dV_{body}}{d\log(I_0)}, \quad (5.52)$$

where S_{tunnel} is given in mV/dec and I_0 is given by Eq. (5.47). S_{tunnel} represents the steepness that results from changing the thickness of the tunneling barrier with a changing bias. Since we are interested in the derivative of the log of I_{tunnel}, any value proportional to it can also be used. To good approximation, we can just consider the tunneling probability, T, instead of I_{tunnel} as we did in Section 5.2

A small sub-threshold swing voltage can be achieved by having either a small S_{tunnel} or a small S_{DOS} and a good gate efficiency.

5.7.3 On-state conductance

Now that we know how to minimize the voltage we need to maximize the on-state conductance. This is discussed thoroughly in two review articles [1, 2] and Chapter 6. Typically there are two methods that are used to increase the conductance. The first is to increase the tunneling area by using a vertical TFET or a bilayer TFET. In these structures, tunneling occurs over a larger overlap region rather than just at the source/channel junction. The second method is to minimize the tunneling barrier. This can be achieved by either reducing the tunneling barrier height or the tunneling barrier width.

Often heavy doping is used to shrink the tunneling barrier width. As we saw in Section 5.3 this causes S_{DOS} to increase and ruins the sub-threshold swing voltage. Because of this, many experimental results with a high conductivity have a terrible sub-threshold swing voltage. Consequently, new methods such as the bilayer are needed to control the tunneling barrier width. The tunneling barrier height can also be reduced by using a smaller bandgap material or by using a heterojunction to create a smaller effective bandgap for tunneling. However, even the minimum effective bandgap is limited by the band edge density of states as we saw from Eq. (5.8). If we want a particular on/off ratio, the barrier height in the off-state, $E_{g,eff}$, must be large enough to suppress the band edge density of states, S_{DOS}, by that on/off ratio.

In Section 5.5 we introduced a new method to increase the conductance. When the tunneling probability is low, quantum confinement in the direction of tunneling will increase the conductance.

5.8   Maximizing the gate efficiency

In addition to having a small S_{tunnel} or a small S_{DOS} we need to maximize the gate efficiency, η_{el}, and η_{quant} to minimize the voltage. Since this is very geometry dependent we will consider the three TFET structures – lateral, vertical, and bilayer – separately.

5.8.1   Lateral TFET gate efficiency

In Fig. 5.12(a) we show the circuit model for the lateral TFET. We have assumed that the body is sufficiently thin that the entire channel will invert. Like a FinFet or nanowire transistor the electrostatics of this device can be very good. If channel is sufficiently long, the gate capacitance, C_{gate}, will be much larger than the source and drain capacitances, C_S and C_D, as only the gate capacitance and the quantum capacitance, $C_{quantum}$, scale with length. This means that we only need to consider C_{gate} and $C_{quantum}$ to compute the gate efficiency. Since there are two gates, the gate capacitance per unit area is given by

$$C_{gate} = 2\frac{\varepsilon_{ox}}{t_{ox}},\tag{5.53}$$

where ε_{ox} is the permittivity of the gate oxide and t_{ox} is the thickness of the gate oxide. The quantum capacitance is simply the voltage needed to add more charge to the channel. The charge in the channel, Q_n, is given by the 2D quantum charge and depends on where the Fermi level is relative to the band edge. The charge is given by the equation for an n-channel device,

$$Q_n = q \times N_{C,2D} \times \ln\left(1 + e^{-\Delta E_{Fn}/k_B T}\right),\tag{5.54}$$

where $N_{C,2D} = (m_{e,t}^*/\pi\hbar^2)k_B T$, ΔE_{Fn} is given by $E_C - E_F$, and $m_{e,t}^*$ is the electron effective mass in the transverse direction. In computing the quantum capacitance, we assumed a small source/drain bias such that we can assume a single Fermi level. We

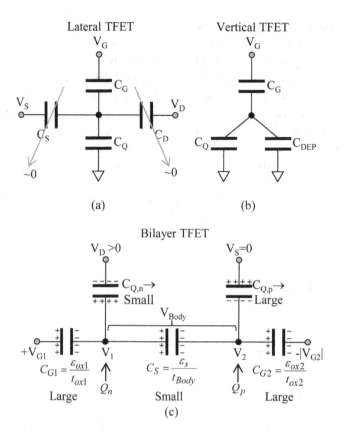

Fig. 5.12 Circuit models for the (a) lateral, (b) vertical, and (c) bilayer TFETs.

make the same assumption for the vertical and bilayer TFETs as well. The quantum capacitance is given by

$$C_Q = \frac{dQ_n}{d\varphi_s} = -\frac{dQ_n}{d\Delta E_{Fn}} = q \times \frac{m^*_{e,t}}{\pi \hbar^2} \frac{1}{1 + e^{\Delta E_{Fn}/k_B T}}. \tag{5.55}$$

The overall gate efficiency is just a voltage divider between C_{gate} and $C_{quantum}$ and is given by $\eta_{el} = C_{gate}/(C_{quantum} + C_{gate})$.

In the sub-threshold regime $C_{quantum} \rightarrow 0$ and so the gate efficiency will approach 1 as long as $C_{gate} \gg C_S$ and C_D. In the overdrive regime, the gate efficiency depends on the Fermi level position and the effective mass. Since the channel needs some charge to conduct current, as given by conventional MOSFET electrostatics, the Fermi level position in the on-state should be set by the minimum required charge to get a given channel conductance. A lower effective mass will also decrease the quantum capacitance and thus increase the gate efficiency in the overdrive regime.

For lateral TFETs the quantum confinement efficiency, η_{quant}, is 1. This is because the confinement energy in the channel is just set by the geometry and does not change with bias.

Overall, we see that a well-designed lateral TFET can have a gate efficiency near 1 in the sub-threshold regime, similar to well-designed FinFet and nanowire transistors. In the overdrive regime, the gate efficiency is limited by the quantum capacitance and can be improved by reducing the effective mass and using the minimum required charge in the channel.

5.8.2 Vertical TFET gate efficiency

The key tunneling junction in a vertical TFET is given by the band diagram in Fig. 5.10(h). It can be modeled as a simple MOS capacitor. In doing so, we are neglecting the details of the 2D electrostatics and focusing on only the essential switching action. A fair bit of engineering is required to ensure that a vertical TFET actually behaves like a 1D MOS capacitor [41, 42]. If the 2D electrostatics are designed incorrectly different regions of the device can turn on at different biases and smear out the sub-threshold swing voltage. Nevertheless, before we even get to that stage of design, we need to understand what the inherent trade-offs in a vertical architecture are. Consequently, we consider the simplest 1D model shown in Fig. 5.10(h). The circuit model is given in Fig. 5.12(b).

First, we find the quantum confinement efficiency, η_{quant} by finding $dE_{\text{OL}}/d\varphi_{\text{s}}$. We choose to measure the potential from the bulk conduction band edge such that the total band bending $= \varphi_{\text{s}}$. The overlap energy E_{OL} is given by

$$E_{\text{OL}} = \varphi_{\text{s}} - E_{\text{G}} - E_{1\text{e}}(\varphi_{\text{s}}), \tag{5.56}$$

where $E_{1\text{e}}$ is the confinement energy in the triangular well and E_{G} is the bandgap. Plugging this into the definition of η_{quant}, Eq. (5.41), we get:

$$\eta_{\text{quant}} = \frac{1}{q}\frac{dE_{\text{OL}}}{d\varphi_{\text{s}}} = 1 - \frac{1}{q}\frac{dE_{1\text{e}}}{d\varphi_{\text{s}}}. \tag{5.57}$$

Thus we need to find $dE_{1\text{e}}/d\varphi_{\text{s}}$. We can approximate the potential well as a triangular quantum well whose slope is set by the peak electric field in the MOS capacitor. Assuming an infinite triangular well will result in an over-estimate of the confinement energy, but it is sufficient for a first approximation. Consequently, the ground state energy is given by

$$E_{1\text{e}} \approx \left(\frac{9\pi}{8}\right)^{2/3} \times \left(\frac{F^2\hbar^2 q^2}{2m^*_{\text{e},z}}\right)^{1/3}, \tag{5.58}$$

where $m^*_{\text{e},z}$ is the electron effective mass in the tunneling direction. The peak electric field is set by the level of band bending, φ_{s}, and by the doping, N_{d}:

$$F = \sqrt{\frac{2qN_{\text{d}}\varphi_{\text{s}}}{\varepsilon_{\text{s}}}}. \tag{5.59}$$

The permittivity of the semiconductor is ε_{S}. Plugging Eq. (5.59) into Eq. (5.58) and evaluating $dE_{1\text{e}}/d(q{\times}\varphi_{\text{s}})$ gives

$$\frac{dE_{1e}}{d(q \times \varphi_s)} = 1 - \eta_{quant} \approx \frac{1}{3}\left(\frac{9\pi}{8}\right)^{2/3} \times \left(\frac{N_d \hbar^2}{m^*_{e,z}\varepsilon_S}\right)^{1/3} \times \varphi_s^{-2/3}. \qquad (5.60)$$

Finally, we can use Eq. (5.60) in Eq. (5.57) to evaluate the quantum efficiency. To estimate η_{quant} we consider silicon ($m^*_{e,z} = 0.92$) with $E_{OL} = 0$ and $N_D = 10^{17}$, 10^{18}, 10^{19}, and $10^{20}/cm^3$. We find $\eta_{quant} = 0.98$, 0.97, 0.93, and 0.88 respectively. In InAs ($m^*_{e,z} = 0.023$), $\eta_{quant} = 0.91$, 0.84, and 0.77 for $N_D = 10^{17}$, 10^{18}, and $10^{19}/cm^3$ respectively. As we can see, lowering the doping will help increase the quantum efficiency, but it will result in a longer depletion width and thus a thicker tunneling barrier and lower current.

Now we can find the electrostatic gate efficiency. The key difference from a lateral TFET is that we have a depletion capacitance in parallel with the quantum capacitance as shown in Fig. 5.12(b). Once again, η_{el} is given by a simple voltage divider: $\eta_{el} = C_{gate}/(C_{quantum} + C_{dep} + C_{gate})$. The gate capacitance is simply given by $C_{gate} = \varepsilon_{ox}/t_{ox}$. The depletion capacitance is given by

$$C_{dep} = \frac{\varepsilon_S}{W_{dep}} = \sqrt{\frac{q\varepsilon_S N_d}{2\varphi_s}}, \qquad (5.61)$$

where W_{dep} is the depletion region width. The quantum capacitance is given by Eq. (5.55) but reduced by η_{quant} as the confinement energy level shifts with bias:

$$C_{quantum} = \frac{dQ_n}{d\varphi_s} = \frac{-dQ_n}{d\Delta E_{Fn}} \times \frac{d\Delta E_{Fn}}{dE_{OL}} \times \frac{dE_{OL}}{d\varphi_s} = q\frac{m^*_{e,t}}{\pi\hbar^2}\frac{1}{1 + e^{\Delta E_{Fn}/k_B T}} \times 1 \times \eta_{quant}. \qquad (5.62)$$

Thus we see that in the sub-threshold regime the electrostatic efficiency is limited by the depletion capacitance, while in the overdrive regime the quantum capacitance will typically limit the efficiency. In addition to improving η_{quant}, minimizing the doping will reduce the depletion capacitance and thus increase η_{el}. In the sub-threshold regime, η_{el} is typically 80–90%. As with the lateral TFET, minimizing the transverse effective mass and the channel charge will result in a lower quantum capacitance and thus a higher overdrive gate efficiency.

Overall, we see that like a conventional planar MOSFET the vertical TFET has a sub-threshold gate efficiency of less than 1, due to the depletion capacitance. Vertical TFETs also suffer slightly from a lower quantum efficiency due to the need for heavy doping to maintain a thin tunneling barrier and a large band bending of at least E_G. Conversely, vertical TFETs do enable a larger tunneling area and thus a higher conductance.

5.8.3   Bilayer TFET gate efficiency

The band diagram for the bilayer TFET is shown in Fig. 5.10(i) and the circuit model is shown in Fig. 5.12(c). As seen in the band diagram the electrons and holes are both quantized, which will result in a lower quantum gate efficiency, η_{quant}, than vertical TFET. Furthermore, since a different voltage is applied to the top and bottom gates, the applied voltage will be split over two gate oxides resulting in a lower electrostatic gate efficiency. Nevertheless, a bilayer structure will allow for the highest on-state conductance. Furthermore, as we will see in Section 5.4, not having doping in the tunneling junction will significantly improve the sub-threshold swing voltage and should more than compensate for the gate efficiency.

To compute the quantum and electrostatic gate efficiencies we first need to redefine the efficiencies in Eq. (5.40) and Eq. (5.41) to refer to the voltage across the body, V_{body}, rather than the surface potential:

$$\eta_{quant} = \frac{1}{q}\frac{dE_{OL}}{dV_{body}}, \tag{5.63}$$

$$\eta_{el} = \frac{dV_{body}}{dV_{G1}}. \tag{5.64}$$

In the lateral and vertical TFETs the potential of the p-side of the junction is fixed by the source and so we only need to know how the surface potential changes. However, in a bilayer TFET, the potential on the p-side, V_2, is not fixed and so it is more convenient to compute the efficiency relative to V_{body}. We also consider the situation where the bias on the n-gate, V_{G1}, is changed while the bias on the p-gate, V_{G2}, is held constant.

We can find η_{quant} by using the definition of E_{OL} to take the derivative of the overlap energy:

$$E_{OL} = qV_{body} - (E_G + E_{1e} + E_{1h}). \tag{5.65}$$

This definition can be seen from Fig. 5.10(i). The triangular well confinement energies are given by:

$$E_{1e} \approx \left(\frac{9\pi}{8}\right)^{2/3} \times \left(\frac{(qV_{body}/t_{body})^2\hbar^2}{2m_{e,z}^*}\right)^{1/3},$$

$$\text{and } E_{1h} \approx \left(\frac{9\pi}{8}\right)^{2/3} \times \left(\frac{(qV_{body}/t_{body})^2\hbar^2}{2m_{h,z}^*}\right)^{1/3}. \tag{5.66}$$

Evaluating Eq. (5.63) using Eq. (5.65) gives

$$\eta_{quant} = 1 - \frac{2}{3}\left(\frac{9\pi}{8}\right)^{2/3} \times \left(\frac{\hbar^2}{2qt_{body}^2}\right)^{1/3} \times \left(\frac{1}{(m_{e,z}^*)^{1/3}} + \frac{1}{(m_{h,z}^*)^{1/3}}\right) \times (V_{body})^{-1/3}. \tag{5.67}$$

Next we consider the electrostatic efficiency, η_{el}. It can be optimized by engineering the bilayer thickness and quantum capacitances. Maximizing the electrostatic efficiency means maximizing dV_{body}/dV_{G1}. As $V_{body} = V_1 - V_2$, we want to maximize dV_1/dV_{G1} while minimizing dV_2/dV_{G1}. The voltages are labeled on the circuit diagram in Fig. 5.12(c). V_1 and V_2 are the n-channel and p-channel potentials, respectively. Consequently, we want the body to be as thick as possible to isolate V_2 from V_{G1} and minimize the body capacitance C_S in the voltage divider. Additionally, we want to minimize the influence of the drain voltage on the n-channel by minimizing the electron quantum capacitance, $C_{quantum,n}$, and thus the electron density. Since we want to fix V_2, we want to maximize the influence of the source on the p-channel by maximizing the hole quantum capacitance, $C_{quantum,p}$, and increase the hole density until the p-side is degenerate. Therefore, choosing the biases/work functions to control the carrier densities will allow us to improve the electrostatic efficiency, especially when thicker gate oxides are used.

Unfortunately, the carrier density and body thickness are constrained by the required on-state conductance. The fewer electrons present in the channel, the lower the channel conductance, and the thicker the body, the lower the tunneling probability. Thus, we need to optimize these trade-offs to maximize the device performance. Furthermore, computing the electrostatic efficiency, η_{el}, needs to be done numerically as the carrier densities depend on the potentials V_1 and V_2 which both change with gate bias. This is done in detail in [43].

As seen in [43], when the bilayer body thickness is optimized for a reasonable on-state current, Si, Ge, and InAs have an overall gate efficiency of 40–50%, with both quantum and electrostatic efficiencies of around 60–70%. While the bilayer has a lower gate efficiency than lateral and vertical TFETs, it will have the highest on-state conductance and has an undoped tunneling junction which will lead to significantly sharper band edges.

5.9   Other design issues to avoid

When designing a TFET there are several additional issues that can prevent a small sub-threshold swing voltage. The first issue that affects many experimental results is trap-assisted tunneling. This process occurs when an electron tunnels to a trap in the bandgap and is then thermally excited out of the trap. This can result in a temperature dependent sub-threshold swing voltage as well as temperature dependent threshold shifts [24, 44–46]. It also increases the sub-threshold swing voltage by preventing the tunneling from turning off. High-quality interfaces and semiconductors are needed to avoid creating states within the bandgap that can lead to trap-assisted tunneling.

Another important design issue to avoid is graded junctions and poor electrostatics. If different regions of the channel start tunneling at different biases, we will get a superposition of I–V curves with different thresholds. This means that the overall sub-threshold swing voltage will be smeared out and will be far worse. This can be seen in a variety of simulation studies [41, 42]. Similarly, spatial inhomogeneity can smear out the sub-threshold swing voltage.

Finally, a short channel length will result in source to drain tunneling or contact broadening which will increase the sub-threshold swing voltage [47]. In order to suppress direct source to drain tunneling, TFETs will need a channel length longer than a corresponding MOSFET in the same material system. This is because TFETs are designed to have a sub-threshold swing voltage less than 60 mV/dec and consequently need a stronger suppression of the direct source to drain tunneling.

5.10   Conclusions

After analyzing the different factors that contribute to the operation of a TFET we find that there are four key design issues to consider:

1. Modulating the tunneling barrier thickness does not work. It cannot give a steep sub-threshold slope at high current density, unless we have an even steeper density of states.
2. We must eliminate doping, spatial inhomogeneity, and preserve material quality to get a steep density of states.
3. Quantum confinement in the tunneling direction increases the tunneling conductance and 2D–2D tunneling (bilayer type) structures have the highest conductance.
4. Lateral tunneling structures tend to have the best gate efficiency, while bilayer structures have the worst gate efficiency.

While it is clear that a steep band edge density of states is needed, to date there have been no electronic measurements of a steep band edge density of states. Fortunately, optical as well as the electronic measurements available indicate that eliminating doping in the tunneling junction may allow us to achieve the steep density of states required. Bilayer-based structures provide an opportunity to achieve this and a high conductance, but unfortunately suffer from a lower gate efficiency. A double gate lateral structure shown in Fig. 5.3 is an alternative that could eliminate doping and potentially have a higher gate efficiency. Unfortunately such a structure will have a lower conductance as it does not take advantage of quantum confinement or the larger tunneling area of the bilayer. To eliminate other forms of spatial inhomogeneity, atomically precise semiconductors such as monolayer semiconductors might be needed.

Overall, there are still some trade-offs that need to be engineered, but by respecting the design principles above it should be possible to make a good TFET with a steep sub-threshold swing voltage at high current densities.

Acknowledgment

This work was supported by the Center for Energy Efficient Electronics Sciences, which receives support from the National Science Foundation (NSF award number ECCS-0939514).

References

[1] A.C. Seabaugh & Q. Zhang, "Low-voltage tunnel transistors for beyond CMOS Logic." *Proceedings of the IEEE*, **98**, 2095–2110 (2010).
[2] A.M. Ionescu & H. Riel, "Tunnel field-effect transistors as energy-efficient electronic switches." *Nature*, **479**, pp. 329–337 (2011).

[3] S. H. Kim, H. Kam, C. Hu, & T.-J. K. Liu, "Germanium-source tunnel field effect transistors with record high I_{on}/I_{off}." In *VLSI Technology, 2009 Symposium on*, pp. 178–179 (2009).

[4] W. Y. Choi, B. G. Park, J. D. Lee, & T. J. K. Liu, "Tunneling field-effect transistors (TFETs) with subthreshold swing (SS) less than 60 mV/dec." *IEEE Electron Device Letters*, **28**, 743–745 (2007).

[5] K. Jeon, W.-Y. Loh, P. Patel *et al.*, "Si tunnel transistors with a novel silicided source and 46 mV/dec swing." In *VLSI Technology, 2010 IEEE Symposium on*, pp. 121–122 (2010).

[6] T. Krishnamohan, K. Donghyun, S. Raghunathan, & K. Saraswat, "Double-gate strained-Ge heterostructure tunneling FET (TFET) with record high drive currents and < 60 mV/dec subthreshold slope." In *IEEE International Electron Devices Meeting, IEDM 2008*, pp. 1–3 (2008).

[7] E. O. Kane, "Theory of tunneling." *Journal of Applied Physics*, **32**, 83–91 (1961).

[8] E. O. Kane, "Zener tunneling in semiconductors." *Journal of Physics and Chemistry of Solids*, **12**, 181–188 (1959).

[9] M. Luisier & G. Klimeck, "Simulation of nanowire tunneling transistors: from the Wentzel–Kramers–Brillouin approximation to full-band phonon-assisted tunneling." *Journal of Applied Physics*, **107**, 084507 (2010).

[10] L. Lattanzio, L. De Michielis, & A. M. Ionescu, "The electron-hole bilayer tunnel FET." *Solid-State Electronics*, **74**, 85–90 (2012).

[11] J. T. Teherani, S. Agarwal, E. Yablonovitch, J. L. Hoyt, & D. A. Antoniadis, "Impact of quantization energy and gate leakage in bilayer tunneling transistors." *IEEE Electron Device Letters*, **34**, 298–300 (2013).

[12] G. Dewey, B. Chu-Kung, J. Boardman *et al.*, "Fabrication, characterization, and physics of III-V heterojunction tunneling field effect transistors (H-TFET) for steep subthreshold swing." In *2011 IEEE International Electron Devices Meeting (IEDM 2011)*, pp. 33.6.1–33.6.4 (2011).

[13] J. Knoch, S. Mantl, & J. Appenzeller, "Impact of the dimensionality on the performance of tunneling FETs: bulk versus one-dimensional devices." *Solid-State Electronics*, **51**, 572–578 (2007).

[14] J. D. Dow & D. Redfield, "Toward a unified theory of Urbach's rule and exponential absorption edges." *Physical Review B*, **5**, 594 (1972).

[15] S. Agarwal & E. Yablonovitch, "Band-edge steepness obtained from Esaki and backward diode current-voltage characteristics." *IEEE Transactions on Electron Devices*, **61**, 1488–1493 (2014).

[16] S. R. Johnson & T. Tiedje, "Temperature dependence of the Urbach edge in GaAs." *Journal of Applied Physics*, **78**, 5609–5613 (1995).

[17] T. Tiedje, E. Yablonovitch, G. D. Cody, & B. G. Brooks, "Limiting efficiency of silicon solar cells." *IEEE Transactions on Electron Devices*, **31**, 711–716 (1984).

[18] J. I. Pankove, "Absorption edge of impure gallium arsenide." *Physical Review*, **140**, A2059–A2065 (1965).

[19] P. Van Mieghem, "Theory of band tails in heavily doped semiconductors." *Reviews of Modern Physics*, **64**, 755–793 (1992).

[20] E. O. Kane, "Band tails in semiconductors." *Solid-State Electronics*, **28**, 3–10 (1985).

[21] Z. Zhang, R. Rajavel, P. Deelman, & P. Fay, "Sub-micron area heterojunction backward diode millimeter-wave detectors with 0.18 pW/Hz 1/2 noise equivalent power." *IEEE Microwave and Wireless Components Letters*, **21**, 267–269 (2011).

[22] J. Karlovsky & A. Marek, "On an Esaki diode having curvature coefficient greater than E/KT." *Czechoslovak Journal of Physics*, **11**, 76–78 (1961).

[23] D. Pawlik, B. Romanczyk, P. Thomas *et al.*, "Benchmarking and improving III-V Esaki diode performance with a record 2.2 MA/cm^2 peak current density to enhance TFET drive current." In *Electron Devices Meeting (IEDM), 2012 IEEE International*, pp. 27.1.1–27.1.3 (2012).

[24] S. Mookerjea, D. Mohata, T. Mayer, V. Narayanan, & S. Datta, "Temperature-dependent I-V characteristics of a vertical In(0.53)Ga(0.47)As tunnel FET." *IEEE Electron Device Letters*, **31**, 564–566 (2010).

[25] S. Agarwal & E. Yablonovitch, "Pronounced effect of pn-junction dimensionality on tunnel switch sharpness." eprint arXiv:1109.0096. (2011). Available online: http://arxiv.org/abs/1109.0096.

[26] K. Tomioka, M. Yoshimura, & T. Fukui, "Steep-slope tunnel field-effect transistors using III-V nanowire/Si heterojunction." In *VLSI Technology, 2012 Symposium on*, pp. 47–48 (2012).

[27] J. Appenzeller, Y.M. Lin, J. Knoch, & P. Avouris, "Band-to-band tunneling in carbon nanotube field-effect transistors." *Physical Review Letters*, **93**, 196805 (2004).

[28] J. Bardeen, "Tunneling from a many particle point of view." *Physical Review Letters*, **6**, 57–59 (1961).

[29] C.B. Duke, *Tunneling in Solids* (New York: Academic Press, Inc, 1969).

[30] W.A. Harrison, "Tunneling from an independent-particle point of view." *Physical Review*, **123**, 85–89 (1961).

[31] J.R. Oppenheimer, "Three notes on the quantum theory of aperiodic effects." *Physical Review*, **31**, 66–81 (1928).

[32] A.F.J. Levi, *Applied Quantum Mechanics* (Cambridge: Cambridge University Press, 2006).

[33] G. Gamow, "Zur Quantentheorie des Atomkernes." *Z. Physik*, **51**, 204, 1928.

[34] M.A. Kastner, "The single-electron transistor." *Reviews of Modern Physics*, **64**, 849 (1992).

[35] S. Datta, *Quantum Transport: Atom to Transistor* (Cambridge: Cambridge University Press, 2005).

[36] P. Patel, "Steep turn on/off "green" tunnel transistors." PhD thesis, Electrical Engineering and Computer Sciences, University of California at Berkeley, Berkeley (2010).

[37] R. Li, Y.Q. Lu, G.L. Zhou *et al.*, "AlGaSb/InAs tunnel field-effect transistor with on-current of 78 uA/um at 0.5 V." *IEEE Electron Device Letters*, **33**, 363–365 (2012).

[38] M. Sweeny & J.M. Xu, "Resonant interband tunnel-diodes." *Applied Physics Letters*, **54**, 546–548 (1989).

[39] S.L. Rommel, T.E. Dillon, M.W. Dashiell *et al.*, "Room temperature operation of epitaxially grown Si/Si$_{0.5}$Ge$_{0.5}$/Si resonant interband tunneling diodes." *Applied Physics Letters*, **73**, 2191–2193 (1998).

[40] S. Krishnamoorthy, P.S. Park, & S. Rajan, "Demonstration of forward inter-band tunneling in GaN by polarization engineering." *Applied Physics Letters*, **99**, 233504–3 (2011).

[41] S.H. Kim, S. Agarwal, Z.A. Jacobson, P. Matheu, H. Chenming, & L. Tsu-Jae King, "Tunnel field effect transistor with raised germanium source." *IEEE Electron Device Letters*, **31**, 1107–1109 (2010).

[42] Y. Lu, G. Zhou, R. Li *et al.*, "Performance of AlGaSb/InAs TFETs with gate electric field and tunneling direction aligned." *IEEE Electron Device Letters*, **33**, 655–657 (2012).

[43] S. Agarwal, J.T. Teherani, J.L. Hoyt, D.A. Antoniadis, & E. Yablonovitch, "Engineering the electron-hole bilayer tunneling field-effect transistor." *IEEE Transactions on Electron Devices*, **61**, 1599–1606 (2014).

[44] G. A. M. Hurkx, D. B. M. Klaassen, & M. P. G. Knuvers, "A new recombination model for device simulation including tunneling." *IEEE Transactions on Electron Devices*, **39**, 331–338 (1992).

[45] G. A. M. Hurkx, D. B. M. Klaassen, M. P. G. Knuvers, & F. G. O'Hara, "A new recombination model describing heavy-doping effects and low-temperature behaviour." In *Electron Devices Meeting, 1989. IEDM '89. Technical Digest, International*, pp. 307–310 (1989).

[46] J. Furlan, "Tunnelling generation-recombination currents in a-Si junctions." *Progress in Quantum Electronics*, **25**, 55–96 (2001).

[47] K. Ganapathi, Y. Yoon, & S. Salahuddin, "Analysis of InAs vertical and lateral band-to-band tunneling transistors: leveraging vertical tunneling for improved performance." *Applied Physics Letters*, **97**, 033504–3 (2010).

6 Tunnel transistors

Alan Seabaugh, Zhengping Jiang, and Gerhard Klimeck

6.1 Introduction

A fundamental characteristic of semiconductor devices is the exponential dependence of current on applied bias. This property of the p-n junction was first revealed by William Shockley in 1948 [1],

$$I = I_0 \left[\exp(qV/kT) - 1 \right], \tag{6.1}$$

where I_0 is the reverse saturation current, V is the applied bias, q is the fundamental charge, k is Boltzmann's constant, and T is temperature. Shockley described the factor kT/q as "the most important single number in semiconductor electronics" [2], with a value at room temperature of approximately one-fortieth of a volt.

The exponential dependence of current on applied bias is a direct result of barrier lowering as illustrated in Fig. 6.1. According to this exponential relationship, to change the current by a factor of 10 in the forward bias direction, the voltage across the junction must increase by $\ln(10)kT/q = 60$ mV, or 60 mV/dec of current at room temperature.

Barrier lowering is the most widely used current control mechanism in semiconductors. In addition to p-n diodes, Schottky diodes, light-emitting diodes, and lasers, the limit of 60 mV/dec bounds almost all commercial semiconductor devices including bipolar junction transistors (BJTs), heterojunction bipolar transistors (HBTs), junction gate field effect transistors (JFETs), metal–semiconductor FETs (MESFETs), metal–oxide–semiconductor FETs (MOSFETs), and high electron mobility transistors (HEMTs).

6.1.1 Need for low-voltage and sub-threshold swing

In transistors (e.g., the MOSFET), the figure of merit that characterizes the exponential dependence of current on voltage is called the inverse sub-threshold slope or sub-threshold swing, S, given by

$$S = \left(\frac{d \log(I_D)}{dV_{GS}} \right)^{-1}, \tag{6.2}$$

where I_D is the drain current and V_{GS} is the gate/source voltage. As with the p-n diode the minimum voltage to effect a change of one order of magnitude in current is 60 mV.

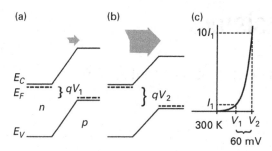

Fig. 6.1 Current control by barrier lowering in a p-n junction: (a) forward bias V_1, (b) $V_2 = V_1 + 60\,\text{mV}$, and (c) current-voltage characteristic. Current control via barrier lowering imposes a fundamental lower limit of 60 mV per decade change in current.

Remarkably, a well-designed 22 nm node trigate MOSFET approaches very near the theoretical limit, achieving a sub-threshold swing of 65 mV/dec [3]. To achieve an on/off current ratio of 10^5 requires at least $5 \times 60\,\text{mV} = 0.3\,\text{V}$ assuming perfect gate control of the channel potential. With the addition of margins to ensure circuit operation in the presence of variability the supply voltage is currently two to three times this lower limit. To continue to decrease the power supply voltage, a current control mechanism is needed that can achieve under 60 mV/dec when averaged over five or more decades. When this is achieved a new basis for ultra-low power circuit development will be enabled.

The power dissipated in complementary MOS (CMOS) logic technology has both static and dynamic components and both are significant when the gate length is reduced below about 50 nm. The static power dissipation is proportional to the off-current multiplied by the supply voltage, while the active power is the product of clocking frequency, node capacitance, and the square of the supply voltage. From a technological perspective, a reduction in voltage directly lowers power dissipation, enabling an increase in transistor density. Significant power reduction at the device level requires new mechanisms for current control, which are not limited to 60 mV/dec.

The realization that electric field control of tunneling can be used to achieve sub-thermal swings is one of the most exciting developments of the past decade [4–6]. Unfortunately Si is not well suited to deliver practical magnitudes of tunnel current because of its indirect bandgap. In indirect bandgap semiconductors, the tunneling process requires a phonon to complete the transition and this dramatically lowers the tunneling probability. Compound semiconductors, with direct bandgaps, are better suited to build a low sub-threshold swing tunnel transistor or tunnel FET (TFET) [7] as are materials like graphene and the transition metal dichalcoginides [8]. The possibility of achieving high tunnel current density and low sub-threshold swing has been a motivating factor in the development of tunnel transistors in III-V materials.

6.1.2   Scope

In this chapter, current understanding of the physics and the design considerations for III-V TFETs is reviewed including material selection, gate arrangement, doping, and heterojunctions. This is followed by a discussion of selected experimental results chosen to illustrate key features of the TFET and the state-of-the-art.

6.2   Tunnel FET

6.2.1   Physical principles

The basic embodiment of the TFET is illustrated in Fig. 6.2. This transistor looks much like a MOSFET with source (S), drain (D), and gate (G) terminals. For the n-channel TFET, the source is p and the channel and drain are n-type. The p-source is doped degenerately to create a large internal junction field of 0.2–0.4 V/nm. In Fig. 6.2(a), the Fermi level is shown to be approximately equal to the valence band maximum. By proper choice of the gate work function, the channel is fully depleted with zero applied gate bias as illustrated in Fig. 6.2(a). A positive gate bias pulls the conduction band minimum below the valence band in the source, opening up a tunneling window, qV_{TW}, through which electrons in the valence band of the source can tunnel into empty states in the channel as indicated in Fig. 6.2(b).

6.2.2   Kane–Sze on-current

In tunneling from the valence band to the conduction band, an electron must conserve both energy and transverse momentum. This selection rule is easiest to satisfy in a direct gap material as illustrated in Fig. 6.3 where no phonon is needed to make the transition.

Sze and Kane [9] derived an analytic expression for the tunneling current (in amperes) for a triangular energy barrier as shown in Fig. 6.3. This equation (see Eq. (6.2)) depends on two material-dependent constants a and b [4], which are controlled by the semiconductor bandgap energy, E_G, and the tunneling effective mass,

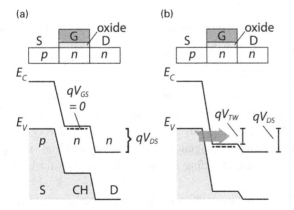

Fig. 6.2 Tunnel field effect transistor schematic cross section of an n-channel device with energy band diagrams drawn for the (a) off-state and (b) on-state. In the off-state current is blocked by the barrier provided by the reverse-biased p-n junction. In the on-state, electrons tunnel from source to the channel within an energy window given by the band overlap.

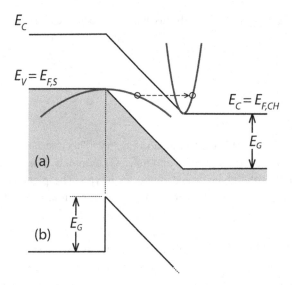

Fig. 6.3 (a) Energy band diagram showing tunneling from valence band to conduction band. The parabolas represent the E–$\hbar k$ (energy–momentum) relationship for the transverse momentum and illustrate a momentum-conserving transition between a heavy-mass valence-band state and light-mass conduction-band state. (b) Triangular potential barrier seen by a valence band electron in the source.

$$I = a V_{TW} \xi \exp\left(-\frac{b}{\xi}\right),$$

$$a = A q^3 \sqrt{2 m_R^*/E_G}/(8\pi^2 \hbar^2), \tag{6.2}$$

$$b = 4 \sqrt{2 m_R^*} E_G^{3/2}/3 q \hbar,$$

where A is the cross-sectional area of the junction; ξ is the electric field in the junction; $m_R^* = (1/m_E^* + 1/m_H^*)^{-1}$ is the reduced effective mass, which is the average of the electron, m_E^*, and hole, m_H^*, effective masses; and $\hbar$ is the reduced Planck's constant.

The Wentzel–Kramers–Brillouin (WKB) approximation, used to obtain Eq. (6.2), has been shown through quantum transport simulations [7] to work reasonably well for direct gap materials such as InAs, but to overestimate the current density by two orders of magnitude for Si and one order of magnitude for Ge. In any materials system, a large electric field in the band-to-band tunneling region reduces the path for least action for tunneling [10].

While the use of bulk material parameters in Fig. 6.4 neglects quantization and band non-parabolicity, the prediction gives a remarkably reasonable estimate in comparison with experiments [11]. Once the material is chosen, the electric field in the junction is set by doping and the ability to form an abrupt junction and this in turn determines the on-current. Detailed numerical simulations discussed later provide more comprehensive

(a)

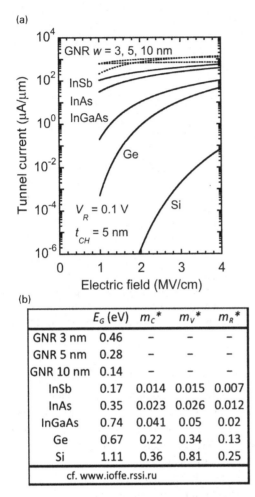

(b)

	E_G (eV)	m_C^*	m_V^*	m_R^*
GNR 3 nm	0.46	–	–	–
GNR 5 nm	0.28	–	–	–
GNR 10 nm	0.14	–	–	–
InSb	0.17	0.014	0.015	0.007
InAs	0.35	0.023	0.026	0.012
InGaAs	0.74	0.041	0.05	0.02
Ge	0.67	0.22	0.34	0.13
Si	1.11	0.36	0.81	0.25
cf. www.ioffe.rssi.ru				

Fig. 6.4 (a) Dependence of Zener tunnel current per unit width on electric field as predicted by the Kane–Sze relation of Eq. (6.2) using the bulk materials parameters in (b), after Zhang [14]. The tunnel current per micron width is computed for a channel thickness (t_{CH}) of 5 nm.

estimates of the current magnitudes, but the analytic approach of Fig. 6.4 gets the trends and provides design guidance. Tools for estimating bandgaps and effective masses are available online at the "*Bandstructure Lab*" at nanoHUB.org [12]. A numerical study that compares ultra-thin body (UTB) TFETs for InAs, GaAs, and InGaAs provides performance trade-offs in these material systems [13].

6.3 Materials and doping trade-offs

6.3.1 Homojunction on-current vs. material

The Kane–Sze expression, Eq. (6.2), for the tunneling current clearly indicates the exponential dependence on the bandgap and invites an exploration of the design space

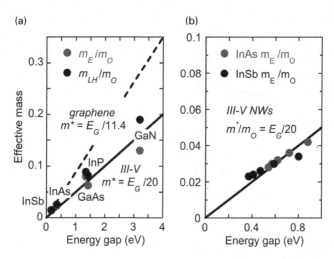

Fig. 6.5 Electron effective mass (gray dots) and light hole effective mass (black dots) vs. energy bandgap for zinc-blende III-V semiconductors. The electron effective mass is given by m_E and the light-hole effective mass, by m_{LH}. (b) Electron effective mass vs. bandgap for nanowires, InAs (gray dots) and InSb (black dots), with diameters in the range from 4 to 12 nm [15]. For III-V semiconductors effective mass is approximately equal to bandgap divided by 20, from InSb to GaN.

with respect to materials as discussed by Zhang [14]. Figure 6.4 compares tunnel currents as a function of electric field for 5 nm ultra-thin body channels. The tunnel currents in the group IV and III-V semiconductors increase monotonically with electric field; current also increases as the bandgap of the semiconductor is decreased. Comparison to graphene nanoribbon (GNR) channels for bandgaps of 0.14, 0.28, and 0.46 eV indicates that graphene can achieve even higher on-currents.

When the material bandgap is selected, effective mass is also determined. In general the two parameters cannot be selected independently. In fact, for the III-V materials, from narrow-bandgap InSb to wide-bandgap GaN, the effective mass is approximately equal to the bandgap of the material divided by 20. Quantization raises both the effective bandgap and effective mass as shown in Fig. 6.5(b), from data given by Khayer and Lake [15]. Using these data the rule that mass equals bandgap divided by 20 is maintained.

6.3.2   Optimum TFET bandgap

While on-current increases as the bandgap decreases, off-current also increases and by a greater exponential factor. This suggests that for a given on/off current ratio there is an optimum bandgap for the TFET which provides the desired on-current while still meeting the off-current requirement. This relationship has been recently explored [16] and the primary conclusion is that the optimum bandgap material depends on three application specifications: on-current, on/off-current ratio, and gate length. Given these specifications, one can choose the bandgap to minimize the supply voltage and power dissipation. These trade-offs have been encapsulated in an analytic model [16] based on the band diagrams

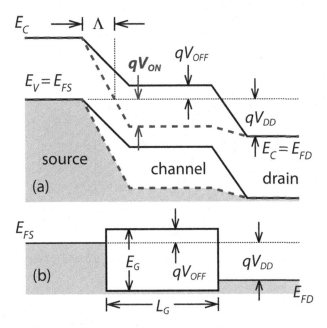

Fig. 6.6 (a) Energy band diagram defining terms used to describe key energies in the on- and off-state: qV_{dd}, qV_{on}, and qV_{off} for a given gate length L_G. (b) The off-current is assumed to be due to direct tunneling and the off-current for this condition is approximated by a rectangular tunnel barrier [16].

in Fig. 6.6. In this model, the on-current is given by the Kane–Sze expression with the particular band relationships defined in Fig. 6.6(a). The off-current is assumed to be due to direct source-to-drain tunneling through a rectangular barrier as shown in Fig. 6.6(b). This simple model is in good agreement with more comprehensive simulations of the on- and off-currents in direct-bandgap nanowire (NW) and two-dimensional channel TFETs, when the gate control of the channel potential is nearly one-to-one.

To select the optimum bandgap material for the TFET, one must first select the gate length since this determines the off-current by direct tunneling. Second, the on-current and on/off-current are set to meet the application requirements. For example, for a 2 nm nanowire TFET in Fig. 6.7, the specification is given that an on-current of $500\,\mu\text{A}/\mu\text{m}$ must be obtained with an off-current of $5\,\text{nA}/\mu\text{m}$, and an on/off ratio of 10^5. The on- and off-voltages, as defined in the band diagrams of Fig. 6.6, are then plotted vs. bandgap. The voltage difference between the two curves in Fig. 6.7(a) then determines the supply voltage according to $V_{\text{dd}} = V_{\text{on}}/0.8 + V_{\text{off}}$, where the factor 0.8 is a gate efficiency factor. Figure 6.7(b) plots the supply voltage and shows that there is a minimum supply voltage and bandgap to meet the performance specifications.

6.3.3   Doping

For the p-n tunnel junction, an abrupt junction and heavy doping are needed to maximize the internal electric field and narrow the depletion width. Figure 6.8 depicts

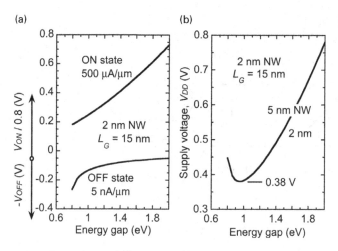

Fig. 6.7 Dependence of (a) on- and off-voltages, as defined in Fig. 6.6, and (b) supply voltage, on energy bandgap to achieve the minimum switching energy.

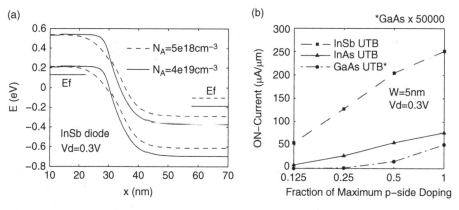

Fig. 6.8 (a) Energy band profiles for two 5 nm UTB InSb p-n junctions with different acceptor doping densities N_A, with a donor doping density, N_D, of $1 \times 10^{19}\,\text{cm}^{-3}$. The InSb bandgap is 0.32 eV due to quantum confinement. (b) Computed tunnel current at 0.3 V through 5 nm UTB p-n junctions based on InAs, InSb, and GaAs as a function of p-doping density with a maximum p-doping of $4 \times 10^{19}\,\text{cm}^{-3}$. InSb delivers the highest current of 250 µA/µm, InAs gives 77 µA/µm, and GaAs has the smallest current of 1 nA/µm.

the strong doping dependence of the on-current for three different p-n junctions with 5 nm body thickness in InAs, InSb, and GaAs. The simulations are based on an atomistic sp^3s^* tight-binding model with spin orbit coupling in the NEMO5 software [17, 18]. The tight-binding parameters are obtained from [19]. An abrupt p-n junction is assumed with the n-side doping fixed at $1 \times 10^{19}\,\text{cm}^{-3}$.

Figure 6.8(a) compares the energy band profiles in the InSb ultra-thin body (UTB) p-n junction for two different p-doping densities. Size quantization increases the overall bandgap of the UTB InSb junction to approximately 0.32 eV from a bulk value of 0.17 eV. If the quantization change was to be estimated from a simple effective mass calculation using the InSb bulk effective masses for electrons, $m_E = 0.014$, and for holes, $m_H = 0.29$, the confinement energies would be 1080 meV and 52 meV, respectively, with an overall bandgap of approximately 1.3 eV. This unrealistic estimate illustrates the inaccuracy of using the parabolic band approximation many kT above the band edge. The sp^3s^* tight-binding model and its parameterization [19] properly captures the non-parabolicities for these direct gap III-V materials. When the X and L bands and strain need to be considered, a more complete $sp^3d^5s^*$ model is needed [20, 21]. Since the tunneling current is exponentially dependent on the tunneling bandgap, it is utterly important to properly model the nonparabolicity of the bands.

A doping change on the p-side from 5 to 40×10^{18} cm^{-3} significantly modifies the band profile as shown in Fig. 6.8(a). As expected, the depletion region width decreases as acceptor doping increases and the interband tunneling distance also decreases. This leads to higher tunnel current as the p-side doping is varied from 4×10^{19} cm^{-3} to 5×10^{18} cm^{-3}, stepping down by factors of 2, Fig. 6.8(b). The tight-binding model predicts quantized bandgaps of 0.32, 0.5, and 1.5 eV for InSb, InAs, and GaAs in this 5 nm UTB configuration.

It is interesting to compare the simple analytic model of Eq. (6.2) as shown in Fig. 6.4 with the atomistic sp^3s^* tight-binding model with spin orbit coupling using NEMO5, Fig. 6.8. From the NEMO5 band diagram of Fig. 6.8(a) the peak electric field in the junction for the bias of 0.3 V is 1 MV/cm and the current is 77 µA/µm. The simple analytic model shown in Fig. 6.4 gives the current for InAs at 1 MV/cm as 31 µA/µm at a bias of 0.1 V. Since from Eq. (6.2) current is directly proportional to bias voltage, the result from Fig. 6.4 should be multiplied by 3 to compare with the NEMO5 result. Thus the simple Kane–Sze formula of Eq. (6.2) using the bulk bandgap and effective masses gives 93 µA/µm vs. 77 µA/µm from the full atomistic simulation accounting for UTB band structure, band non-parabolicity, and quantization.

6.3.4   Heterojunctions – broken-gap benefit

The GaSb/InAs heterojunction has a broken-gap band alignment, where the GaSb valence band and InAs conduction band overlap in energy [22, 23]. Heterojunctions with this broken band alignment are of significant interest for TFETs because they decrease the tunneling distance for band-to-band tunneling (BTBT) and thereby increase the tunnel current. In addition, the broken band alignment means that less heavy doping densities can be used, which should lessen the effects of doping-induced band tails and defect-assisted tunneling, both of which degrade the sub-threshold swing as discussed later.

In TFETs, UTB and NW configurations are needed to provide the best gate control and quantum confinement is an important element in the broken-gap device design. In general for III-V semiconductors, the conduction band edge is more sensitive to confinement than the valence band edge due to the lighter conduction-band mass. With quantization the broken-gap alignment is diminished or even converted into a staggered alignment. Alloys of AlGaSb can also be used to produce the staggered band alignment. Equilibrium band diagrams for the p-n GaSb/InAs broken-gap heterojunction, the AlGaSb/InAs staggered heterojunction, and the InAs homojunction are shown in Fig. 6.9(a–c).

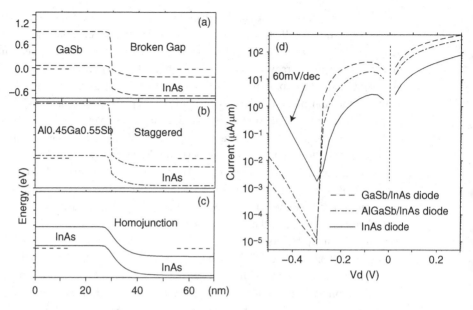

Fig. 6.9 (a–c) Computed energy band profiles for three 5 nm UTB p-n junction diodes with donor doping $N_D = 1 \times 10^{19}\,cm^{-3}$, acceptor doping $N_A = 4 \times 10^{19}\,cm^{-3}$ at zero bias: (a) broken-gap GaSb/InAs, (b) staggered-gap AlGaSb/InAs, (c) homojunction InAs. (d) Computed absolute value of the current–voltage characteristic for all three UTB diodes with positive voltage corresponding to the Zener tunneling direction.

Figure 6.9(d) compares the diode characteristics of the corresponding broken gap, staggered gap, and homojunction UTB devices. The positive bias corresponds to the Zener tunneling direction and successively stronger tunneling currents are obtained with the introduction of staggered and broken band alignments. Indeed the broken-gap structure delivers the highest current in either bias direction (445 µA/µm at 0.3 V) showing an improvement of approximately 6× in current over the homojunction InAs diode.

For the negative bias polarity, the Esaki tunneling behavior is observed with the strongest peak current obtained in the broken-gap arrangement. In the negative differential resistance (NDR) region the current turns off abruptly at less than 5 mV/dec over roughly five orders of magnitude in current showing current control which is much less than the thermal limit imposed by barrier lowering. The current increases again for more negative biases and here the slope for the InAs homojunction is 60 mV/dec due to thermionic emission over a voltage-dependent barrier. Further studies of the GaSb/InAs TFET are available [24]. In a nanowire geometry, tunnel current in the GaSb/InAs TFET has been predicted to reach 1900 µA/µm at 0.4 V [25].

6.4   Geometrical considerations and gate electrostatics

Like the MOSFET, the TFET can be designed for operation as an inversion-mode or accumulation-mode FET or it can be operated as an enhancement-mode FET with a fully depleted channel. These two styles have substantially different vertical field profiles and doping arrangements. In the inversion accumulation style, the vertical electric field is highest in the on-state. In contrast in the enhancement-mode style, the highest vertical field occurs in the off-state when the channel is fully depleted. When the transistor is on the vertical field is tends toward a flat band condition. In the enhancement-mode configuration, higher channel mobility may be expected for transport from channel to drain.

The doping arrangements differ for the inversion accumulation-mode vs. the enhancement-mode transistor styles. The inversion-mode n-channel TFET uses a p-i-n source/channel/drain doping arrangement. The i-layer means intrinsic, but in practice the intrinsic designation means that the region is undoped and then depends on the background impurity type and density. The enhancement-mode n-channel TFET uses a p-n-n doping arrangement, meaning the doping in the channel is intentionally controlled. Both inversion accumulation-mode and enhancement-mode transistor arrangements can be fabricated in vertical or lateral configurations and each can be formed in single-gate (SG), double-gate (DG), trigate, or gate-all-around (GAA) geometries from UTB sheets, fins, or wires. The vertical and lateral arrangements are not fundamental aspects of the carrier transport.

Within these two operation styles, the gate can be placed in two orthogonal directions with respect to the tunnel junction. The gate can be substantially perpendicular to the tunnel junction as in a conventional n-channel MOSFET or the gate can be arranged so that the tunnel junction and gate planes are in parallel. In this case the gate electric field is substantially in line with the tunnel current direction. If the gate bias increases the junction electric field, this enhances the current, Eq. (6.2) [4].

6.4.1   Tunnel junction perpendicular to the gate

The InSb accumulation-mode TFET, illustrated schematically in Fig. 6.10, is simulated using atomistic tight-binding calculations with an sp^3s^* basis to show the performance vs. gate geometry: SG, DG, and GAA. Similar geometries have been explored in InAs [26]. In the SG configuration, increasing the body thickness increases the on-current but also diminishes the sub-threshold swing, see Fig. 6.10(d). Sub-threshold swing is less than 60 mV/dec only for body thicknesses less than 2 nm at a gate length of 15 nm and increases as body thickness increases [26]. A 10 nm thick InSb body with appropriate doping ($N_D = N_A = 5 \times 10^{19} \text{cm}^{-3}$) can deliver an on-current in excess of 200 μA/μm, but only with large gate/source voltage, $V_{GS} = 0.9$ V at $V_{DS} = 0.2$ V. The off-current, however, increases to an impractical level as the body thickness increases. A 2 nm thick body achieves sub-threshold swing under 60 mV/dec, but the on-current is limited to about 0.1 μA/μm, again at this same large gate drive.

TFET performance depends to a large extent on establishing excellent gate control as shown in Fig. 6.10(e). On-current, off-current, and sub-threshold swing are all dramatically improved as the gate electrostatics are improved, with the best performance gate control obtained using a nanowire GAA geometry.

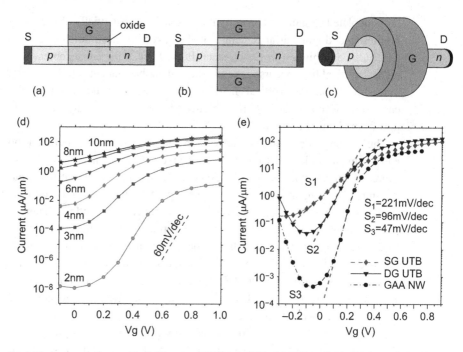

Fig. 6.10 Accumulation-mode InSb p-i-n TFETs: (a) SG, (b) DG, and (c) GAA configurations. (d) Simulated drain current vs. gate/source voltage at a drain-source bias of 0.2 V for SG UTB TFETs. (e) Simulated transfer characteristics vs. gate geometry in SG, DG, and GAA arrangements. The UTB is 6 nm and the NW diameter is also 6 nm. The gate length is 15 nm and the equivalent oxide thickness (EOT) is 0.3 nm.

6.4.2   Tunnel junction parallel to the gate

This geometry was first proposed by Hu *et al.* [27] and called the green transistor. The origin of this transistor concept and a variation on its design are given in Fig. 6.11 after [28]. In the enhancement-mode TFET of Fig. 6.11(a) carriers are injected from the source into the channel at the source end of the gate/oxide interface. Hu reasoned that the area for tunneling could be increased by having carriers injected from the body into a doped n-type pocket under the source side of the gate, Fig. 6.11(b). The n-pocket is fully depleted at zero gate bias and turned on with positive gate bias. The positive gate bias also accumulates the n-region between the pocket connecting the n-channel to the drain. In this configuration the tunneling current can be substantially directed toward the gate. The introduction of the thin heavily doped pocket and the underlying p+ also raises the leakage current between source and drain, Fig. 6.11(b). Luisier [29] proposed to geometrically terminate the parasitic leakage as shown in Fig. 6.11(c).

Quantum transport simulations in Fig. 6.11(c) for a gate length of 40 nm show that on-current increased by two orders of magnitude in the enhancement-mode TFET compared to the p-i-n geometry TFET. By eliminating the parasitic leakage path, low off-state leakage is obtained. The high on-currents in the enhancement-mode TFET

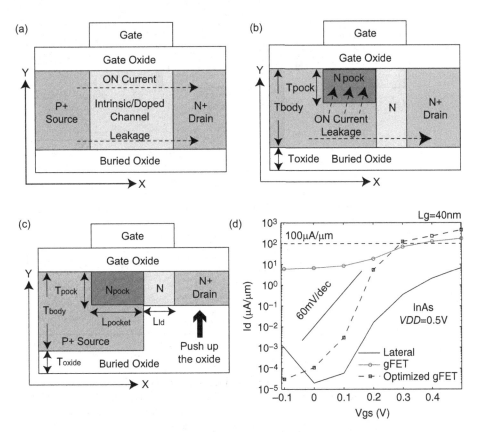

Fig. 6.11 Evolution of TFET geometries: (a) inversion-mode p-i-n InAs TFET, and (b) enhancement-mode p-n InAs TFET, so-called green TFET (gFET) [27], where electrons tunnel substantially toward the gate. The lightly doped *n*-region between the *n*-pocket (Npock) and the drain is incorporated to lower the off-state leakage. (c) Enhancement-mode TFET eliminating the parasitic off-state leakage path. (d) Simulated drain current vs. gate/source voltage comparing the approaches [28, 29].

relative to the accumulation-mode TFET are diminished with gate length scaling, because the area of the tunneling injection is reduced, ultimately limited by the transfer length at the source/channel junction.

The enhancement mode TFET can also be envisioned in SG, DG, and GAA geometries as illustrated in Fig. 6.12. A blocking layer is needed at the drain end of the source in both the DG and GAA geometries. This is indicated as an oxide in Fig. 6.12(b); the blocking layer in the GAA geometry, Fig. 6.12(c), is not visible.

There are several new geometrical parameters which must be specified and controlled in the enhancement-mode TFET. The full list of parameters is provided in Fig. 6.12(d). The additional parameters add a new source of variability, but also provide knobs for controlling the transistor behavior, particularly the off-current. NEMO5 quantum transport simulations are shown in Fig. 6.12(e) to illustrate the effects of some of these parameters. A wider investigation of these parameters is provided by Lu *et al.* [30]. Figure 6.12 shows the influence of drain extension on the off-current in the GaSb/InAs

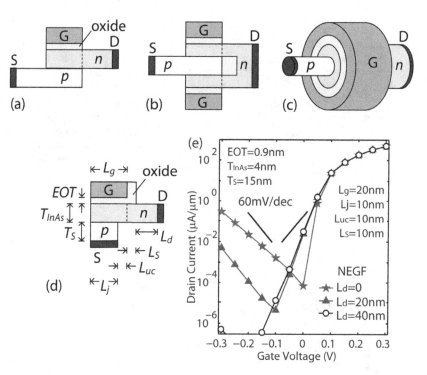

Fig. 6.12 Enhancement-mode TFETS in (a) SG, (b) DG, and (c) GAA geometries. (d) Simplified geometry used for simulations with design parameters indicated. (e) NEMO5 simulations for the p-GaSb/n-InAs TFET vs. drain extension, for a source doping of $4 \times 10^{18}\,\mathrm{cm}^{-3}$ (a 1 nm thick p-type δ-doping layer, $6 \times 10^{19}\,\mathrm{cm}^{-3}$, is located 2 nm below the tunnel junction). Channel and drain are homogeneously doped at $5 \times 10^{17}\,\mathrm{cm}^{-3}$.

TFET, for a drain doping of $5 \times 10^{17}\,\mathrm{cm}^{-3}$. For this doping the gate voltage controls the potential in the drain extension. Increasing the length of the drain extension reduces the off-current.

Figure 6.13 shows that the undercut L_{uc} of the source under the gate improves both the sub-threshold swing and off-current. Increasing the undercut improves the electrostatic control of the channel and provides better shielding of the gate field without significant penalty on the on-current.

6.4.3   p-TFET

A p-channel TFET is highly desirable to implement a complementary logic technology. In comparison to the effort on n-channel TFETs the p-TFET has received little attention. Knoch and Appenzeller [31] simulated the current–voltage (I–V) characteristics for the 20 nm gate length DG n-InAs/p-AlGaSb p-TFET and found the achievable currents to be less than 1 μA/μm at $V_{DS} = 0.4\,\mathrm{V}$. Avci *et al.* [32] simulated the DG InAs p-TFET and found that even lower currents are achieved in the homojunction arrangement. The reason is the low density of states in the n-InAs source. To achieve a low sub-threshold

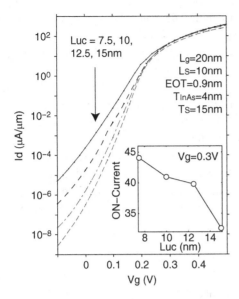

Fig. 6.13 Role of the undercut L_{uc} on the GaSb/InAs broken-gap TFET (10 nm L_d with $5 \times 10^{17}\,\mathrm{cm}^{-3}$ doping density and a 10 nm heavy-doping region, $4 \times 10^{18}\,\mathrm{cm}^{-3}$, at the drain contact to minimize the gate/drain coupling. Other parameters are the same as in Fig. 6.12). Increasing the undercut increases the gate control over the GaSb/InAs interface and reduces the sub-threshold swing. The inset shows the effect of undercut on the on-current.

swing the source doping needs to be low so that the Fermi level in the source is near the conduction band edge. This leads to low drive current because the junction electric field is then low. If the source doping is raised, the current can be increased, but if the Fermi level pushes into the band much beyond $3kT$ the sub-threshold swing tends toward thermionic due to the thermal tail in the expanded tunneling window.

6.5 Non-idealities

The motivating factors for development of the TFET for logic are its sub-thermal sub-threshold swing, low off-state leakage, sub-0.4 V operation, high on-current relative to the MOSFET at low voltage, and high integration density. Loss of any one of these attributes places an impediment to adoption of the TFET for computing and logic. This section outlines current understanding of the main factors which act to degrade TFET performance with a focus on III-Vs. These are traps, band tails, interface roughness, and alloy disorder. It is also important to consider what may be expected in terms of variability in comparison to the MOSFET. In keeping with the treatment thus far, the focus here is on fundamental aspects.

6.5.1 Traps

Traps are known to introduce discrete states in the bandgap. Trap-assisted tunneling couples the conduction and the valence bands when the bands are not overlapping. As is well known from the excess current that degrades the peak-to-valley current in the tunnel diode [33–36], tunneling paths through defects can provide leakage when the conduction and valence bands are misaligned. In the TFET this leakage degrades both the sub-threshold swing and raises the off-current.

Quantum transport models based on the non-equilibrium Green's function (NEGF) and that include trap-assisted-tunneling have been developed relatively recently for optical devices [37] and for InAs TFETs [38, 39]. These models are clarifying the role of traps in the tunneling process and how the traps degrade the sub-threshold swing. Pala and Esseni [38] show that even a single energy trap in the bandgap degrades the sub-threshold swing and that shallow traps are more important than deeper traps. The trap-assisted tunneling process requires phonons and this leads to a temperature-dependent I–V characteristic in the sub-threshold region, which is not expected from direct tunneling.

6.5.2   Band tails/phonons

It is often assumed that the semiconductor band edges are abrupt with zero density of states in the bandgap. However, the band edges are not abrupt and tails into the bandgap due to lattice disorder due to impurities, dopants, and phonons [40]. Interface roughness and alloy disorder also degrade TFET performance since they modify the local band edges.

The effects of heavy doping and bandgap narrowing have been simulated [41] through an empirical model in NEGF. The modeling of quantum transport subject to incoherent scattering of phonons has been demonstrated [42] in the NEGF formalism. This approach is computationally taxing for multiband models in one-dimension [43] and even more taxing for three-dimensional devices [44]. A computationally more-efficient model is needed.

Khayer and Lake [45] devised an empirical model to mimic the effects of band tails and studied their effect on the off-state of the TFET. As shown in Fig. 6.14 the empirical model introduces a finite density of states (DOS) into the bandgap region through an empirical parameter. An increase in the band tails and the DOS significantly reduces the sub-threshold swing of the corresponding TFET over several orders of magnitude. This work indicates the critical importance of the band tails in TFET off-state performance and the need to model these band tails with a physics-based model. Such a model needs to include phonon-scattering induced virtual states in the bandgap, bandgap narrowing due to heavy doping, and tunneling through spatially localized defects.

6.5.3   Interface roughness

Quantum transport modeling has shown that line edge roughness is of critical import-ance in graphene nanoribbons [46] as is interface roughness in resonant tunneling diodes [47]. Interface roughness is also of critical importance in the mobility of MOSFETs [48, 49], in NWs [50], and in NW TFETs [51]. Here we analyze interface roughness in III-V TFETs.

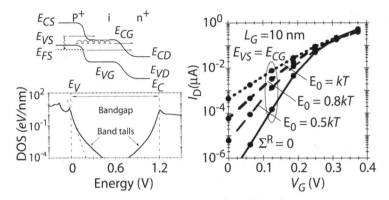

Fig. 6.14 Modeling of the role of traps in the bandgap on the off-current of a 10 nm TFET after Khayer and Lake [45]. The presence of a finite density of states in the gap enables tunneling in the nominally forbidden range and increases sub-threshold swing.

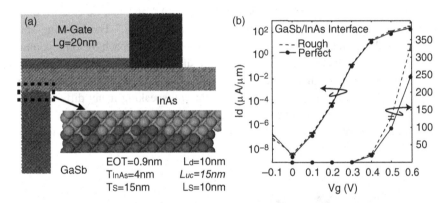

Fig. 6.15 Modeling of interface roughness in a GaSb/InAs broken-gap TFET with 10 nm L_d and a 10 nm heavy-doping region at the drain as in Fig. 6.12. (a) Schematic drawing of the TFET cross section and interface and (b) simulated transfer characteristic. Interface roughness does not significantly degrade the sub-threshold swing and increases the on-current slightly.

Figure 6.15 shows the numerical representation of an experimentally realized [52] TFET in the GaSb/InAs material system. The NEMO5 software [17, 18] used for these simulations can explicitly model interface roughness through an atomistic representation of the interface. The simulation assumes a finite width of the repeated periodic cell for the structure that is assumed to be infinitely deep in the transverse direction. Therefore the interface roughness modeled here has a periodicity of one unit cell of width 0.61 nm.

The NEMO5 simulations show that interface roughness at the GaSb/InAs interface does not have a detrimental effect on the sub-threshold swing. The off-state characteristic for the rough and the perfect interface TFET are virtually identical. The on-state is slightly increased by the interface roughness.

6.5.4   Alloy disorder

Alloys are widely accepted as having a well-defined band edge and effective mass. Strictly speaking, however, there is no repeated unit cell in an alloy that warrants the definition of a band edge or an effective mass. As such the band edge and effective mass of an alloy are defined only as mean quantities [53, 54]. This raises the question as to how small a device would have to be until the local order of a constituent alloy makes a significant difference on the overall device performance [55, 56]. Alloy disorder in SiGe buffer material around a slanted Si quantum well has been studied to understand the electronic structure valley splitting in the Si quantum well [57]. Simulations of alloy disorder in AlGaAs NWs [58] show that the transmission is significantly reduced relative to a smoothed alloy assumption.

Alloy disorder can be modeled in AlGaSb/InAs TFETs through an explicit atomistic representation with a unit cell thickness of 0.61 nm as depicted in Fig. 6.16. A perfect binary GaSb/InAs TFET is here compared to a AlGaSb/InAs TFET. The AlGaSb is represented in two fundamentally different ways. The virtual crystal approximation (VCA) averages the Hamiltonian elements of "Al" and "Ga" to create a smooth material which effectively creates a new "atom" called "AlGa." The resulting Hamiltonian is spatially homogeneous and smooth. The VCA approach is very common in alloy electronic structure and carrier transport simulations [42]. A more realistic representation considers explicitly the different species of "Al" and "Ga" atoms and places them explicitly into the random alloy [58]. Strictly speaking many different alloy configurations need to be considered to obtain an average device behavior. In fact for the results to be presented, 20 different simulations were made to provide the average.

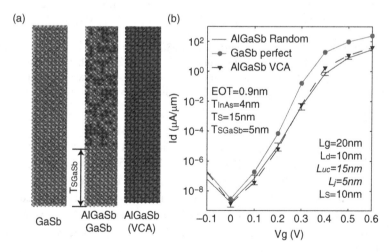

Fig. 6.16 Influence of alloy disorder scattering on n-TFET performance. (a) Graphical representation of GaSb, a p-GaSb/n-AlGaSb atomistically disordered random alloy channel source heterojunction, and a smoothed out AlGaSb channel represented in the VCA approximation. (b) Drain current vs. gate/source voltage using the same δ-doping layer and heavy drain doping as discussed in Fig. 6.12.

To consider the effect of alloy disorder an enhancement-mode TFET based on the one shown in Fig. 6.12(d) is used. The pure GaSb p-type section is replaced by an $Al_{0.45}Ga_{0.55}Sb$ staggered gap device with carriers injected from a pure GaSb source, Fig. 6.16(a). The "perfect" GaSb device is compared to the AlGaSb device performances computed in the two different alloy models in Fig. 6.16(a). In contrast to the previous numerical studies in NWs [58] the UTB TFET appears to be weakly dependent on the explicit alloy disorder scattering.

Previous explicit random alloy disorder studies that found significant alloy disorder effects have focused on NWs rather than UTB structures. The UTB device analyzed here is represented with a single unit cell in the periodic direction, thus creating a periodic order in the depth of the device and atomistic disorder along the principal transport direction. The limitation to a single unit cell may be critical in this study here. A significantly increased computational burden is required to expand the unit cell depth.

6.5.5   Variability

A significant question in the development of TFETs is whether variability will be greater or less than in MOSFETs. This question has been explored by Esseni and Pala [39] who conclude that the location of traps leads to greater variability in InAs NW TFETs than in InAs NW MOSFETs. The effects of traps in the MOSFET can be mitigated by greater EOT scaling in the TFET; however, since the traps are providing new tunneling paths, their location with respect to the source cause variations in the off-current. Depending on location the traps can cause the off-current to vary about an order of magnitude and sub-threshold swing can be increased by approximately $2\times$.

Avci et $al.$ [59] have utilized the atomistic tight-binding approach of NEMO5 to explore and compare the 20 nm gate length InAs TFET with the MOSFET. Using known process variations, simulations were performed to explore changes in on-current and off-current resulting from variations in gate length, channel random dopant fluctuation, oxide thickness, source random dopant fluctuation, channel thickness, and work function. This analysis shows that TFETs are generally more sensitive to variations than the MOSFET, but that the differences are not decidely significant. These studies have been recently extended to the GaSb/InAs TFET with similar findings [60]. The dominant source of variation in the TFET is the dependence on gate-metal work function.

6.6   Experimental results

There are a growing number of experimental reports on III-V TFETs. The first demonstration of a III-V TFET was reported by Mookerjea et $al.$ [61, 62] who used a p-i-n InGaAs homojunction and a sidewall gate. This was followed by improvements in on-current and gate dielectrics by Zhao et $al.$ [63, 64], also using an InGaAs channel. The introduction of heterojunctions by Mohata et $al.$ [65–68] resulted in steady improvements in on-current.

InAs TFETs, with the gate-field oriented in-line with the tunneling direction, were first demonstrated in InGaAs on Si by Ford et $al.$ [69]. Heterojunction TFETs

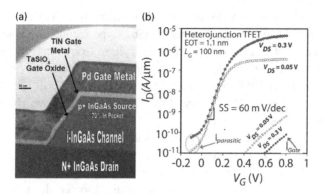

Fig. 6.17 InGaAs TFET with low sub-threshold after Dewey *et al.* [80]. (a) Transmission electron micrograph of an InGaAs heterojunction TFET showing a 6 nm $In_{0.7}Ga_{0.3}As$ pocket between the intrinsic layer and the source. (b) Drain current and gate leakage vs. gate/source voltage for a gate length of 100 nm with an EOT of 1.1 nm.

with the in-line-gate geometry appeared simultaneously by Li *et al.* in p-AlGaSb/n-InAs [70, 71] and Zhou *et al.* in p-InP/n-InGaAs [72–74], followed by Zhou *et al.* in p-AlGaSb/n-InAs [75]. On-currents have continued to advance steadily since the first reports, see Zhou [76], Dey *et al.* [77, 78], and Bijesh *et al.* [79]. The best result to date is by Bijesh *et al.* [79], 240 $\mu A/\mu m$ at $V_{DS} = V_{GS} = 0.5$ V, but these results should not be expected to stand. Currents as high as 740 $\mu A/\mu m$ have been reported [79], but these and other reported high on-currents are achieved by high gate overdrives with gate/source voltage much greater than drain/source voltage.

In the III-V semiconductors, sub-threshold swings of less than 60 mV/dec have been reported by Dewey *et al.* [80, 81] using InGaAs in a single-gate geometry with p-i-n channel doping. Shown in Fig. 6.17 is the device cross section and transfer characteristics. Subsequent comparison of the experiments results against atomistic simulations showed that there are no unaccountable parameters and that with scaling and improvement in electrostatics the TFET is a feasible low-voltage device candidate [82].

One other report of low sub-threshold swing has appeared. Tomioka *et al.* [83] reported sub-60 mV/dec sub-threshold swing in a p-Si/n-InAs NW TFET. Noguchi *et al.* [84] achieved a sub-threshold swing as low as 64 mV/dec in an InGaAs channel with a Zn-diffused source, using a Ta/Al_2O_3 gate stack with an EOT of 1.4 nm.

The best electrostatic control in the TFET should be obtained in the GAA geometry and for this reason NW TFETs are of particular importance. Progress in nanowire development for TFETs is advancing quickly: InAs/GaSb by Bjorg *et al.* [85], Si/InAs by Tomioka and Fukui [86], InAs/Si by Moselund *et al.* [87] and Riel *et al.* [88], and GaSb/InAsSb by Borg *et al.* [89] and Dey *et al.* [77, 78]. Representative progress is shown in Fig. 6.18 from Dey *et al.* [78]. In this n-channel TFET the source is GaSb and the channel is InAsSb. The transistor characteristics are shown for both positive and negative drain biases. For negative gate biases the channel/source junction becomes forward biased and in this bias polarity clear Esaki tunneling is observed. This is a

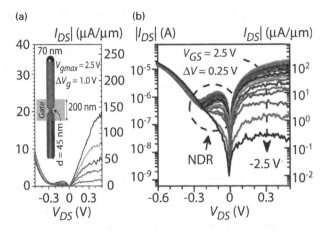

Fig. 6.18 p-GaSb/n-InAsSb nanowire TFET after Dey *et al.* [77]. (a) Common source characteristics. (b) Log plot of the same device showing clear Esaki tunneling with negative differential resistance (NDR) when the source junction is forward biased.

signature characteristic apparent in all III-V TFETs and an excellent proof of the tunneling current mechanism.

At the time of writing only one III-V *p*-TFET has been demonstrated, this is by Rajamohanan *et al.* [90] using an n-In$_{0.7}$Ga$_{0.3}$As/p-GaAs$_{0.35}$Sb$_{0.65}$ staggered-gap heterostructure. This is an area where much more work is needed.

6.7   Conclusions

Moore's law device scaling is approaching its end at atomistic device scales. The reduction of supply voltage for MOSFETs is fundamentally limited to the 60 mV/dec sub-threshold swing. Tunnel FETs provide a current mechanism which can be controlled at less than the 60 mV/dec limit, opening a way to reduce the system operation voltage.

The materials, doping, gate configurations, and layer structures of the TFET offer opportunities for end-of-roadmap device design. TFET device optimization requires new rules and design methodologies. This chapter highlights the fundamental design choices and device limitations for TFETs. The field of TFETs is evolving rapidly [91]. Future computing systems will become much more energy efficient if the materials and process challenges of the TFET can be solved.

Acknowledgments

This research on TFETs was principally funded by the Semiconductor Research Corporation's Nanoelectronics Research Initiative, through the Midwest Institute for Nanoelectronics Discovery (MIND). The previously unpublished simulation results were enabled by the development of NEMO5 and we acknowledge the work of Professor Michael Povolotskyi, Professor Tillmann Kubis, Dr. James Fonseca,

Dr. Bozidar Novakovic, Dr. Hong-Yun Park, Mr. Yu He, and Mr. Daniel Mejia for their help and support in this effort. NEMO5 was funded under an NSF Peta-Apps grant OCI-0749140, and nanoHUB.org computational resources were operated by the Network for Computational Nanotechnology funded by the NSF under grants EEC-1227110, EEC-0228390, EEC-0634750, OCI-0438246, OCI-0721680.

References

[1] W. Shockley, "The theory of *p-n* junctions in semiconductors and *p-n* junction transistors." *Bell Systems Technical Journal*, **28**, 439–489 (1949).

[2] J. R. Haynes & W. Shockley, "Minority carriers in semiconductors." Semiconductor Electronics Education Committee film, www.youtube.com/watch?v=zYGHt-TLTl4.

[3] C. H. Jan, U. Bhattacharya, R. Brain *et al.*, "A 22nm SoC platform technology featuring 3-D tri-gate and high-k/metal gate, optimized for ultra low power, high performance and high density SoC applications." In *Electron Devices Meeting (IEDM), 2012 IEEE International*, pp. 3.1.1–3.1.4 (2012).

[4] A. C. Seabaugh & Q. Zhang, "Low-voltage tunnel transistors for beyond CMOS logic." *Proceedings of the IEEE*, **98**, 2095–2110 (2010).

[5] A. M. Ionescu & H. Riel, "Tunnel field-effect transistors as energy-efficient electronic switches." *Nature*, **479**, 329–337 (2011).

[6] A. Seabaugh, "The tunneling transistor." *IEEE Spectrum*, **50**, 35–62 (2013).

[7] M. Luisier & G. Klimeck, "Simulation of nanowire tunneling transistors: From the Wentzel-Kramers-Brillouin approximation to full-band phonon-assisted tunneling." *Journal of Applied Physics*, **107**, 084507 (2010).

[8] D. Jena, "Tunneling transistors based on graphene and 2-D crystals." *Proceedings of the IEEE*, **101**, 1585–1602 (2013).

[9] S. M. Sze & K. K. Ng, *Physics of Semiconductor Devices*, 3rd edn. (New York: Wiley-Interscience, 2007), p. 103.

[10] S. E. Laux, "Computation of complex band structures in bulk and confined structures." In *2009 International Workshop on Computational Electronics*, pp. 1–4 (2009).

[11] Q. Zhang, S. Sutar, T. Kosel, & A. Seabaugh, "Fully-depleted Ge interband tunnel transistor: modeling and junction formation." *Solid-State Electronics*, **53**, 30–35 (2009).

[12] S. Mukherjee, A. Paul, N. Neophytou *et al.*, "Band structure lab." (2013). https://nanohub.org/resources/bandstrlab http://dx.doi.org/10.4231D3HD7NS6N.

[13] M. Luisier & G. Klimeck, "Investigation of In$_x$Ga$_{1-x}$As ultra-thin-body tunneling FETs using a full-band and atomistic approach." In *2009 Simulation of Semiconductor Processes and Devices (SISPAD)*, pp. 1–4 (2009).

[14] Q. Zhang, T. Fang, H. Xing, A. Seabaugh, & D. Jena, "Graphene nanoribbon tunnel transistors." *IEEE Electron Device Letters*, **29**, 1344–1346 (2008).

[15] M. A. Khayer & R. K. Lake, "Performance off *n*-type InSb and InAs nanowire field-effect transistors." *IEEE Transactions on Electron Devices*, **55**, 2939–2945 (2008).

[16] Q. Zhang, Y. Lu, C. Richter, D. Jena, & A. Seabaugh, "Optimum band gap and supply voltage in tunnel FETs." *IEEE Transactions on Electron Devices*, **61**, 2719–2724 (2014).

[17] NEMO5 is available under an academic open source license at: https://nanohub.org/groups/nemo5distribution.

[18] S. Steiger, M. Povolotskyi, H.-H. Park, T. Kubis, & G. Klimeck, "NEMO5: a parallel multiscale nanoelectronics modeling tool." *IEEE Transactions on Nanotechnology*, **10**, 1464 (2011).

[19] G. Klimeck, R. Bowen, T. Boykin, & T. Cwik, "sp3s* tight-binding parameters for transport simulations in compound semiconductors." *Superlattices and Microstructures*, **27**, 519–524 (2000).

[20] T. Boykin, G. Klimeck, R. Bowen, & F. Oyafuso, "Diagonal parameter shifts due to nearest-neighbor displacements in empirical tight-binding theory." *Physics Reviews B*, **66**, 125207 (2002).

[21] G. Klimeck, F. Oyafuso, T. Boykin, R. Bowen, & P. Allmen, "Development of a nanoelectronic 3-D (NEMO 3-D) simulator for multimillion atom simulations and its application to alloyed quantum dots." *Computer Modeling in Engineering and Science (CMES)*, **3**(5), 601–642 (2002).

[22] D. L. Smith & C. Maihiot, "Proposal for strained type II superlattice infrared detectors." *Journal of Applied Physics*, **62**, 2545 (1987).

[23] G. A. Sai-Halasz, R. Tsu, & L. Esaki, "A new semiconductor superlattice." *Applied Physics Letters* **30**, 651 (1977).

[24] M. Luisier & G. Klimeck, "Performance comparisons of tunneling field-effect transistors made of InSb, carbon, and GaSb-InAs broken gap heterostructures." In *Electron Devices Meeting (IEDM), 2009 IEEE International*, pp. 37.6.1–37.6.4 (2009).

[25] S. O. Koswatta, S. J. Koester, & W. Haensch, "On the possibility of obtaining MOSFET-like performance and sub-60-mV/dec swing in 1-D broken-gap tunnel transistors." *IEEE Transactions on Electron Devices*, **57**, 3222–3230 (2010).

[26] M. Luisier & G. Klimeck, "Atomistic, full-band design study of InAs band-to-band tunneling field-effect transistors." *IEEE Electron Device Letters*, **30**, 602–604 (2009).

[27] C. Hu, D. Chou, P. Patel, & A. Bowonder, "Green transistor – a V_{DD} scaling path for future low power ICs." In *VLSI Technology, Systems and Applications, 2008. VLSI-TSA 2008. International Symposium on*, pp. 14–15 (2008).

[28] S. Agarwal, G. Klimeck, & M. Luisier, "Leakage reduction design concepts for low power vertical tunneling field-effect transistors." *IEEE Electronic Device Letters*, **31**, 621–623 (2010).

[29] M. Luisier, S. Agarwal, & G. Klimeck, "Tunneling field-effect transistor with low leakage current." *US Patent No. 8,309,989*, November 13 (2012).

[30] Y. Lu, G. Zhou, R. Li *et al.*, "Performance of AlGaSb/InAs TFETs with gate electric field and tunneling direction aligned." *IEEE Electron Device Letters*, **33**, 655–657 (2012).

[31] J. Knoch & J. Appenzeller, "Modeling of high-performance p-Type III-V heterojunction tunnel FETs." *IEEE Electron Device Letters*, **31**, 305–307, (2010).

[32] U. E. Avci, R. Rios, K. J. Kuhn, & I. A. Young, "Comparison of power and performance for the TFET and MOSFET and considerations for P-TFET." In *Nanotechnology (IEEE-NANO), 2011 IEEE Conference on*, pp. 869–872 (2011).

[33] R. P. Nanavati & C. A. M. De Andrade, "Excess current in gallium arsenide tunnel diodes." *Proceedings of the IEEE*, **52**, 869–870 (1964).

[34] R. P. Nanavati & M. Eisencraft, "On thermal and excess currents in GaSb tunnel diodes." *IEEE Transactions on Electron Devices*, **15**, 796–797 (1968).

[35] A. Seabaugh & R. Lake, "Tunnel diodes." In *Encyclopedia of Applied Physics*, vol. 22 (New York: Wiley, 1998), pp. 335–359.

[36] C. D. Bessire, M. T. Björk, H. Schmid, A. Schenk, K. B. Reuter, & H. Riel, "Trap-assisted tunneling in Si-InAs nanowire heterojunction tunnel diodes." *Nano Letters*, **11**, 4195–4199 (2011).

[37] S. Steiger, R. G. Veprek, & B. Witzigmann, "Electroluminescence from a quantum-well LED using NEGF." In *2009 International Workshop on Computational Electronics, (IWCE)*, pp. 1–4 (2009).

[38] M. G. Pala & D. Esseni, "Interface traps in InAs nanowire tunnel-FETs and MOSFETs – part I: model description and single trap analysis in tunnel-FETs." *IEEE Transactions on Electron Devices*, **60**, 2795–2801 (2013).

[39] D. Esseni & M. G. Pala, "Interface traps in InAs nanowire tunnel FETs and MOSFETs – part II: comparative analysis and trap-induced variability." *IEEE Transactions on Electron Devices*, **60**, 2802–2807 (2013).

[40] M. H. Cohen, M. Y. Chou, E. N. Economou, S. John, & C. M. Soukoulis, "Band tails, path integrals, instantons, polarons, and all that." *IBM Journal of Research and Development*, **32**, 82–92 (1988).

[41] W.-S. Cho, M. Luisier, D. Mohata *et al.*, "Full band modeling of homo-junction InGaAs band-to-band tunneling diodes including band gap narrowing." *Applied Physics Letters*, **100**, 063504 (2012).

[42] R. Lake, G. Klimeck, R. Bowen, & D. Jovanovic, "Single and multiband modeling of quantum electron transport through layered semiconductor devices." *Journal of Applied Physics*, **81**, 7845–7869 (1997).

[43] C. Rivas, R. Lake, G. Klimeck *et al.*, "Full band simulation of indirect phonon-assisted tunneling in a silicon tunnel diode with delta-doped contacts." *Applied Physics Letters*, **78**, 814–816 (2001).

[44] M. Luisier & G. Klimeck, "Atomistic full-band simulations of Si nanowire transistors: effects of electron phonon scattering." *Physics Reviews B*, **80**, 155430 (2009).

[45] M. A. Khayer and R. K. Lake, "Effects of band-tails on the subthreshold characteristics of nanowire band-to-band tunneling transistors." *Journal of Applied Physics*, **110**, 074508 (2011).

[46] M. Luisier & G. Klimeck, "Performance analysis of statistical samples of graphene nanoribbon tunneling transistors with line edge roughness." *Applied Physics Letters*, **94**, 223505 (2009).

[47] G. Klimeck, R. Lake, & D. Blanks, "Numerical approximations to the treatment of interface roughness scattering in resonant tunneling diodes." *Semiconductor Science and Technology*, **13**, A165 (1998).

[48] S. Takagi, A. Toriumi, M. Iwase, & H. Tango, "On the universality of inversion layer mobility in Si MOSFET's: Part II – effects of surface orientation." *IEEE Transactions on Electron Devices*, **41**, 2363–2368 (1994).

[49] S. Jin, M. V. Fischetti, & T.-W. Tang, "Modeling of electron mobility in gated silicon nanowires at room temperature: surface roughness scattering, dielectric screening, and band nonparabolicity." *Journal of Applied Physics*, **102**, 083715 (2007).

[50] S. G. Kim, A. Paul, M. Luisier, T. Boykin, & G. Klimeck, "Full three-dimensional quantum transport simulation of atomistic interface roughness in silicon nanowire FETs." *IEEE Transactions on Electron Devices*, **58**, 1371–1380 (2011).

[51] F. Conzatti, M. G. Pala, & D. Esseni, "Surface-roughness-induced variability in nanowire InAs tunnel FETs." *IEEE Electron Device Letters*, **33**, 806–808 (2012).

[52] R. Li, Y. Lu, G. Zhou *et al.*, "AlGaSb/InAs tunnel field-effect transistor with on-current of 78 μA/μm at 0.5 V." *IEEE Electron Device Letters*, **33**, 363–365 (2012).

[53] T. Boykin, N. Kharche, & G. Klimeck, "Brillouin-zone unfolding of perfect supercells having nonequivalent primitive cells illustrated with a Si/Ge tight-binding parameterization." *Physics Reviews B* **76**, 035310 (2007).

[54] T. Boykin, N. Kharche, G. Klimeck, & M. Korkusinski, "Approximate bandstructures of semiconductor alloys from tight-binding supercell calculations." *Journal of Physics: Condensed Matter*, **19**, 036203 (2007).

[55] F. Oyafuso, G. Klimeck, R. Bowen, & T. Boykin, "Atomistic electronic structure calculations of unstrained alloyed systems consisting of a million atoms." *Journal of Computational Electronics*, **1**, 317–321 (2002).

[56] F. Oyafuso, G. Klimeck, R. Bowen, T. Boykin, & P. Allmen, "Disorder induced broadening in multimillion atom alloyed quantum dot systems." *Physica Status Solidi (C)*, **0**, 1149–1152 (2003).

[57] N. Kharche, M. Prada, T. Boykin, & G. Klimeck, "Valley-splitting in strained silicon quantum wells modeled with 2 miscuts, step disorder, and alloy disorder." *Applied Physics Letters* **90**, 092109 (2007).

[58] T. Boykin, M. Luisier, A. Schenk, N. Kharche, & G. Klimeck, "The electronic structure and transmission characteristics of AlGaAs nanowires." *IEEE Transactions on Nanotechnology*, **6**, 43–47 (2007).

[59] U. E. Avci, R. Rios, K. Kuhn, & I. A. Young, "6B-5 comparison of performance, switching energy and process variations for the TFET and MOSFET in logic." In *VLSI Technology (VLSIT), 2011 Symposium on*, pp. 124–125 (2011).

[60] U. E. Avci, D. H. Morris, S. Hasan, & R. Kotlyar, "Energy efficiency comparison of nanowire heterojunction TFET and Si MOSFET at Lg = 13 nm, including P-TFET and variation considerations." In *Electron Devices Meeting (IEDM), 2013 IEEE International*, pp. 33.4.1–33.4.4 (2013).

[61] S. Mookerjea, D. Mohata, R. Krishnan *et al.*, "Experimental demonstration of 100nm channel length $In_{0.53}Ga_{0.47}As$-based vertical inter-band tunnel field effect transistors (TFETs) for ultra low-power logic and SRAM applications." In *Electron Devices Meeting (IEDM), 2009 IEEE International*, pp. 13.7.1–13.7.3 (2009).

[62] S. Mookerjea, D. Mohata, T. Mayer, V. Narayanan, & S. Datta, "Temperature-dependent I–V characteristics of a vertical $In_{0.53}Ga_{0.47}As$ tunnel FET." *IEEE Electron Device Letters*, **31**, 564–566, (2010).

[63] H. Zhao, Y. Chen, Y. Wang, F. Zhou, F. Xue, & J. Lee, "$In_{0.7}Ga_{0.3}As$ tunneling field-effect transistors with an I_{on} of 50 µA/µm and a subthreshold swing of 86 mV/dec using HfO_2 gate oxide." *IEEE Electron Devices Letters*, **31**, 1392–1394 (2010).

[64] H. Zhao, Y. Chen, Y. Wang, F. Zhou, F. Xue, and J. Lee, "InGaAs tunneling field-effect-transistors with atomic-layer-deposited gate oxides." *IEEE Transactions on Electron Devices*, **58**, 2990–2995 (2011).

[65] D. K. Mohata, R. Bijesh, S. Mujumdar *et al.*, "Demonstration of MOSFET-like on-current performance in arsenide/antimonide tunnel FETs with staggered hetero-junctions for 300 mV logic applications." In *Electron Devices Meeting (IEDM), 2011 IEEE International*, pp. 33.5.1–4 (2011).

[66] D. K. Mohata, R. Bijesh, V. Saripalli, T. Mayer, & S. Datta, "Self-aligned gate nanopillar $In_{0.53}Ga_{0.47}As$ vertical tunnel transistor." In *2011 Device Research Conference (DRC)*, pp. 203–204 (2011).

[67] D. Mohata, S. Mookerjea, A. Agrawal *et al.*, "Experimental staggered-source and n+ pocket-doped channel III-V tunnel field-effect transistors and their scalabilities." *Applied Physics Express*, **4**, 024105 (2011).

[68] D. Mohata, B. Rajamohanan, T. Mayer *et al.*, "Barrier-engineered arsenide–antimonide heterojunction tunnel FETs with enhanced drive current." *IEEE Electron Device Letters*, **33**, 1568–1570 (2012).

[69] A. C. Ford, C. W. Yeung, S. Chuang *et al.*, "Ultrathin body InAs tunneling field-effect transistors on Si substrates." *Applied Physics Letters*, **98**, 113105 (2011).

[70] R. Li, Y. Lu, S. D. Chae *et al.*, "InAs/AlGaSb heterojunction tunnel field-effect transistor with tunnelling in-line with the gate field." *Physica Status Solidi (C)*, **9**, 389–392 (2011).

[71] R. Li, Y. Lu, G. Zhou *et al.*, "InAs/AlGaSb heterojunction tunnel FET with InAs airbridge drain." In *Compound Semiconductors (ISCS), 2011 International Symposium on*, pp. 189–190 (2011).

[72] G. Zhou, Y. Lu, R. Li *et al.*, "Self-aligned $In_{0.53}Ga_{0.47}As/InP$ vertical tunnel FET." In *Compound Semiconductor Manufacturing Technology (CS ManTech), 2011 International Conference on*, (2011).

[73] G. Zhou, Y. Lu, R. Li *et al.*, "Vertical InGaAs/InP tunnel FETs with tunneling normal to the gate." *IEEE Electron Devices Letters*, **32**, 1516–1518 (2011).

[74] G. Zhou, Y. Lu, R. Li *et al.*, "InGaAs/InP tunnel FETs with a subthreshold swing of 93 mV/dec and I_{on}/I_{off} ratio near 10^6." *IEEE Electron Devices Letters*, **33**, 782–84 (2012).

[75] G. Zhou, Y. Lu, R. Li *et al.*, "Self-aligned $InAs/Al_{0.45}Ga_{0.55}Sb$ vertical tunnel FETs." In *Device Research Conference (DRC), 2011 69th Annual*, pp. 205–206 (2011).

[76] G. Zhou, R. Li, T. Vasen *et al.*, "Novel gate-recessed vertical InAs/GaSb TFETs with record high I_{on} of $180\,\mu A/\mu m$ at $V_{DS} = 0.5$ V." In *Electron Devices Meeting (IEDM), 2012 IEEE International*, pp. 32.6.1–32.6.4 (2012).

[77] A. W. Dey, B. M. Borg, B. Ganjipour *et al.*, "High current density InAsSb/GaSb tunnel field effect transistors." In *Devices Research Conference (DRC), 2012 IEEE*, pp. 205–206 (2012).

[78] A. W. Dey, B. M. Borg, B. Ganjipour *et al.*, "High-current GaSb/InAs(Sb) nanowire tunnel field-effect transistors." *IEEE Electron Device Letters*, **34**, 211–213 (2013).

[79] R. Bijesh, H. Liu, H. Madan *et al.*, "Demonstration of $In_{0.9}Ga_{0.1}As/GaAs_{0.18}Sb_{0.82}$ near broken-gap tunnel FET with $I_{on} = 740\,\mu A/\mu m$, $G_M = 700\,\mu S/\mu m$ and gigahertz switching performance at $V_{DS} = 0.5$V." In *Electron Devices Meeting (IEDM), 2013 IEEE International*, pp. 28.2.1–28.2.4 (2013).

[80] G. Dewey, B. Chu-Kung, J. Boardman *et al.*, "Fabrication, characterization, and physics of III-V heterojunction tunneling field effect transistors (H-TFET) for steep sub-threshold swing." *Electron Devices Meeting (IEDM), 2011 IEEE International*, pp. 33.6.1–33.6.4 (2011).

[81] G. Dewey, B. Chu-Kung, R. Kotlyar, M. Metz, N. Mukherjee, & M. Radosavljevic, "III-V field effect transistors for future ultra-low power applications." In *VLSI Technology (VLSIT), 2012 Symposium on*, pp. 45–46 (2012).

[82] U. E. Avci, S. Hasan, D. E. Nikonov, R. Rios, K. Kuhn, & I. A. Young, "Understanding the feasibility of scaled III-V TFET for logic by bridging atomistic simulations and experimental results." In *VLSI Technology (VLSIT), 2012 Symposium on*, pp. 183–184 (2012).

[83] K. Tomioka, M. Yoshimura, & T. Fukui, "Steep-slope tunnel field-effect transistors using III-V nanowire/Si heterojunction." In *VLSI Technology (VLSIT), 2012 Symposium on*, pp. 47–48 (2012).

[84] M. Noguchi, S. Kim, M. Yokoyama *et al.*, "High I_{on}/I_{off} and low subthreshold slope planar-type InGaAs tunnel FETs with Zn-diffused source junctions." In *Electron Devices Meeting (IEDM), 2013 IEEE International*, pp. 28.1.1–4 (2013).

[85] B. M. Borg, K. A. Dick, B. Ganjipour, M.-E. Pistol, L.-E. Wernersson, & C. Thelander, "InAs/GaSb heterostructure nanowires for tunnel field-effect transistors." *Nano Letters*, **10**, 4080–4085 (2010).

[86] K. Tomioka & T. Fukui, "Tunnel field-effect transistor using InAs nanowire/Si heterojunction." *Applied Physics Letters*, **98**, 083114 (2011).

[87] K. E. Moselund, H. Schmid, C. Bessire, M. T. Bjork, H. Ghoneim, & H. Riel, "InAs–Si nanowire heterojunction tunnel FETs." *IEEE Electron Device Letters*, **33**, 1453–1455 (2012).

[88] H. Riel, K. E. Moselund, C. Bessire *et al.*, "InAs-Si heterojunction nanowire tunnel diodes and tunnel FETs." In *Electron Devices Meeting (IEDM), 2012 IEEE International*, pp. 16.6.1–16.6.4 (2012).

[89] B. Mattias Borg, M. Ek, K. A. Dick *et al.*, "Diameter reduction of nanowire tunnel heterojunctions using in situ annealing." *Applied Physics Letters*, **99**, 203101 (2011).

[90] B. Rajamohanan, D. Mohata, D. Zhernokletov *et al.*, "Low-temperature atomic-layer-deposited high-κ dielectric for p-channel $In_{0.7}Ga_{0.3}As/GaAs_{0.35}Sb_{0.65}$ heterojunction tunneling field-effect transistor." *Applied Physics Express*, **6**, 101201 (2013).

[91] H. Lu & A. Seabaugh, "Tunnel field-effect transistor: state-of-the-art." *IEEE Journal of the Electron Devices Society*, **2**, 44–49 (2014).

7 Graphene and 2D crystal tunnel transistors

Qin Zhang, Pei Zhao, Nan Ma, Grace (Huili) Xing, and Debdeep Jena

7.1 What is a low-power switch?

Transistors in the traditional field effect geometry operate by the injection of mobile carriers – electrons or holes – from a source reservoir to the drain reservoir through a conducting channel region. The carriers enter the channel region by surmounting an electrostatic potential barrier. The gate electrode controls the height of this barrier capacitively. The carriers in the source reservoir are in thermal equilibrium with the source contact. This means that the carriers, say electrons, are distributed in energy in the conduction band according to the Fermi–Dirac distribution $f(E) = 1/1[1+\exp((E - E_F)/kT)]$. The Maxwell–Boltzmann approximation $f(E) \sim \exp[- E/kT]$ of the Fermi–Dirac distribution for large energies represents the high-energy tail of the distribution. There are electrons in this tail with energy higher than the potential barrier; the gate cannot stop them from being injected into the channel. This leads to a sub-threshold "leakage" drain current $I_D \sim \exp[qV_{GS} / kT]$, which leads to the well-known sub-threshold slope (S) requirement of $S \sim (kT / q)\ln 10$ ~ 60 mV/dec change of current. Methods to make the SS steeper than the 300 K value of 60 mV/dec value are expected to substantially lower the power dissipation in digital logic and computation [1, 2]. The methods must explore novel mechanisms of charge transport, or of electrostatic gating. This chapter focuses on transport.

The high-energy tail of electrons exists because of the available density of states (DOS) $D_C(E)$ of the conduction band; the electron distribution in energy is $n(E) = D_C(E)f(E)$. If the DOS were cut off, there would be no tail, and it is possible to obtain S less than 60 mV/dec. This sort of energy filtering is possible if we replace the n-type source for electrons by a p-type source, which has a valence band maximum and zero DOS above. For injection into the channel of the n-FET, the electrons cannot undergo the traditional drift/diffusion process, but have to quantum mechanically *tunnel* through the bandgap. This energy-filtering scheme to achieve sub-60 mV/dec switching is the central idea behind the tunneling FET (or TFET). The TFET is expected to be a low-power switch compared to traditional FETs because the steep S enabled by tunneling will facilitate a reduction of the voltage necessary for switching. The challenge is to maintain a high current when the device has switched on. TFETs have been demonstrated with traditional 3D crystal semiconductors and their heterostructures. To date, Si, Ge, and SiGe material-based TFETs have been realized, as have III-V embodiments based on InAlGaAs and InAlGaSb materials that use the rich range of homojunction and heterojunctions with staggered and broken gap alignments. Rapid progress is being made on the design, control, and performance of such TFETs based on 3D crystals. Their performance is also being actively benchmarked, and new avenues for their application are being sought.

Compared to 3D crystal semiconductors and their heterostructures, 2D crystals such as graphene, BN, and the transition metal dichalcogenides such as MoS_2 are relatively new materials whose intrinsic electronic, structural, and optical properties are in the process of being understood. Many building blocks of devices such as controlled growth of the crystals, chemical doping, low-resistance contacts, dielectric integration, etc. are in an embryonic stage for this material family. However, because of their intrinsic scalability, they present an exciting alternative to traditional 3D semiconductors for realizing TFETs in the future. Because of the relative material immaturity and the consequent lack of substantial TFET experimental data for 2D crystal semiconductors, a large part of this chapter will focus on modeling and projections. The experimental advances are also reviewed.

7.2   Brief review of 2D crystal materials and devices

Traditional semiconductors such as Si, Ge, GaAs, GaN, etc. are formed by covalent bonding between the constituent atoms. The chemical nature of the bonds is sp^3, formed by the hybridization of an s orbital with three p orbitals. The resulting covalent bonds are tetrahedral, and are thus intrinsically three-dimensional. In contrast, 2D crystals are sheets – with sp^2 chemical bonds for graphene and BN. For the transition metal dichalcogenide family semiconductors such as MoS_2, WS_2, etc., the covalent bonds involve d-orbitals in addition to s and p. The chemical bonds are *saturated*, imparting these layered materials their characteristic chemical inertness. For example, to form a chemical bond to a carbon atom in graphene, we have to convert an sp^2 bond to sp^3 – this requires a buckling of the crystal structure and consequently large formation energy. Multiple layers are weakly bonded by van der Waals interactions. The electronic band structure evolves with the number of layers: typically single layers are direct bandgap with their conduction band minima and valence band maxima at the K and K' points – the vertices – of a planar hexagonal Brillouin zone. The conduction band minimum moves into the hexagonal Brillouin zone from the vertices for multilayer counterparts, making them indirect bandgap. A single layer of graphene has an energy bandgap of 0 eV, BN in excess of 5 eV, and the TMD 2D crystals have gaps in the 1–2 eV range. Consequently, traditional FETs made of graphene exhibit very low on/off ratios, and forming ohmic contacts to BN has remained very challenging. On the other hand, the TMD 2D crystals behave for the most part as traditional semiconductors. Due to the existence of a manageable bandgap, devices made from them exhibit the two central FET characteristics: on/off ratio of several orders of magnitude and current saturation. Due to the crystal structure and the lack of dangling bonds, such FETs also exhibit low S values of close to 60 mV/dec. The carrier mobilities in the TMD materials are in the 100s of cm^2/V s, and are currently under intense scrutiny. Because of their sub-1 nm thicknesses with no dangling bonds, these materials are being explored for the possibility of the near complete elimination of thickness variations in thin body transistors. Integration of dielectrics poses challenges because of the chemical inertness, but substantial progress is being made by atomic layer deposition. The possibility of 2D crystal dielectrics such as BN also exists. Chemical doping and low-resistance ohmic contacts pose challenges at this point, but should be achievable in the near future.

7.3   Carbon nanotubes and graphene nanoribbons

An energy bandgap can be opened in graphene by quantum confinement. Carbon nanotubes are cylinders formed by the seamless wrapping of ribbons of the graphene crystal framework. They exhibit bandgaps depending on the chirality of the wrapping and diameter of the cylinder. Similarly, open graphene ribbons exhibit bandgap depending on the nature of the edges. In this section, tunneling transport and TFETs formed from these sp^2 nanostructructures of carbon are discussed.

7.3.1   First demonstration of less than 60 mV/dec sub-threshold swing in TFETs with carbon nanotubes

The first TFET with a sub-threshold swing less than 60 mV/dec was demonstrated in 2004 with carbon nanotubes [3]. Figure 7.1(a) shows the double gate structure of this transistor: the Si gate with a negative voltage dopes the source and drain to be p$^+$, while the Al gate controls the channel. The drain current as a function of the Al gate voltage is shown in Fig. 7.1(b), where the sub-threshold swing of the positive branch is measured to be 40 mV/dec, less than the MOSFET's fundamental limit at room temperature. The energy band diagram in Fig. 7.1(c) explains this behavior. The positive Al gate pushes the conduction band in the channel down below the valence band in the source, turning on the transistor by enabling band-to-band tunneling.

7.3.2   Zener tunneling in carbon nanotubes and graphene nanoribbons

The band-to-band tunneling probability in carbon nanotubes (CNTs) or graphene nanoribbons (GNRs) can be calculated with the Wentzel–Kramer–Brillouin (WKB) approximation in a manner similar to other direct bandgap semiconductors, but by accounting for the different band structure. The E–k dispersion for the conduction band and valence band of a semiconducting zigzag CNT or armchair GNR is approximated by

$$E_{C,V} = \pm \hbar v_F \sqrt{k_x^2 + k_1^2}, \qquad (7.1)$$

where $\hbar$ is the reduced Planck's constant, $v_F \sim 10^6$ m/s is the Fermi velocity of graphene, E_C (with "+") and E_V (with "−") are the conduction band and valence band energy respectively, k_x is the wavevector along the transport, and k_1 is the transverse wavevector of the first subband quantized by the ribbon width w or the tube diameter d,

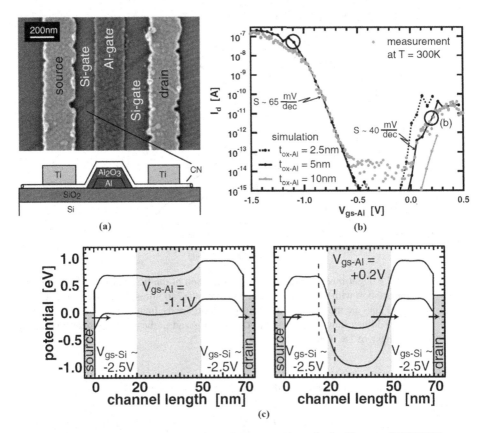

Fig. 7.1 (a) SEM top-view image and schematic cross section of a double-gate CNT TFET. (b) Transfer characteristic and (c) band diagrams of this CNT TFET, the first time a sub-threshold swing of less than 60 mV/dec was demonstrated in TFETs [3].

which is approximated by $k_1 = \pi/3w$ for GNRs and $k_1 = 2/3d$ for CNTs. Figure 7.2 shows that the E–k relation calculated by Eq. (7.1) is in good agreement with tight-binding simulations inside the bandgap where k_x is imaginary for a GNR. Using Eq. (7.1) with the WKB approximation, the energy-dependent tunneling transmission coefficient for semiconducting GNRs or CNTs is given by

$$T_{\text{WKB}}(E) = \exp\left(-2\int_{x_i}^{x_f} \frac{1}{\hbar v_F}\sqrt{(E_G/2)^2 - (E - E_C(x) - E_G/2)^2}\,dx\right), \qquad (7.2)$$

where E_G is the energy bandgap, and x_i and x_f are the initial and final position of the tunneling, the classical turning points.

Applying Eq. (7.2) to a reverse-biased p^+n^+ junction, the tunneling probability becomes energy independent if the electric field F is constant in the junction, given by [4], where q is the electron charge:

$$T_{\text{WKB}} = \exp\left(-\frac{\pi E_G^2}{4q\hbar v_F F}\right). \qquad (7.3)$$

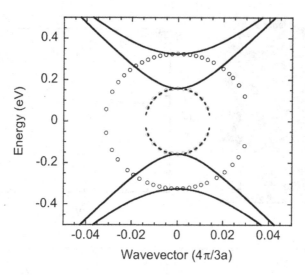

Fig. 7.2 Tight-binding energy dispersion relation of the first and second subband of a graphene nanoribbon with the width of $15a$, where $a = 0.246$ nm is the lattice constant of graphene. The solid lines are the real E–k relation and the dots are the imaginary E–k relation in the bandgap. The dashed lines show Eq. (7.1) fit to the first subband in the bandgap (tight-binding simulations courtesy of Dr. Tian Fang).

The tunneling current is then calculated assuming 1D ballistic transport:

$$I = \frac{g_v q}{\pi \hbar} \int (f_p(E) - f_n(F)) T_{WKB} dE, \qquad (7.4)$$

where g_v is the valley degeneracy ($g_v = 1$ for GNRs and $g_v = 2$ for CNTs), and f_p and f_n are the Fermi–Dirac function at the p- and n-sides of the junction respectively. Figure 7.3 shows the calculated tunneling current for GNR p^+n^+ junctions for Fermi levels at the valence band edge in the p^+ side and at the conduction band edge in the n^+ side. The plots show currents at different reverse biases, temperatures, ribbon widths, and electric fields using Eqs. (7.3) and (7.4). It is seen that the tunneling current increases at lower temperature due to more abrupt Fermi–Dirac functions, and increases with higher electric field due to larger tunneling probability. The tunneling current density can reach 1 mA/μm at a bias of 0.1 V at room temperature with $w = 3$ nm and $F = 5$ MV/cm. Such values are attractive for high-performance TFET operation.

7.3.3   TFET device structure and semi-classical modeling

Consider a p-type double-gate (DG) GNR TFET [5], where the source is heavily doped to be n^+ and the drain is p^+. The complementary n-type TFET has the same structure but with a heavily doped p^+ source and n^+ drain.

The electrostatics is solved using a quasi-1D Poisson equation [3, 5]. The scaling length λ for double-gate GNR transistors can be extracted from the analytical solution to

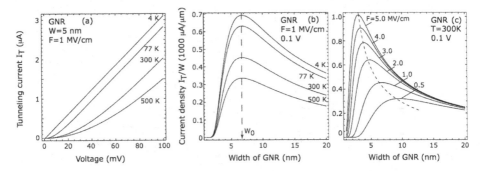

Fig. 7.3 Tunneling current in GNRs with constant electric field: (a) ribbon width $w = 5$ nm, voltage dependence at various temperatures. Current density as a function of ribbon width for (b) different temperatures at electric field $F = 1$ MV/cm and (c) different electric field at temperature $T = 300$K [4].

2D Poisson equations [6]. Charge injections into the channel from the source and drain reservoirs are calculated separately by

$$Q_S = q \int \rho_{GNR}(E)(1 - f_S(E)) T_{WKB}(E) dE \qquad (7.5)$$

and

$$Q_D = q \int \rho_{GNR}(E)[(f_D(E) - 1) T_{WKB}(E) + 2(1 - f_D(E))] dE, \qquad (7.6)$$

respectively, where ρ_{GNR} is the density of states for the first subband of the GNR, and f_S, f_D are the source and drain Fermi–Dirac distributions. The energy-dependent tunneling probability from the source $T_{WKB}(E)$ is calculated by Eq. (7.2), using $E_C(x)$ obtained from the electrostatic solution. Therefore the electrostatics and the charge in the channel can be solved self-consistently. Figure 7.4(a) shows the calculated off- and on-state energy band diagrams for a p-type GNR TFET at $V_{DS} = -0.1$ V. The transfer characteristics are then calculated using Eq. (7.4) and shown in Fig. 7.4(b), where $f_{p/n}$ is substituted by $f_{D/S}$ and T_{WKB} is energy-dependent, given by Eq. (7.3). With a supply voltage of 0.1 V, the GNR TFET is able to achieve an I_{on}/I_{off} ratio of more than 10^3 with an average sub-threshold swing of less than 30 mV/dec, and an on-state current density (normalized with the ribbon width) larger than 100 μA/μm. Due to the symmetry of the GNR band structure for electrons and holes near the Dirac point, the transfer characteristics of complementary n- and p-type GNR TFETs are also symmetric, as seen in Fig. 7.4(b).

The WKB approximation used in semi-classical TFET modeling has been compared with non-equilibrium Green's function (NEGF) quantum simulations. From the same band diagrams of an n-type GNR TFET self-consistently obtained by a 3D atomistic quantum transport solver [7], the transmission coefficients are calculated by WKB approximation (dots) and NEGF method (solid line) for both the off- and on-states, shown in Fig. 7.5(a) and (b). It is seen that the WKB approximation and NEGF are in good agreement in the off-state, while the resonant effect in the on-state observed in the NEGF simulations is not captured by the WKB approximation. Therefore, the on-state

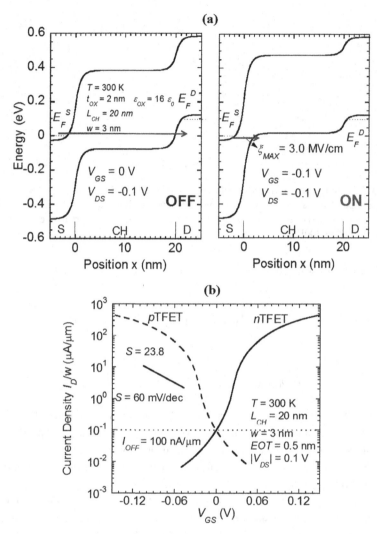

Fig. 7.4 (a) Band diagrams for a p-type GNR TFET in the off- and on-state with a supply voltage of 0.1 V. The calculated gate oxide is 2 nm thick with a relative dielectric constant of 16, the GNR width is 3 nm, and the gate length is 20 nm. (b) Transfer characteristics of this p-type GNR TFET (dashed line) and its complementary n-type TFET (solid line).

current is usually overestimated by the WKB approximation, and the overestimation is about 30% for the calculated geometry with $L_G = 40$ nm and ribbon width = 5.15 nm, shown in Fig. 7.5(c) and (d). Nevertheless, the WKB method offers crucial insights to the design of the device and material parameters.

7.3.4   Geometry and doping dependence, optimization, and benchmarks

The geometry- and doping-dependent performance of GNR TFETs has been studied by Chin *et al.* in detail by self-consistently solving NEGF and quasi-2D Poisson

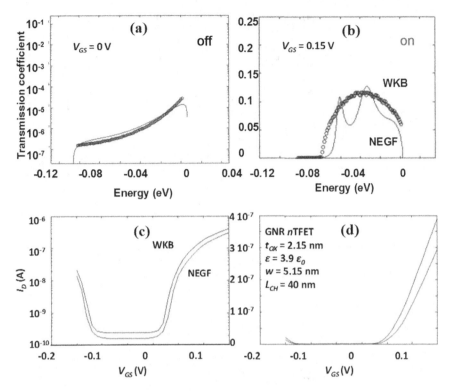

Fig. 7.5 WKB approximation vs. NEGF simulation for transmission coefficients in the (a) off- and (b) on-state of an n-type GNR TFET at $V_{DS} = 0.1$ V. The oxide is 2.15 nm SiO$_2$, the GNR width is 5.15 nm, and the gate length is 40 nm. (c), (d) Log and linear scale transfer characteristic of this GNR TFET calculated from WKB approximation and NEGF simulation (courtesy of Dr. Mathieu Luisier for the NEGF simulations).

equations [8]. Figure 7.6 shows the transfer characteristics of the DG GNR TFETs with (a) different ribbon widths (different E_G), (b) different gate lengths, and (c) different source/drain doping concentrations. Here, W12, W15, W23, and W27 denote the ribbon widths of 1.2, 1.5, 2.3, and 2.7 nm respectively, and correspond to bandgaps of 1.22, 0.953, 0.661, and 0.573 eV respectively. The off-state current is found to be strongly dependent on the bandgap. It is also strongly dependent on the gate length when scaled below 20 nm, due to strong direct source-to-drain tunneling. Narrow ribbons with a large bandgap are needed to achieve the required low off-state current at ultra-scaled gate lengths. However, the on-state current is degraded for narrower ribbons since the tunneling probability is lowered by the larger bandgaps and larger effective masses.

One way to improve the on-state current is to increase the source doping concentration, but the sub-threshold swing degrades in doing so, as seen in Fig. 7.6(c). There is a trade-off between the on-state current and the sub-threshold swing in engineering the source doping concentration. It is also shown by calculating the intrinsic delay and power delay product (PDP) that the W12 ribbon can meet the ITRS 2009 MOSFET requirements for low-power (LP) logic with gate length down to 12 nm.

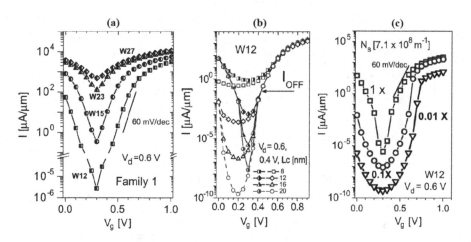

Fig. 7.6 Transfer characteristics of GNR TFETs with (a) different ribbon widths, (b) different gate lengths, and (c) different source/drain doping. The default is W12 ribbon, 16 nm gate length, and 7.1×10^8 /m source/drain doping [8].

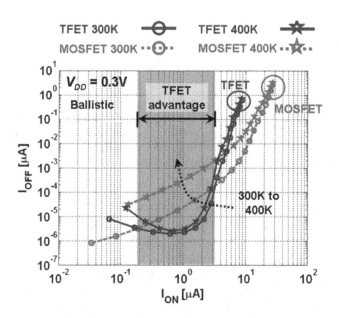

Fig. 7.7 I_{off} vs. I_{on} comparison for a zigzag (13, 0) CNT TFET and MOSFET at a supply voltage of 0.3 V and a gate length of 15 nm. The shaded area shows where TFET can outperform the MOSFET by having higher I_{on} for the same I_{off} at room temperature

A similar quantum transport study on CNT TFETs by Koswatta *et al.* [9] has shown that the TFETs "might be better suited for low-power applications with moderate drive-current requirements." Figure 7.7 compares I_{off} vs. I_{on} for a zigzag (13, 0) CNT TFET and MOSFET at a supply voltage of 0.3 V and a gate length of 15 nm. The shaded region at room temperature $T = 300\,\text{K}$ shows where the TFET can outperform the

MOSFET in the sense of having higher on-current but with the same off-current. It is seen that the on-current in this "advantage" region of the TFET is up to 3 μA per CNT. Considering the CNTs have a diameter of 1 nm, a high density of CNTs is necessary to achieve the ITRS drive current requirements.

A more general approach to optimize the TFET by engineering the bandgap, gate length, on/off-currents, and supply voltage has been given by Zhang et al. [10], using a physics-based analytic model described in Sections 7.3.2 and 7.3.3. The schematic band diagrams for a n-type homojunction p-i-n TFET in the on- and off-states have been shown in Fig. 6.6, where the energy qV_{on} is defined as the tunneling window below the source Fermi level in the on-state, and qV_{off} is defined as the energy barrier between the source Fermi level and the channel conduction band in the off-state. The sum of qV_{on} and qV_{off} is determined by the supply voltage and gate control efficiency. The current is calculated as a function of the bandgap, gate length, and the voltage V_{on} (on-state) or V_{off} (off-state) with Eqs. (7.2)–(7.4), assuming the direct source to drain tunneling dominates off-current for short gate lengths. The design space is then summarized in the constant current plot, where the optimized supply voltage V_{dd} can be determined according to the required gate length, on-current, off-current, or on/off-current ratio. Figure 7.8 shows the example of optimization for GNR TFETs, where the current is normalized by twice the ribbon width. The constant current plot, Fig. 7.8(a), is generated for gate lengths of 10 and 15 nm, where two off-currents of 5 nA/μm and 15 nA/μm with an on/off ratio of 10^5 are designed to meet the LP and HP (high-performance) requirements respectively. It is shown in Fig. 7.8 that the

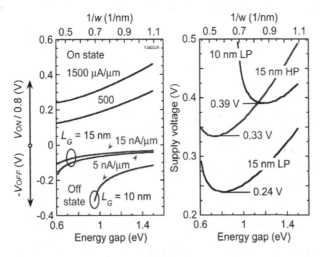

Fig. 7.8 Constant current density contour plot for GNR TFETs for $L_G = 10$ and 15 nm, with varied energy bandgap E_G and potential qV_{off} or $qV_{on}/0.8$ in the channel, so the distance between the on-state (upper) curves and the off-state (lower) curves represents $qV_{dd} - I_{off} = 5$ nA/μm for low-power (LP) application and $I_{on} = 1500$ μA/μm for high-performance (HP) application and $I_{on}/I_{off} = 10^5$. (b) Supply voltage V_{dd} vs. E_G and ribbon width w, where the minimum V_{dd} and the corresponding optimum E_G are extracted [10].

optimized supply voltage increases with shorter gate length and the bandgap must also increase to meet the off-current requirement.

7.3.5   Non-ideal effects

It has been shown by simulations that ideally the TFET with carbon-based materials is a promising candidate for low-power logic applications. But so far many simulations have assumed ballistic transport. Koswatta *et al.* [9] have studied the influence of phonon scattering for CNT TFETs. Figure 7.9 shows the simulated transfer characteristics with and without phonon scattering effects for temperature $T = 300$ K and 400 K. It is observed that phonon-assisted inelastic tunneling plays an important role in the off-state and degrades the sub-threshold swing. This degradation is worse at higher temperatures when there is a larger phonon occupation and absorption processes are favored.

For graphene nanoribbons sub-10 nm ribbon widths are necessary to obtain significant energy bandgaps. It is a considerable experimental challenge to fabricate perfect ribbons with minimal line edge roughness (LER) at these dimensions. The deleterious effects of LER have been studied with NEGF simulations by Luisier and Klimeck [7]. LER is randomly generated for an armchair GNR of width 5.1 nm, and the mean transfer characteristics of the GNR TFETs are shown in Fig. 7.10. Both the on-state and off-state current are increased with higher LER probability, due to the extra states in the bandgap introduced by the LER. However, the off-state current increases faster than the on-state current so that the sub-threshold swing and the on/off-current ratio are degraded. It is perceivable that advances in lithography from FinFET scaling will be used for controlling LER within acceptable limits to enable GNR TFETs in the future.

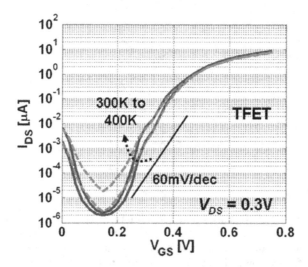

Fig. 7.9 Temperature-dependent transfer characteristics for a CNT TFET under ballistic (solid lines) and dissipative (dashed lines) transport. Phonon-assisted tunneling can degrade the sub-threshold swing.

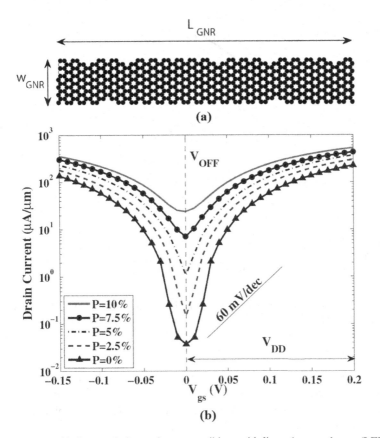

Fig. 7.10 (a) An armchair graphene nanoribbon with line edge roughness (LER). (b) Mean transfer characteristics of GNR TFETs with various LER probabilities, where the LER degrades the sub-threshold swing of the device.

7.4   Atomically thin body transistors

On the ITRS roadmap, the physical gate length, L_G, has been rapidly scaled down. It will reach below 10 nm beyond 2020. The electrostatic integrity of sub-10 nm channels will be a great challenge. This requires an electrostatic scaling length λ smaller than ~2 nm as the sub-threshold swing degrades significantly with excessive thermal leakage if the ratio $L_G/(\pi\lambda)$ is below 1.5 ~ 2 [11, 12]. The scaling length for graphene, with a double gate structure (the top and bottom oxides are not necessarily symmetric), has been extracted from a general analytic solution to the 2D Laplace equation [6]. The same theory can also be applied to other atomically thin materials. Figure 7.11 compares the scaling length for single-gate (SG), double-gate (DG) extremely thin SOI (ETSOI), and gate-all-around (GAA) Si nanowire MOSFETs with a channel thickness of 5 nm, to SG and DG graphene nanoribbon Schottky-barrier FETs [13]. From such considerations, it appears that to maintain good short-channel control for $L_G \leq 10$ nm, only GAA-Si nanowire and graphene (or other atomically thin materials) can meet the

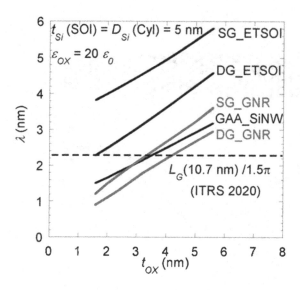

Fig. 7.11 Comparison of electrostatic scaling length λ for SG- DG- ETSOI FETs, GAA- SiNW FETs, and SG- DG- GNR FETs. λ is obtained from evanescent-mode analysis for Si, and from 2D Laplace solution for graphene [13].

requirement (the region below the dashed line in Fig. 7.11). Furthermore, the atomically thin channel materials show better scalability. Two-dimensional semiconductor crystals such as the transition-metal dichalcogenides are currently being explored as atomically thin channel materials.

7.4.1   In-plane tunneling transport in 2D crystals

The ultra-thin nature of 2D semiconductors offers the opportunity to scale electronic devices to much smaller dimensions than conventional 3D semiconductors, such as Si-MOSFETs and III-V HEMTs. For FET devices, the current in the active channel is dominated by drift-diffusion transport. For very short channels, the performance approaches the limiting case of ballistic transport. For TFETs, interband carrier tunneling to a large extent determines the device performance for electrostatically well-designed devices. To find out the feasibility of 2D crystal semiconductors for TFETs, it is important to evaluate the interband tunneling current in an in-plane 2D tunnel junction.

Figure 7.12(a) shows a 2D crystal p-i-n junction. The corresponding energy band-diagram is shown in Fig. 7.12(b) under the application of a reverse bias voltage, where a finite tunneling window is opened for electrons from the p-side to the n-side [14]. The doping in the p- and n-sides is assumed to align the Fermi levels to the respective band edges. The tunneling probability is obtained by the Wentzel–Kramer–Brillouin (WKB) approximation [15]. Each state in the 2D k-space, their respective group velocities and their interband tunneling probabilities are tracked. Figure 7.13 shows the zero-temperature interband tunneling current spectrum resolved in the k-space for 2D $MoTe_2$ [16].

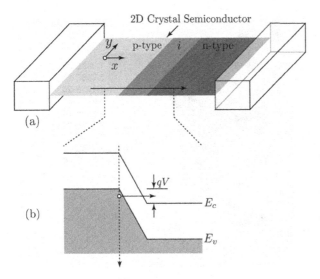

Fig. 7.12 Schematic depiction of a 2D crystal p-i-n junction (a), and (b) the corresponding energy band-diagram [14].

The total tunneling current is evaluated by summing the contributions from the individual k-states. At zero temperature, the interband tunneling current per unit width ($\mu A/\mu m$) in a 2D crystal p-i-n junction is given by

$$J_T^{2D} = \frac{q^2}{h}\left(\frac{g_s g_v T_0}{2\pi}\right)\sqrt{\frac{2m_v^* \bar{E}}{\hbar^2}} \times \left[\sqrt{\pi}\left(V - \frac{V_0}{2}\right)\mathrm{Erf}\left(\sqrt{\frac{V}{V_0}}\right) + \sqrt{V \times V_0}\exp\left(-\frac{V}{V_0}\right)\right],$$

(7.7)

where g_s, g_v are spin and valley degeneracy, respectively; and T_0, V_0, and $\bar{E}$ are parameters determined by the electric field F, the effective mass (m^*), and the energy bandgap E_G of the 2D crystal. With this expression, the tunneling current for various 2D crystals are calculated. When T is high, the Fermi–Dirac distributions of electrons are considered. However, this only introduces weak temperature dependence. The interband tunneling current densities of various 2D crystals at $T = 4$ K and 300 K are plotted as solid and dashed lines in Fig. 7.14(a) respectively. The tunneling current of $MoTe_2$ and a 2D semiconductor crystal with $E_G = 1.0$ eV and $m^* = 0.1m_0$ as a function of the voltage at different T is shown in Fig. 7.14(b) and (c), respectively.

The tunneling current densities for MoS_2 and the family of TMDs are low for in-plane fields less than 4 MV/cm owing to their large bandgaps. Methods to increase the in-plane electric field by electrostatic and geometric design can allow for much higher current densities, as may easily be calculated from Eq. (7.7). For TFET applications, 2D crystals with smaller bandgaps are necessary for boosting the current. For example, tunneling currents for 2D crystals with bandgaps of 0.5 eV and 1.0 eV with corresponding lower effective masses are plotted in Fig. 7.14(a). The currents for such crystals are attractive for high-performance TFET applications. To show the relative importance of effective masses and bandgaps, these are treated as independent

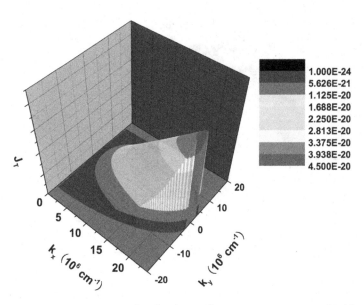

Fig. 7.13 Zero-temperature interband tunneling current spectrum resolved in the k-space in MoTe$_2$ [16].

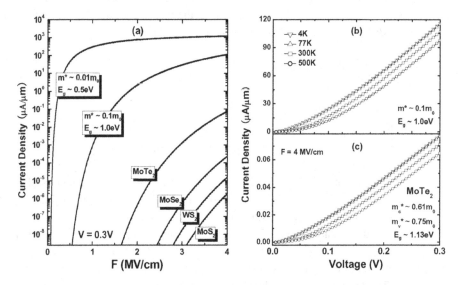

Fig. 7.14 (a) Interband tunneling current density for various 2D crystal semiconductors. The solid lines are at $T = 4$ K, and the dashed lines at $T = 300$ K, the temperature dependence is weak. (b) Current–voltage curves at various temperatures at a junction field $F = 4$ MV/cm for a 2D crystal semiconductor with band parameters indicated. (c) Same as (b), but for the 2D crystal MoTe$_2$ [14].

material parameters, and the interband tunneling currents in 2D crystals at a high junction field are shown in Fig. 7.15.

As is evident from Fig. 7.15(a), there is a trade-off in the choice of effective mass and bandgap for maximizing the tunneling current. Figure 7.15(b) zooms in to highlight this

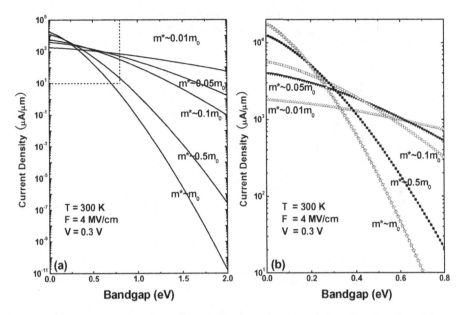

Fig. 7.15 (a) Interband tunneling currents for 2D crystal semiconductors as a function of the energy bandgap for various effective mass parameters. (b) The high-current part of (a), zoomed in for more details [14].

trade-off for a fixed in-plane electric field of 4 MV/cm. For high-performance TFETs for digital switching applications, currents exceeding 100 μA/μm are highly desirable.

Figure 7.16 plots interband tunneling currents for 2D crystal semiconductors as a function of the electron mass for energy bandgap of 0.1 and 1.0 eV [16].

7.4.2   In-plane 2D crystal TFETs

Following the discussion in Section 7.4.1 for high-performance TFETs, 2D crystals with bandgaps in the ~0.6–0.7 eV range and effective masses of 0.1–0.5m_0 are desirable for boosting the tunneling current. Such small-bandgap materials could be intrinsic 2D crystals. As new 2D crystals come to the fore, it is possible to have small-bandgap intrinsic 2D crystal, for example, ScS_2 (E_G ~ 0.44 eV), CrO_2 (E_G ~ 0.5 eV), and $CrTe_2$ (E_G ~ 0.6 eV) [1]. These 2D crystals need further investigation to explore their potential for device applications. Besides the TMD family of 2D crystal semiconductors, 2D in-plane TFETs can be realized with bilayer graphene, where an energy bandgap can be opened by means of an applied vertical electric field. Fiori and Iannaccone proposed the realization of an ultralow-power bilayer graphene TFET based on the solution of the coupled Poisson and Schrödinger in 3D with the non-equilibrium Green's function approach to conductivity [17]. The device structure and the band edge profile of the device in the off-state are shown in Fig. 7.17. The black and grey arrows indicate carrier fluxes from source and drain, including tunneling and thermal emission components respectively. When V_{diff} ($= V_{top} - V_{bottom}$) was varied from 6 to 8 V, the capability of

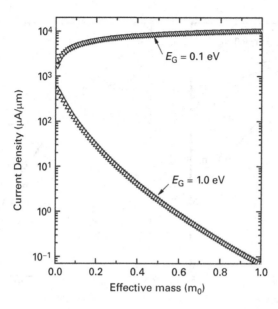

Fig. 7.16 shows the current density as a function of m^* with E_G of 0.1 eV and 1.0 eV, respectively. For 2D crystals semiconductors with bandgaps smaller than ~0.3–0.4 eV, a choice of a *higher* effective mass will maximize the interband tunneling current, far exceeding typical transistor on-currents for high-performance switching. But for larger bandgaps, a *lower* effective mass is more desirable. For high-performance TFETs, 2D crystals with bandgaps in the ~0.6–0.7 eV range and effective masses of 0.1–0.5m_0 can thus be potentially very attractive.

electrostatically tuning the energy gap resulted in an increase of on/off ratio by a factor of 33. Bilayer graphene presents an interesting new material property of a *field-tunable energy gap*. This feature has not been used directly in FETs before, and the paradigm may find increased interest for switching devices in the future.

Figures 7.18(a) and (b) show the transfer characteristics of the bilayer graphene TFETs for different dopant molar fraction in the drain and source leads and for different gate overlaps respectively. It is observed that the on/off ratio increases as the mole fraction decreases, and as the overlap increases. Both of them change the on/off ratio by manipulating the off-state current. Figure 7.18(c) shows the transfer characteristic for the device with an asymmetric oxide thickness, where the bottom gate voltage is fixed and the top gate is used as the control. This structure is more process-compatible for circuit integration and still has an on/off ratio of a few hundreds. Bilayer graphene TFETs can benefit from the low quantum capacitance, and combine the advantages of GNRs and CNTs, but also with an easily manufacturable planar geometry.

For in-plane TFETs, the capability to grow heterojunctions within a single layer 2D crystal can considerably enhance the possibilities for high performance. This paradigm is identical to the choice of staggered or broken-gap III-V heterostructure TFETs, that have shown significant improvements in on-currents. However, the scalability of the 2D crystal materials can extend the paradigm to much smaller channel thicknesses and gate lengths than conventional III-V 3D crystal heterostructures. The choice of

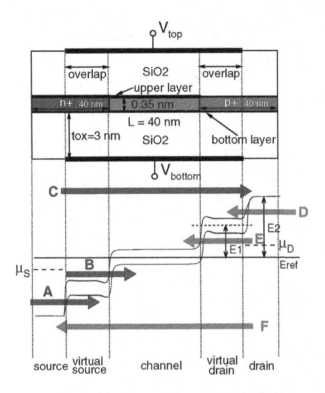

Fig. 7.17 Sketch of the bilayer graphene TFET with double-gate: the channel length is 40 nm, and the n^+ and p^+ reservoirs are 40 nm long. The device is embedded in a 3 nm thick SiO_2 dielectric. The (black) arrows pointing to the right are from the source; the arrows pointing to the left are from the drain (grey). V_{top} and V_{bottom} are the voltages applied to the top and bottom gate, respectively. The bottom shows the band edge profile of the device in the off-state [17].

heterojunctions must take advantage of the necessary band alignments by choosing the proper single-layer TMD materials, as shown in Fig. 7.19 [18]. The on-state current can be improved by an order of magnitude for both n- and p-type TFETs by introducing a common-X heterojunction at the source/channel interface. The performance enhancement is due to the reduced transition distance as a result of the abrupt quasi-electric field resulting from the band offset between common-X MX_2. However, the sub-threshold swing still needs further improvement. Whether the band edges will change as abruptly as predicted by theory remain to be experimentally verified. Another degree of freedom in the design is the engineering of the built-in electric field by proper choice of the surrounding dielectric materials, which can be used to boost the tunneling currents.

Fiori *et al.* have calculated that planar hBCN (hexagonal boron carbon nitrogen)-graphene heterostructures present a viable option to obtain truly planar nanoscale FETs [19]. They explore the behavior of a resonant tunneling FET using a channel with a graphene-BN-graphene-BN-graphene heterostructure with two metals. The advantage of this structure is the extremely good electrostatic control of the device due to the flat channel and the efficient gate coupling to the channel, which is unachievable with bipolar operation and vertical heterostructures.

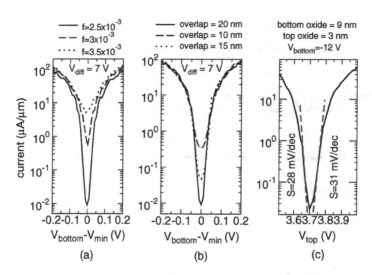

Fig. 7.18 Transfer characteristics of bilayer graphene TFETs with (a) different dopant molar fractions f, (b) different gate overlap, where $V_{\text{diff}} = 7\,\text{V}$, $V_{\text{DS}} = 0.1\,\text{V}$, and $V_{\text{min}} = -3.55\,\text{V}$; and (c) asymmetric oxide thickness, $V_{\text{DS}} = 0.1\,\text{V}$ [17].

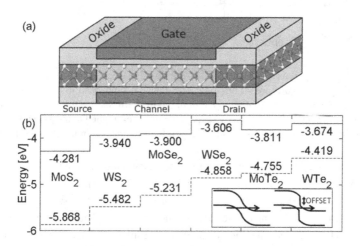

Fig. 7.19 (a) Device schematic of the double-gated MX_2 TFET. (b) Band offset from DFT calculation aligned at vacuum level. Solid and dashed lines correspond to the lowest conduction and highest valence bands, respectively. Inset: modification to band-to-band tunneling because of the introduction of band offset [18].

7.5   Interlayer tunneling transistors

The tunnel-junction of in-plane 2D crystal TFETs is a *line*, which restricts the current drive. Moving from line tunneling to *area* tunneling can substantially boost the current drive, and the performance. This requires a change in the geometry of the devices. In this section, such interlayer tunneling devices taking advantage of 2D crystals is discussed.

7.5.1   Interlayer tunneling between graphene layers

A major novel feature of graphene is the perfect symmetry of the band structure, which can lead to enhanced functionality. Motivated by the above question, the single-particle interlayer tunneling current–voltage curves were calculated explicitly for finite area 2-terminal graphene–insulator–graphene (GIG) heterostructures [20]. At most interlayer bias voltages, energy and momentum conservation force a small tunneling current to flow at one particular energy halfway between the two Dirac points (Fig. 7.20(a) and (b)). However, at a particular interlayer voltage when the Dirac points of the p- and n-type graphene layers align, a very large interlayer tunneling current flows. This is because energy and momentum are conserved in this process for all electron energies between the quasi-Fermi levels of the n- and p-type graphene layers (Fig. 7.20(c)). The I–V curve is dominated by a Dirac-delta function-like peak at the critical interlayer voltage, and smaller currents at all other voltages (Fig. 7.20 (d)). The calculation of the tunneling current also showed that the effect is robust to temperature, but less robust to rotational misalignment of the two graphene layers [20]. This sort of resonant tunneling leads to a strong negative differential resistance (NDR), commonly observed in III-V resonant tunneling diodes. An advantage of the GIG 2D crystal building block is that it can be gated with higher electrostatic control than III-V RTD structures. Furthermore, the tunneling mechanism is rather different from a III-V RTD.

7.5.2   Symmetric tunneling field effect transistor (SymFET)

The resonant tunneling behavior between two graphene layers in the GIG heterostructure can be incorporated into an FET geometry in a novel device called the SymFET [21]. The device structure is shown in Fig. 7.21. An insulator separates two graphene

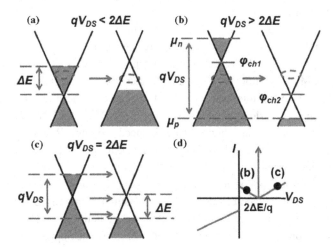

Fig. 7.20 The energy-band diagrams for a doped graphene/insulator/graphene junction, at voltages of (a) $qV_{DS} < 2\Delta E$, (b) $qV_{DS} > 2\Delta E$, and (c) $qV_{DS} = 2\Delta E$. A qualitative current–voltage (I–V_{DS}) characteristic is shown in (d) [21].

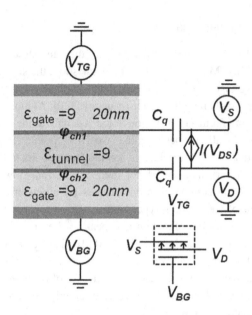

Fig. 7.21 Sketch of the SymFET. The inset shows the symbol defined for the SymFET [21].

layers, and this GIG structure is sandwiched between a top and bottom gate. Ohmic contacts are formed to the two graphene layers individually representing the source (S) and the drain (D). The top and bottom gate voltages, V_{TG} and V_{BG}, control the quasi-Fermi levels μ_n and μ_p in the top and bottom layers of graphene. The gate insulator thicknesses of both gates are assumed to be the same, t_g. The quasi-Fermi level is ΔE above the Dirac point in the n-type graphene layer and below the Dirac point in the p-type graphene layer. The top and back gates are symmetric $V_{TG} = -V_{BG}$, and the drain/source voltage is $V_{DS} = V_D - V_S$. The insert in Fig. 7.21 shows a proposed device symbol for the SymFET.

The tunneling insulator thickness t_t plays a similar role as the tunneling barrier thickness in double quantum well heterostructures. As t_t increases the resonant peak current decreases, as expected. The gate insulator can be a high-k material similar to that employed in Si CMOS technology. Alternatively, 2D materials such as BN might be a better choice to reduce the interface trap density because of the absence of dangling bonds. The measured breakdown field is as high as ~7.9 MV/cm for BN [22]. Thinner t_g offers better gate control and higher gate induced doping. When t_g decreases, ΔE becomes larger at same gate bias. The resonant peak moves to a higher bias and the peak current increases.

The I_D–V_{DS} characteristics at a fixed V_G are shown in Fig. 7.22(a). The resonant behavior shows clearly on- and off-states without a saturation region. Due to the finite chemical doping of graphene, the SymFET with a resonant current peak can operate at $V_G = 0$ V. As mentioned above, the gate will induce electrostatic doping in the graphene layer. With larger V_G, ΔE increases, the resonant condition $qV_{DS} = 2 \Delta E$ occurs at larger drain bias, and the resonant current peak moves to the right. Higher V_G induces more doping and thus increases the on-state current. In the 2-terminal GIG device, the

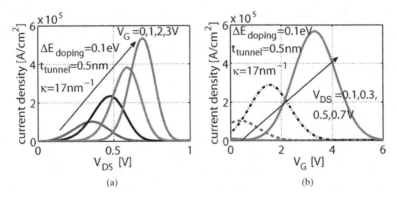

Fig. 7.22 (a) I_D vs. V_{DS} curves at different V_G, and (b) I_D vs. V_G at different V_{DS}.

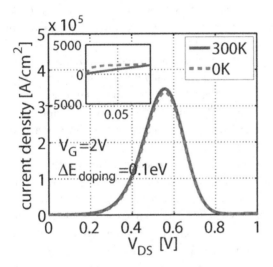

Fig. 7.23 Comparison of I_D–V_{DS} characteristics at $T = 300\,\mathrm{K}$ and $T = 0\,\mathrm{K}$.

resonant current peak is proportional to the coherence length L of the grain size of graphene, and the peak width is proportional to $1/L$ [20]. In the gated SymFET, since the gate bias electrostatically dopes the graphene, it offers the additional flexibility to adjust the on- and off-states. In Fig. 7.22(b), I_D–V_G curves are shown with a strong non-linear and resonant behavior, but with wider peaks. When V_G is small and outside the resonant peak, the transconductance is small, but it is large in the peak condition.

Because tunneling is the main current transport mechanism, the I_D–V_{DS} curve is quite insensitive to temperature, as shown in Fig. 7.23. But since the Fermi Dirac distribution smears out the state occupancy at a finite temperature, slight differences can still be observed between $T = 300\,\mathrm{K}$ and $T = 0\,\mathrm{K}$. At low V_{DS}, the transport energy window between the quasi-Fermi levels μ_n and μ_p is small. Then, the Fermi distribution smearing reduces the carrier density at higher temperature and the current decreases. The increase of the resonant peak current at room temperature is because the Fermi distribution tail extends to higher energy with more states. When the Dirac points are aligned, states at all energies conserve lateral momentum upon tunneling, and thus are allowed to tunnel. The predicted gated NDR behavior of the SymFET has already been experimentally confirmed at room temperature, and is discussed later in this chapter.

7.5.3   Bilayer pseudospin FETs (BiSFETs)

The studies of out-of-plane charge transport in 2D crystals have been motivated by the proposal of the bilayer pseudospin FET (BiSFET) [23]. The BiSFET exploits the fact that two graphene layers can be placed in close proximity and if populated by electrons and holes, the strong Coulomb attraction between them can lead to strong electron–hole Coulomb attraction, and the formation of excitons. Since excitons are bosonic quasiparticles, they can effectively undergo a Bose–Einstein condensation below a certain critical temperature. Since the Fermi degeneracy in graphene is tunable over a large energy window, the critical temperature for the excitonic condensate has been calculated to be higher than room temperature. The formation of the excitonic condensate is expected to lead to a macroscopic correlated tunneling current between the layers, similar to the coherent many-electron transport in superconductors. The BiSFET thus has the potential to realize many-body excitonic tunneling phenomena at room temperature. The power dissipation in computation using the functionality of BiSFET is predicted to be many orders lower than conventional CMOS switching (Fig. 7.24(c)).

The proposed energy consumption in logic operations in a BiSFET has been evaluated with SPICE simulation for an inverter including two BiSFETs (Fig. 7.25(a)). Figure 7.25(b) shows the SPICE-simulated response for such an inverter. A 20 nm gate width was assumed for each device. The clock signal is pulsed with a frequency of 100 GHz, and a peak V_{clock} of 25 mV. Both input and output signals were subject to a fan-in and fan-out of four inverters, respectively. The average energy consumed per clock cycle per BiSFET was 0.008 aJ at 100 GHz. For comparison, current MOSFETs consume ∼100 aJ per switching on average at a ∼5 GHz clock frequency, and the projected power consumption is about 5 aJ per switching at 100 GHz by 2020. The BiSFET serves as an exciting point of departure for truly novel device ideas that exploit the unique properties of 2D crystal semiconductors. This device proposal has also motivated the exploration of similar ideas in conventional 3D crystal materials, as described next.

7.5.4   Experimental progress of interlayer tunneling transistors

Significant prior investigation of 2D–2D tunneling has been carried out on coupled electron gas systems in closely placed quantum wells in AlGaAs/GaAs heterostructures [24–26]. The parallel 2DEGs (two-dimensional electron gases) are controlled by the top contact and back contact separately (Fig. 7.26). Considering the case of unequal doping between the 2DEGs, at a voltage bias corresponding to aligned bands of the 2D systems, a large, sharp peak in the tunnel current has been experimentally observed.

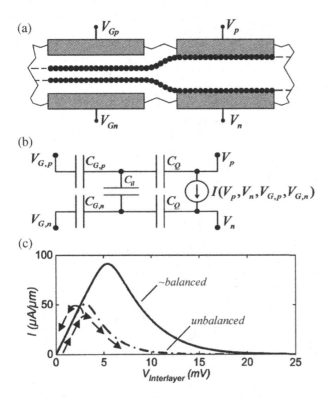

Fig. 7.24 (a) Sketch of a BiSFET structure, (b) equivalent circuit model of BiSFET, (c) the interlayer I–V characteristics of BiSFET with a low operation bias.

The width of this peak was found to be temperature independent (except possibly from inelastic effects) as shown in Fig. 7.27. The tunneling is between the same bands, and therefore is expected to be limited by the Boltzmann limit.

The observation of resonant tunneling in graphene interlayer tunneling transistor with a peak to valley ratio larger than 2 has been reported at room temperature and at 7 K [27]. A four atomic-layer thick h-BN was used as the tunneling insulator between two graphene layers. The doping of the bottom and top graphene layers is controlled by the bottom gate and contact on top graphene layer V_b separately. Repeatable NDR can be observed from 7 K to 300 K (Fig. 7.28). Comparing the I–V relationship at 7 K and 300 K, the peak current is almost the same as the interlayer tunneling is insensitive to temperature. The expected barrier height at the graphene/BN interface is about 2.6 eV (approximately half of the bandgap of BN). The thermionic emission current is low with such a high barrier. The peak current increases with higher doping (larger gate voltage) in the graphene layer. This is similar to the peak-to-valley ratio (PVR) increase at larger V_G (inset of Fig. 7.28). The gate-controlled resonant tunneling vertical graphene transistor with high PVR and stable NDR has some advantages over other resonant tunneling devices. The position of the resonant peak and the PVR can be controlled by the gate, which is absent in 2-terminal devices such as RTDs, but can be obtained in gated RTDs. Due to atomically thin graphene, the dwell time of carriers in graphene at

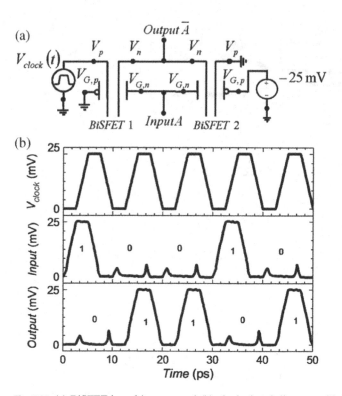

Fig. 7.25 (a) BiSFET-based inverter and (b) clock signal diagram with input and output wave forms based on SPICE simulation.

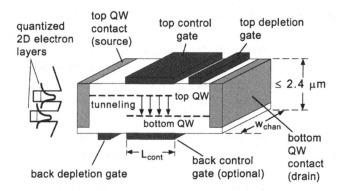

Fig. 7.26 Schematic diagram showing the 2D–2D tunneling device structure indicating the positions of front and back gates. The left contact makes electrical contact to the top quantum well only, while the right contact controls the bottom quantum well.

the resonant bias is expected to be shorter than that of traditional quantum well-based resonant structures, which could improve the switching speed for potential applications in high-speed electronics. The demonstration also proves the feasibility of the SymFET, as was discussed earlier.

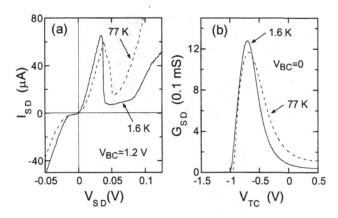

Fig. 7.27 (a) Graph of the $I–V$ curve, and (b) the small-signal G_{SD} vs. V_{TC} (top control gate voltage), at both 1.6 K and 77 K. The characteristics change little between 1.6 K and 77 K.

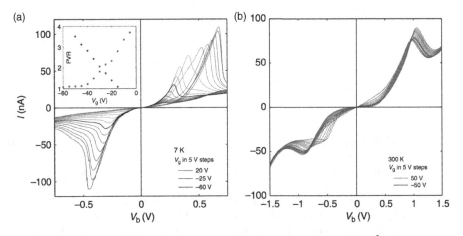

Fig. 7.28 The h-BN tunneling barrier is five atomic layers thick with a 0.6 μm² active region. (a) T = 7 K; V_G values range, in 5 V steps, from +20 V to −60 V. The inset shows the V_G dependence of the PVR. (b) Room temperature $I–V_b$ characteristics for the same device; V_G ranges from +50 to −50 V in 5 V steps.

Electron transport out of the plane of 2D materials has attracted strong experimental interest. For example, a graphene/BN/graphene sandwich heterostructure was recently reported [28], and interlayer electron transport was measured. This stacked graphene/BN/graphene heterostructure showed a room temperature switching ratio of 50 and 10 000 for a similar graphene/MoS₂/graphene structure. The on/off ratio is six orders of magnitude in graphene/WS₂ heterostructure, which is among the highest in a graphene-based device [29]. The output conductance of the devices, however, was not suitable for logic devices in the reported device topology. It is conceivable that low output conductance FETs could be realized in this fashion.

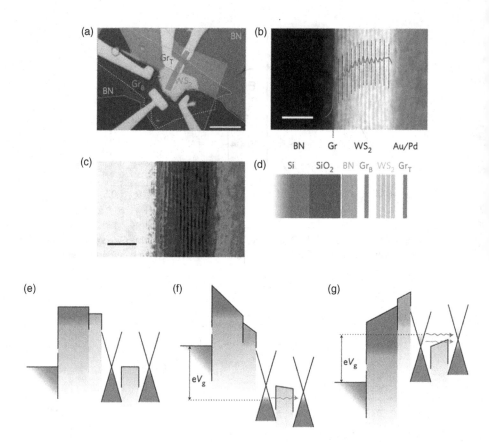

Fig. 7.29 Graphene/WS$_2$ heterotransistor. (a) Optical image (scale bar, 10 μm). (b) Cross-section of high-resolution, high-angle, annular dark-field, scanning transmission electron microscopy (HAADF STEM) image (scale bar, 5 nm). (c) Bright-field STEM image (scale bar, 5 nm). (d) Schematic representation of vertical architecture of transistor. (e) Band diagram corresponding to no V_G and applied V_B. (f) Negative V_G shifts the Fermi level of the two graphene layers down from the neutrality point, increasing the potential barrier and switching the transistor off. (g) Applying positive V_G results in an increased current between Gr_B and Gr_T due to both thermionic and tunneling contributions [29].

7.5.5   Challenges

Significant challenges exist in the realization of interlayer tunneling devices based on 2D crystals of various forms. For example, for the SymFET, the greatest amount of non-linearity in the I–V characteristics is achieved with a perfect rotational orientation of graphene layers. As the misalignment angle increases, the intensity of the resonant peak rapidly decreases; the peak shifts to higher voltages and a related peak appears at lower (negative voltages) [20]. Similar I–V characteristics appear in the experimental data (Fig. 7.28). The realization of PVR ~ 100 with perfect angle alignment between two layers of graphene presents a challenge in the fabrication of such devices based on layer transfer technology. The epitaxial growth of graphene, BN, or other 2D materials promises to overcome this problem. At the time of writing, epitaxial single-layer graphene on SiC wafers of only a few inches in diameter are available [30, 31]. Chemical-vapor deposition (CVD) grown graphene has been realized on metals, and transferred to other substrates [32]. The crystal quality is not yet perfect, but similar to the development of 3D crystals, it will undergo drastic improvements in the near future. Similarly, BN 2D crystals have been grown by CVD [33], as have electronic-grade MoS_2- and WS_2-based layered materials [34, 35]. Due to the many-particle nature of the excitonic condensate, the BiSFET is expected to be insensitive to rotational misalignment. However, BiSFET is sensitive to variations in insulator thickness.

The tunneling-based devices are expected to be robust to temperature and can potentially enable lower than 60 mV/dec sub-threshold swings. For high-performance applications, a high on-current density is required. In III-V TFETs, a high on-current can be obtained with staggered-band-offset heterostructures as AlGaAsSb/InGaAs, and broken-gap heterostructures such as GaSb/InAs heterojunctions [36]. Based on the calculated band structure of monolayer TMD materials, the staggered-gap heterostructure also exists, among which MoS_2/WTe_2 heterojunction gives the smallest energy overlap. This would favor a high on-current in interlayer tunneling FETs [37]. In-plane interband tunneling was discussed in an earlier section. However, little is known for interlayer tunneling in 2D material heterostructures. The calculation of the carriers transport can follow the Bardeen transfer Hamiltonian matrix formalism described in [20]. Due to the asymmetry of the conduction valence bands in 2D TMD materials, novel features distinct from graphene can be expected in them, and will provide new ways of realizing high-performance TFETs.

7.6   Internal charge and voltage gain steep devices

The discussion in this chapter was restricted to sub-Boltzmann steep switching device realizations with 2D crystal semiconductor materials by exploiting tunneling transport. Several complementary ideas exist for steep devices that do not depend on tunneling, and 2D crystal channel materials can take advantage of many such ideas because of the favorable electrostatics, geometry, and band structure in these materials. For example, impact-ionization-based internal charge gain can amplify the charge modulated by the gate and result in sub-60 mV/dec switching. This internal charge gain mechanism can be utilized in a 2D crystal geometry to explore the differences, and possible advantages over their 3D crystal counterparts. Similarly, several ideas related to active (or smart) gate dielectrics exist, which propose internal voltage amplification for achieving sub-60 mV/dec switching in a conventional FET without tunneling. Integration of such active gate materials with 2D crystals can be explored for steep switching devices.

7.7   Conclusions

In this chapter, the electronic properties of 2D crystal semiconductors were discussed, with the aim of using these new materials for tunneling transistors for low-power logic devices in a beyond-CMOS scenario. These materials, ranging from zero-gap graphene, large-bandgap BN, to semiconducting TMD crystals, collectively offer an exciting alternative to conventional semiconductors for extending the scaling of conventional transistors. They also offer a range of possibilities for ultrascale steep switching devices by exploiting tunneling. Currently, their level of technological maturity is very low, but rapid progress is being made in crystal growth, doping, ohmic contacts, and integration with dielectrics, and we hope that they will live up to the promise of the range of proposed devices. But more importantly, as they are explored further, we expect new physical phenomena to emerge that we have not yet foreseen. It is such new phenomena that will truly determine the future of these materials and the devices that result from them.

References

[1] G. E. Moore, "Cramming more components onto integrated circuits." *Electronics*, 114–117 (1965).
[2] G. G. Shahidi, "Device scaling for 15 nm node and beyond." In *Device Research Conference, 2009. DRC 2009*, pp. 247–250 (2009).
[3] J. Appenzeller, Y.-M. Lin, J. Knoch, & Ph. Avouris, "Band-to-band tunneling in carbon nanotube field-effect transistors." *Physics Review Letters*, **93**(19), 196805 (2004).
[4] D. Jena, T. Fang, Q. Zhang, & H. Xing, "Zener tunneling in semiconducting nanotube and graphene nanoribbon p-n junctions." *Applied Physics Letters*, **93**, 112106 (2008).
[5] Q. Zhang, T. Fang, H. L. Xing, A. Seabaugh, & D. Jena, "Graphene nanoribbon tunnel transistors." *IEEE Electron Device Letters*, **29**, 1344–1346 (2008).
[6] Q. Zhang, Y. Q. Lu, H. G. Xing, S. J. Koester, & S. O. Koswatta, "Scalability of atomic-thin-body (ATB) transistors based on graphene nanoribbons." *IEEE Electron Device Letters*, **31**, 531–533 (2010).
[7] M. Luisier & G. Klimeck, "Performance analysis of statistical samples of graphene nano-ribbon tunneling transistors with line edge roughness." *Applied Physics Letters*, **94**, 223505 (2009).
[8] S.-K. Chin, D. Seah, K.-T. Lam, G. S. Samudra, & G. Liang, "Device physics and characteristics of graphene nanoribbon tunneling FETs." *IEEE Transactions on Electron Devices*, **57**, 3144–3152 (2010).
[9] S. O. Koswatta, M. S. Lundstrom, & D. E. Nikonov, "Performance comparison between p-i-n tunneling transistors and conventional MOSFETs." *IEEE Transactions on Electron Devices*, **56**, 456–464 (2009).
[10] Q. Zhang, Y. Lu, C. A. Richter, D. Jena, & A. Seabaugh, "Optimum band gap and supply voltage in tunnel FETs." submitted to *IEEE Transactions on Electron Devices* (2014).
[11] D. J. Frank, Y. Taur, & H.-S. P. Wong, "Generalized scale length for two dimensional effects in MOSFETs." *IEEE Electron Device Letters*, **19**, 385–387 (1998).
[12] X. Liang & Y. Taur, "A 2-D analytical solution for SCEs in DG MOSFETs." *IEEE Transactions on Electron Devices*, **51**, 1385–1391 (2004).
[13] Q. Zhang, L. Ye, G. H. Xing, C. A. Richter, S. J. Koester, & S. O. Koswatta, "Graphene nanoribbon Schottky-barrier FETs for end-of-the-roadmap CMOS: challenges and opportunities." In *Device Research Conference, 2010. DRC 2010*, pp. 75–76 (2010).

[14] N. Ma & D. Jena, "Interband tunneling in two-dimensional crystal semiconductors." *Applied Physics Letters* **102**, 132102 (2013).

[15] A. C. Seabaugh & Q. Zhang, "Low-voltage tunnel transistors for beyond CMOS logic." *Proceedings of the IEEE* **98**, 2095 (2010).

[16] N. Ma & D. Jena, "Interband tunneling transport in 2-dimensional crystal semiconductors." In *Device Research Conference, 2013. DRC 2013*, pp. 103–104 (2013).

[17] G. Fiori & G. Iannaccone, "Ultralow-voltage bilayer graphene tunnel FET." *IEEE Electron Device Letters*, **30**, 1096 (2009).

[18] K.-T. Lam, X. Cao, & J. Guo, "Device performance of heterojunction tunneling field-effect transistors based on transition metal dichalcogenide monolayer." *IEEE Electron Device Letters*, **34**, 1331–1333 (2013).

[19] G. Fiori, A. Betti, S. Bruzzone, & G. Iannaccone, "Lateral graphene-hBCN heterostructures as a platform for fully two-dimensional transistors." *ACS Nano*, **6**, 2642–2648 (2012).

[20] R. M. Feenstra, D. Jena, & G. Gu, "Single-particle tunneling in doped graphene-insulator-graphene junctions." *Journal of Applied Physics*, **111**, 043711 (2012).

[21] Pei Zhao, R.M. Feenstra, Gong Gu, & D. Jena. "SymFET: a proposed symmetric graphene tunneling field effect transistor." *IEEE Transactions on Electron Devices*, **60**(3), 951–957 (2013).

[22] G.-H. Lee *et al.*, "Electron tunneling through atomically flat and ultrathin hexagonal boron nitride." *Applied Physics Letters*, **99**, 243114 (2011).

[23] S. K. Banerjee, L. F. Register, E. Tutuc, D. Reddy, & A. H. MacDonald, "Bilayer pseudo-spin field-effect transistor (BiSFET): a proposed new logic device." *IEEE Electron Device Letters*, **30**, 158 (2009).

[24] J. P. Eisenstein, L. N. Pfeiffer, & K. W. West, "Field-induced resonant tunneling between parallel two-dimensional electron systems." *Applied Physics Letters* **58**, 1497 (1991).

[25] J. P. Eisenstein, T. J. Gramila, L. N. Pfeiffer, & K. W. West, "Probing a two-dimensional Fermi surface by tunneling." *Physics Reviews B* **44**, 6511 (1991).

[26] J. A. Simmons *et al.*, "Planar quantum transistor based on 2D–2D tunneling in double quantum well heterostructures." *Journal of Applied Physics* **84**(10), 5626–5634 (1998).

[27] Britnell, L. *et al.*, "Resonant tunnelling and negative differential conductance in graphene transistors." *Nature Communications* **4**, 1794 (2013).

[28] L. Britnell *et al.*, "Field-effect tunneling transistor based on vertical graphene heterostructures." *Science*, **335**, 947 (2012).

[29] T. Georgiou, R. Jalil, B. D. Belle *et al.*, "Vertical field-effect transistor based on graphene–WS_2 heterostructures for flexible and transparent electronics." *Nature Nanotechnology* **8**, 100–103 (2013).

[30] G. Gu *et al.*, "Field effect in epitaxial graphene on a silicon carbide substrate." *Applied Physics Letters*, **90**, 253507 (2007).

[31] J. S. Moon *et al.*, "Epitaxial-graphene RF field-effect transistors on Si-face 6H-SiC substrates." *IEEE Electron Device Letters*, **30**, 650–652 (2009).

[32] X. Li *et al.*, "Large-area synthesis of high-quality and uniform graphene films on copper foils." *Science*, **324**, 1312–1314 (2009).

[33] L. Song *et al.*, "Large scale growth and characterization of atomic hexagonal boron nitride layers." *Nano Letters*, **10**, 3209–3215 (2010).

[34] Y. H. Lee *et al.*, "Synthesis of large-area MoS_2 atomic layers with chemical vapor deposition." *Advanced Materials.*, **24**, 2320–2325 (2012).

[35] W. S. Hwang *et al.*, "Transistors with chemically synthesized layered semiconductor WS_2 exhibiting 10^5 room temperature modulation and ambipolar behavior." *Applied Physics Letters*, **101**, 013107 (2012).

[36] R. Li *et al.* "AlGaSb/InAs tunnel field-effect transistor with on-current of 78μA/μm at 0.5V." *IEEE Electron Device Letters*, **33**(3), 363–365 (2012).

[37] J. Kang, S. Tongay, J. Zhou, J. Li, & J. Wu. "Band offsets and heterostructures of two-dimensional semiconductors." *Applied Physics Letters*, **102**(1), 012111 (2013).

8 Bilayer pseudospin field effect transistor

Dharmendar Reddy, Leonard F. Register, and Sanjay K. Bannerjee

8.1 Introduction

The bilayer pseudospin field effect transistor (BiSFET) is intended to enable much lower voltage and power operation than possible with complementary metal–oxide–semiconductor (CMOS) field effect transistor (FET)-based logic [1, 2]. The ultimate limits of CMOS are not due to fabrication technology limitations. Rather, they are intrinsic to its operating principles, defined by basic physics such as charge carrier thermionic emission over the channel barrier and quantum mechanical tunneling through it. New operating principles are required. The BiSFET relies on the possibility of room temperature excitonic (electron-hole) superfluid condensation in two dielectrically separated graphene layers [3, 4]. While the physics is interesting in its own right, from the device point of view this many-body physics brings with it the possibility of a strong sensitivity to sub-thermal voltages (sub-$k_{\mathrm{B}}T/q$ voltages, where k_{B} is Boltzmann's constant, T is the temperature in Kelvin, and q is the magnitude of electron charge) in the current–voltage (I–V) characteristics [5–7]. With power consumption proportional to the square of voltage, use of voltages on the scale of or less than room temperature $k_{\mathrm{B}}T/q \approx 26\,\mathrm{mV}$ offers order of magnitude reductions in switching energies as compared to even end-of-the roadmap CMOS [8]. Circuit simulation with 25 mV power supplies show switching energies on the scale of 10 zeptojoules (zJ) per BiSFET (where $1\,\mathrm{zJ} = 10^{-21}\,\mathrm{J} = 10^{-3}\,\mathrm{aJ}$)! However, with this potential for voltage reduction also come I–V characteristics much different from those of MOSFETs that must be worked around at worst, and may provide new circuit opportunities at best. In terms of interconnects, information would continue to be passed via charge among devices. In this way, BiSFETs would also be compatible with existing electronic devices after voltage level shifts.

There are substantial challenges to realizing BiSFETs, however. At the time of writing, the BiSFET remains a novel transistor concept based on novel physics in a novel material system. Such a room temperature condensate has yet to be observed, although it is also the case that the expected necessary experimental conditions have yet to be realized. Moreover, if theory holds, there remain significant technological challenges to realization of BiSFETs. However, experimental results in mostly III-V semiconductor double quantum well heterostructures have exhibited much of the essential transport physics required, if only at very low temperatures and under high magnetic fields [9–14]. It is "only" the latter requirements that the use of graphene layers is intended to eliminate [3]. Moreover, other analogous two-dimensional (2D) material systems for which the requirements may be qualitatively similar but quantitatively different, such as silicene, germanene, or transition metal dichalcogenides, represent potential alternatives to graphene.

8.2   Overview

Figure 8.1 illustrates the transition from essential physics to low-power circuits, and, roughly, an outline for the body of this chapter and overview section. First, we discuss the essential physics that underlies condensate formation (a) and the resulting potentially very low-voltage onset of negative differential resistance for interlayer transport (b). We then discuss the associated BiSFET physical layout requirements (c) and the compact modeling thereof (d). We note that the BiSFET layout and, therefore, compact model have evolved as our understanding of the condensate physics has improved. However, the expected I–V characteristics remain much the same. Based on these output characteristics, we present compatible logic circuits (e) which we have used on SPICE-level simulations to estimate power consumption – very low power consumption (f). In addition, not represented in Figure 8.1, we also consider critical technology needs for the BiSFET. In this overview section, we focus on essential information (of which there is a significant amount for this novel device concept). The reader will then find more detail and support on each topic in Sections 8.3–8.6.

8.2.1   Condensate formation and low-voltage NDR

As noted, the BiSFET concept relies upon formation of an excitonic superfluid condensate between the electron/n-type and hole/p-type graphene bilayer system. Therefore, like classical exciton formation, anything that increases the Coulomb interactions between layers increases the strength of the excitonic condensate. Moreover, it is also optimal that the charge densities on the two layers be equal in magnitude; increasing charge imbalance has an analogous effect to increasing temperature.

The excitonic condensate, however, differs from classical excitons in critical ways. When calculated in terms of electronic states as done here, the excitonic condensate does not result directly from the interlayer Coulomb interactions among electrons so much as from the many-body exchange correction to these Coulomb interactions when and only when there is interlayer quantum coherence. The exchange correction recognizes that the electrons of the same spin are further away from each other in a coherent many-body quantum mechanical state than otherwise would be expected. (The Pauli Exclusion Principle can be considered a limiting example of this behavior.) Equivalently, electrons and holes are closer to each other than otherwise would be expected. For bilayer systems, this means that a coherent interlayer state, the condensate, is energetically selected for even when there is no single-particle/bare interlayer coupling mechanism.

In the energy band structure, this reduction in energy is associated with the forming of an energy band/anti-crossing gap about where the cone-shaped valence and

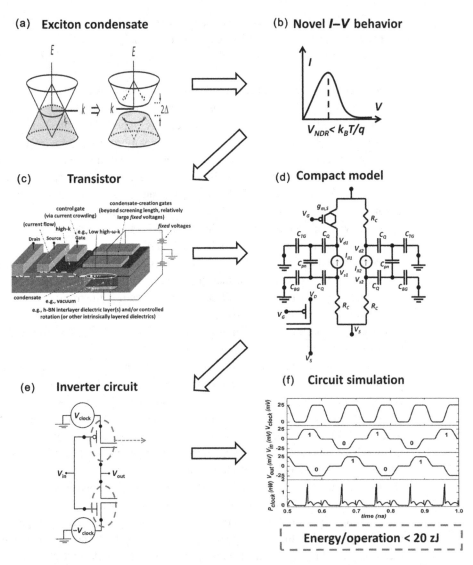

Fig. 8.1 From essential physics to low-power circuits: (a) exciton superfluid condensate formation in dielectrically separated graphene layers also showing nominally occupied states in gray, and (b) associated novel low-voltage negative differential characteristics for interlayer tunneling current; (c) BiSFET layout, and (d) compact models thereof; and (e) BiSFET compatible circuits, and (d) circuit simulation results. Each topic will be addressed in more detail in the body of the text. (Part (a) adapted with permission from [15], copyright 2012 by the American Physical Society.)

conduction bands of the two layers would otherwise have crossed in energy (Figure 8.1(a)), and which encompasses the Fermi level. The latter requirement also implies near-balanced charge densities between the layers. The size of this gap is directly related to the maximum temperature for condensate formation, the critical temperature T_c. There also is a positive feedback loop in which the stronger the exchange interaction between layers, the more coherence there is, the stronger the exchange interaction, etc., and vice versa. This feedback leads to an abrupt transition with temperature about T_c, as well as strong sensitivity to interlayer separation and dielectric environment.

Once the condensate forms, the I–V characteristics of such a bilayer system are altered drastically. As just one consequence, condensate formation can nearly short the two layers together (in the sense of closely approaching the Landauer–Büttiker ballistic limit of conduction) via interlayer tunneling near the edge of the condensate region. However, only a limited amount of current, the critical current I_c, can be carried in this way. With further increases in interlayer voltage, the steady-state current drops drastically (Figure 8.1(b)) [5–7, 14, 16]. This transition is characterized by a region of negative differential resistance (NDR). However, it is the change from high conductance to low conductance that is of importance for the logic applications as discussed here, not the region of NDR over which it occurs.

The more important point for the proposed BiSFET is not the general shape of the I–V characteristic, however, but that in principle the NDR onset voltage V_{NDR} can be scaled below the thermal voltage $k_B T/q$ in this system with strong collective behavior with reductions in the single-particle/bare interlayer coupling (Figure 8.1(b)), as recently illustrated in quantum transport simulations in [6, 7]. This low voltage of the NDR onset is the basis for low-power computation.

We emphasize that excitonic superfluid condensate formation and associated enhanced interlayer conductance up to a critical current have been observed experimentally in III-V double quantum well heterostructures [10, 13, 14]. These I–V characteristics are also much like those of the better known superconducting Josephson junctions [17], and for much the same reasons (although the superfluid is not excitonic in Josephsen junctions). These superconductive systems, however, are limited to low temperatures that severely limit their applications.

Graphene (and/or perhaps other 2D materials systems as previously noted) may provide a synergy of properties that will allow the condensate formation above room temperature, screening considerations notwithstanding [15]. These properties include the ability to bring the two layers much closer together and within a much lower dielectric environment, both of which strengthen the interlayer Coulomb interactions and, thus, the condensate. It also is believed to be necessary to push the Fermi energy roughly $8k_B T$ or more above (below) the conduction (valence) band edge to allow condensate formation in this graphene system for reasons discussed in [3], or to or above roughly 200 meV at room temperature via layer sheet carrier densities of about 5×10^{12} carrier/cm^2 or more in graphene.

8.2.2   Device layout and compact modeling

To create BiSFETs, one must first create regions of condensation. Both initial estimates [1] and more recent quantum transport simulations [6, 7] suggest that the length of these regions need be only on the scale of ten to a few tens of nanometers. The regions of condensation can be defined by creation of the required layer charge densities in the required low dielectric constant (low-k) dielectric environment. Inclusion of air gaps above and/or below the graphene layers may even be necessary to obtain the required low-k environment. We had initially presumed the required charge densities to be produced via gating above and below the graphene bilayer system. We had also considered the possibility of achieving the required gate fields via gate work function engineering alone for one or both of the gates. However, it now seems more likely that use of substantial but fixed voltages applied to gates a few nm or more away from the condensate in a low dielectric constant (low-k) environment will be required to produce the condensate due to screening considerations (Fig. 8.1(c)).

At this point one could (and should for some applications) "just" apply separate contacts to the graphene layers to form what could be called a "bilayer pseudospin junction" (BiS junction). Ideally, but perhaps not necessarily for short condensate regions, the contacts should be on the same side of the region of condensation for either a BiS junction or a BiSFET; again, tunneling occurs near the edge of the condensate region. A BiS junction would be expected to behave much like a Josephson junction [17] – producing a symmetric low-voltage-onset DC NDR characteristic – and for similar reasons, but potentially at room temperature.

Unlike a Josephson junction, a BiSFET also needs gate control for switching. However, the gating is intended only to vary the critical current of the BiSFET. Moreover, essentially any amount of critical current variation would work in principle. In circuits (to be elaborated on below) the on/off condition is defined by which side of the NDR onset voltage the interlayer voltage falls; the gate need only control which device reaches its critical current first. In practice, the gate control will need to overcome variations in critical current in and among devices. As a result, switching voltages may well be defined by fabrication technology limitations, not theoretical limitations (much as for CMOS until recently).

While gate work function engineering could be used to create the condensate absent any applied voltage; these gates could also be used as "control gates," with small voltages applied to these gates to imbalance (balance) the charge distributions between layers and, thus, weaken (strengthen) the condensate and reduce (increase) the critical current. This approach was the first that we considered, in what we will call BiSFET 1.

However, with the gates that are used to create the condensate moved further out, substantial gate voltages are required for that purpose. Therefore, these gates cannot also be used as control gates for low-voltage switching. Another gate and gating approach is required to provide low-voltage control. To this end, we have considered gate-induced current shunting between parallel conduction paths, in what we will call BiSFET 2 (Fig. 8.1(c)). For example, to the extent that the current can be shunted to one side in two identical parallel paths (identical perhaps not being optimal) the combined critical current can be effectively halved. Once one condensate region falls into the NDR regime, current is shunted back and the other is forced into the NDR regime as well. However, halving of the critical current is not nominally necessary; again, gate control need only overcome unintended variations in critical current within and among devices.

For circuit modeling purposes, we have created simple device models intended to capture essential device elements and behavior (Fig. 8.1(d)). These models are not meant to be precise. We are not concerned with fractional changes in performance at this stage of development. We include a capacitance between each gate and the adjacent graphene layers, a capacitance between the two graphene layers, and an intra-layer quantum capacitance (associated with moving the Fermi level with respect to the band edge) for each layer. We include contact resistances. We include a model of interlayer current as a function of the interlayer voltage (Fermi level difference) and charge balance/imbalance. Note that while we now actually expect an abrupt NDR onset with a large on/off ratio [6, 7], neither is actually necessary for the circuits we consider, and we have been conservative with both to reinforce the point. For BiSFET 2, we also considered two parallel conduction paths, essentially two parallel BiS junctions, with gate-controlled resistance along (at least) one of the paths. This variation is treated as an effective variation of the contact resistance of one (or more) leads, and has been modeled after a graphene FET with thermal and inhomogeneous smearing of the conductance minimum.

8.2.3   BiSFET compatible circuits and circuit simulation results

With a strong NDR characteristic with increasing interlayer voltage, rather than current saturation with increasing source-to-drain voltage, BiSFETs do not make for simple low-voltage drop-in replacements for MOSFETs in CMOS-like circuits. Indeed, if we drop BiSFETs into CMOS-like circuits, the input to the BiSFET control gate(s) cannot change the output.

Instead, alternative circuit layouts have been found, including the introduction of clocked power supply voltages to our logic circuits (Fig. 8.1(e)) with a quarter period delay between basic logic elements (NOT, NAND, etc.) in series. This basic approach is much like what has been proposed previously for gated resonant tunneling diode circuits that also have gate-controlled NDR [18] but with the voltage scale much reduced. For a given logic element, the inputs to the control gates is first set by a preceding logic element. The supply voltage is then ramped up causing one (or more) BiSFETs to reach their critical current and fall into its (their) low conductance regime, i.e., switch off, with an interlayer voltage greater than V_{NDR}. Current is thereby blocked to the remaining BiSFET or BiSFETs, and they fall back to essentially zero interlayer voltage along the sub-V_{NDR} portion of their current–voltage characteristics. Output signals/voltages are taken between devices in series – in this case like CMOS – to set the input for the gates of the next logic element or elements. The supply voltages are then ramped on for the next logic element(s), and so forth.

A disadvantage of this approach is that the activity factor for each logic element is effectively unity. An advantage is that each logic element acts as its own latch; once set it will hold its output until its clock is ramped back down, even while its inputs change. Therefore, each logic element is free to consider new inputs once the next logic element in the series is set, no matter how many logic elements there are in series.

Time-dependent SPICE-level circuit simulations have exhibited remarkably low switching energies (Fig. 8.1(f)). Assuming clock voltages of 25 mV for illustrative purposes (approximately the room temperature thermal voltage $k_B T/q$), power consumption per BiSFET per computation have been obtained on the scale of ten to a few tens of zeptojoules ($1 \text{ zJ} = 10^{-21}$ J) depending on the logic element. Note that these energies are the energies consumed per BiSFET over the entire clock cycle; there is no real quiescent state. With the clock voltage about an order of magnitude or more below end-of-the-roadmap CMOS and power scaling with voltage squared, these switching energies are predictably two or more orders of magnitude below end-of-the-roadmap CMOS switching energies [8].

We also note that we have only considered application to logic circuits in simulation, and by only one basic approach; there should be more opportunities if BiSFETs can be realized. Some of these opportunities are suggested by the above-discussed similarities of the *I–V* characteristics of the proposed BiSFETs to those of gated resonant tunneling diodes but at lower voltages with stronger NDR characteristics, and to those of Josephson junctions but with gating and potentially room temperature operation. The possibility of multiple connections to the same condensate region, all cumulatively subject to the same critical current, also suggests the possibility of current controlled logic rather than voltage controlled gating, which we are just now beginning to explore. The possibility of multiple contacts to the same coherent region of condensation, or coupling or gate-controlled coupling and decoupling of coherent regions of condensation even if at higher voltages, also suggests the possibility of quantum computing applications.

8.2.4   BiSFET technology

If theory holds and creation of condensates and even sub-thermal switching voltages are possible, then realization of the BiSFET and its switching voltage and power consumption well may be defined by fabrication technology limitations, not theoretical limitations (much as for CMOS previously). The greatest technology challenges for the BiSFET are providing the dielectric environment necessary for creation of the condensate, controlling of the single-particle/bare interlayer tunneling via the interlayer dielectric, and providing only limited but reliable gate control.

The dielectric requirements for creating a condensate between dielectrically separated graphene layers are fundamentally different from those required for gate control in a MOSFET. In the region of condensation, the goal is to maximize the Coulomb interactions among individual charged particles, not minimize the Coulomb interactions (thus, requiring more charge for the same gate field) between gate and channel sheets of charge as in a MOSFET. As a result, low-*k* dielectrics are required. Moreover, the dielectric constants of the material above and below the dielectrically separated graphene layers generally will be more important than that of the dielectric between. On the positive side, preliminary theory suggests that it is the high-frequency dielectric constant – often substantially smaller than the low-frequency dielectric constant in polar materials – that is most relevant. Nevertheless, air gaps in the vicinity of the condensate may be necessary. While technologically challenging, the use of air gaps between the gate and the source and the drain has already been proposed to reduce parasitic gate-to-source and gate-to-drain capacitances in conventional MOSFETs [19], and top and bottom gated suspended graphene bilayers have already been reported, although without the interlayer dielectric required for the BiSFET at the time of writing [20].

In contrast, the dielectric requirements for the control gate region of BiSFET 2 are much the same as for conventional MOSFETs, or at least "conventional" graphene FETs. Moreover, the amount of gate control required is perhaps significantly smaller. The challenge will be reproducibility.

The interlayer dielectric, while also preferably low-k, must also be thin and provide both limited and reproducible interlayer single particle/bare coupling. Remember that the critical current varies with the interlayer bare coupling. This requirement suggests naturally layered systems such as hexagonal boron-nitride (hBN), transition metal dichalcogenides (TMD), or graphene intercalants, all of which colleagues and we are currently exploring.

8.3   Essential physics

Much of the essential physics of exciton condensate formation and interlayer current flow is not new to the system under consideration here, as per the experimental works [9–14]. Moreover, exciton condensation in bilayer systems was theoretically first predicted by Lozovik and Yudson in 1975 [21]. Most of these experimental investigations have focused on closely spaced bilayers in predominately III-V systems, although silicon-based systems also have been considered. The recent work of [14] exhibits enhanced but critical current-limited interlayer currents that are of importance here, along with perfect Coulomb drag within an alternative "drag-counterflow" biasing arrangement, both as predicted by theory [16].

However, the exciton condensate in III-V systems is observed only at very low temperatures and high magnetic fields. It is due to a synergy of unique electronic properties of graphene that it has been theoretically predicted that the superfluid state might occur above room temperature in this latter system [3, 4, 15]. In this section, we will look in more detail at the essential physics of the condensate formation in bilayer graphene, the source of the NDR characteristic with potentially sub-thermal voltage onset, and the various factors that govern both.

8.3.1   Condensate formation in bilayer graphene

The exciton condensate in bilayer systems results from the many-body exchange correction to the interlayer Coulomb interactions. The exchange interaction both self-consistently leads to and results from formation of an energy band anti-crossing/gap formed where the electron and hole bands would otherwise cross and necessarily encompassing the Fermi level. If the anti-crossing occurs about the Fermi level, this implies both that the Fermi level contour in k-space must be much the same for both electrons and holes – i.e., the electron and hole Fermi contours must be "nested" – and that the charge density magnitudes must be nearly balanced between the layers. In typical semiconductors, such Fermi contour nesting is not possible between the conduction and valence bands given the highly anisotropic heavy-hole valence band. Therefore, in the III-V systems, high magnetic fields have been used to form Landau levels in the conduction band, with a partially (optimally half) full Landau level in one layer corresponding to an electron band, and a partially (optimally half) empty Landau level in the other layer corresponding to a hole band. Because of the nearly symmetric conduction and valence bands of graphene, a dielectrically separated bilayer graphene system allows Fermi contour nesting with equal electron and hole densities in opposite layers.

For graphene layers, in mean-field theory on a p_z/π orbital-based atomistic tight binding lattice, the Fock many-body exchange interactions between layers, V_F, can be approximated as [5, 22]

$$V_F(\mathbf{R}_T, \mathbf{R}_B) \approx \frac{q^2}{4\pi\varepsilon_{eff}\sqrt{|\mathbf{R}_T - \mathbf{R}_B|^2 + d^2}} \rho(\mathbf{R}_T, \mathbf{R}_B), \quad (8.1)$$

where $\rho(\mathbf{R}_T, \mathbf{R}_B)$ is the interlayer portion of the quantum mechanical density matrix characterizing the interlayer quantum coherence; $\mathbf{R}_T$ and $\mathbf{R}_B$ are the two-dimensional (2D) in-plane location vectors for the atoms in the top and bottom graphene layers, respectively; d is the separation between the two layers; and q is the electron charge. Here we have neglected the detailed screening models considered in [15] and assumed a constant effective dielectric permittivity ε_{eff} to simplify discussion. (V_F remains proportional to ρ in a more careful analysis, but the dependencies of the Coulomb proportionality constant on $\mathbf{R}_T$, $\mathbf{R}_B$ and d become more complicated.)

We note that the interlayer density matrix, ρ, can also be described as the collective "pseudospin" matrix of the condensate, where "pseudospin" refers to the "which layer" degree of freedom, and wherein lies the BiSFET moniker [1, 3]. (This pseudospin should not be confused with the "which sublattice" degree of freedom for an individual graphene layer system that also can be described as a pseudospin.)

Now consider n-type and p-type graphene layers with their symmetric conduction and valence bands with equal charge densities under equilibrium conditions. The conduction band Dirac cone and the valence band Dirac cone will intersect at their common Fermi level. No magnetic fields are required to achieve the desired Fermi contour nesting. Assuming some coupling between the layers, there will be an energy band anti-crossing/gap formed about the Fermi level with interlayer-coherent quantum mechanical energy eigenstates near the anti-crossing bandgap edges, with the states below the bandgap preferentially occupied according to Fermi statistics. Therefore, the interlayer density matrix ρ and Fock exchange potential V_F will be non-zero. Therefore, there will be coupling between the layers, which will produce a bandgap and interlayer coherence, which will produce a non-zero Fock interaction, and so on. Moreover, the formation of the bandgap will lower the energy of the preferentially occupied states near but below the bandgap. This lowering will reduce the overall energy of the system, making gap formation energetically favorable. While some single-particle/bare coupling can strengthen the condensate, it is possible to find self-consistent solutions for the many-body condensate without any bare coupling, the so-called "spontaneous" condensates.

Graphene being a nearly perfectly two-dimensional material means both that the electronic states of even dielectrically separated layers can be brought much closer together than is conceivable for quantum wells in III-V systems, and that the effective dielectric permittivity is largely an extrinsic property, governed more by the surrounding dielectrics than the graphene itself, allowing the incorporation of low-k dielectrics. Both of these properties enhance the exchange interaction, Eq. (8.1), and thus the opportunity for condensate formation.

Even with other conditions being satisfied, the predicted maximum temperature for condensate formation, the critical temperature T_c, is expected to be approximately $E_F/(8k_BT)$ for reasons associated with the Kosterlitz–Thouless temperature as discussed in [3], where E_F is the magnitude of the Fermi energy relative to the band edge. The zero bandgap and low density of states make it relatively easy to move the Fermi level well into the bands with limited carrier concentrations and, thus, manageable gating fields. $T_c = E_F/(8k_BT)$ translates to electron and hole densities of only $n_o \approx p_o \approx 5 \times 10^{12}$ cm^{-2} for condensation at 300 K [1]. The carrier density in graphene layers has been electrostatically modulated to as high as 10^{13} cm^{-2} using independent gates [23].

Some of the basic properties of the condensate (obtained within a nearest neighbor atomistic tight-binding calculations except for the long-range Fock interaction of Eq. (8.1)) are illustrated for the spontaneous condensate in Fig. 8.2 (adapted from [22]). The solid lines in Fig. 8.2(a) show the low energy band structure of two weakly coupled graphene layers in the absence of interlayer exchange coupling. Absent interlayer coupling, there are two ("top" and "bottom") offset graphene band structures, indicated by solid lines, intersecting at the Fermi level E_F, which is taken to be the zero energy reference. The various dashed lines show the zero temperature low-energy dispersion for the bilayer graphene in presence of the Fock exchange interactions for different values of interlayer separation. A zero temperature bandgap E_{G0} opens due to the Fock interaction indicating condensate formation, and the condensate bandgap increases with decreasing interlayer separation due to strengthening interlayer Coulomb interactions and, thus, strengthening Fock exchange corrections to those Coulomb interactions. Figure 8.2(b) shows the temperature dependence of the condensate bandgap for three values of dielectric constant. The transition from the superfluid state to normal state is evident in the collapse of the condensate bandgap at the "critical temperature," T_c. Contributions to the density matrix and, thus, exchange interaction from quantum mechanical energy eigenstates above the bandgap are opposite in phase to those from states below the bandgap. (This behavior is analogous to formation of symmetric and anti-symmetric states when two otherwise isolated degenerate states couple.) The abruptness of condensate collapse with temperature is due to a positive feedback loop in which increasing temperature and the associated occupation of states above the gap and emptying of those below, weakens the exchange interaction, which reduces the bandgap, which leads to more occupation of states above the bandgap and less below, which weakens the exchange interaction, and so forth. As a result, the form of the temperature dependencies when normalized to the zero temperature bandgap E_{G0} are quite similar independent of the dielectric constant, with $k_BT_c \approx 0.25E_{G0}$. However,

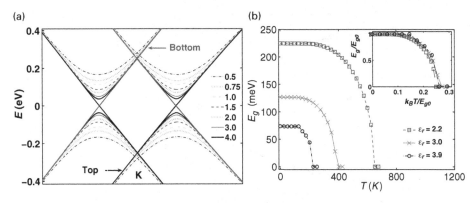

Fig. 8.2 Illustration of condensate formation, and dependencies on various parameters, taken from [22]. Basic properties of the graphene interlayer condensate obtained via the Fock interaction of Eq. (8.1). (a) Low-energy dispersion of the graphene bilayer system with a potential energy difference between layers of $\Delta E_\Gamma = 0.5$ eV, a relative dielectric constant $\varepsilon_r = 2.2$ at 0 K, and balanced charge distributions, as a function of layer separation d shown in the legend in units of nm. The solid lines are the band structures of the top and bottom layer graphene, respectively, in absence of interlayer exchange coupling. (b) Temperature dependence of the bandgap for three different dielectric constants with $\Delta E_\Gamma = 0.5$ eV, $d = 1$ nm, and balanced charge distributions. Lower ε_r result in a stronger Coulomb interaction and, thus, larger Foch correction potential, which leads to larger 0 K bandgaps that are also more robust at higher temperatures. The top right insert shows the same data, illustrating the similarity of the T dependence of bandgap for different ε_r when normalized by the 0 K bandgap, E_{G0}. We note, however, with screening considered (see Fig. 8.4 later) the required dielectric constant would be smaller and the abruptness of condensate formation likely greater. (Note that the nominal stacking, Bernal or otherwise, is essentially irrelevant for spontaneous condensates.) (Parts (a) and (b) reprinted with permission from [22], copyright (2010) by the American Physical Society.)

E_{G0} and, thus, T_c increase even more strongly with decreases in the dielectric constant than with layer separation.

Charge imbalance between layers has much the same effect on the condensate as raising the temperature. The imbalance shifts the Fermi level E_F toward one edge or the other of the condensate bandgap, either increasing occupancy of states above the bandgap or decreasing occupancy of states below the bandgap, weakening the condensate and leading to collapse of the condensate for large imbalance. Figure 8.3 (also adapted from the work [22]) illustrates the effect of the carrier density imbalance on the condensate bandgap, where an imbalance (n − p)/(n + p) in the range of about 20% to 25% leads to collapse of the condensate in these simulations. As shall be seen, in our initial BiSFET designs, gate-controlled charge imbalance was proposed as the switching mechanism, although concerns over screening (see below) have led us to consider alternative gating mechanisms.

The essential effects on the equilibrium condensate of adding interlayer single particle coupling can include strengthening it somewhat and smoothing out the transition about the otherwise critical temperature T_c depending on the strength of the interlayer coupling [5]. However, for bare interlayer coupling strengths of likely interest

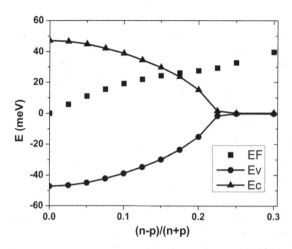

Fig. 8.3 300 K simulated energy band edges and Fermi level as a function of carrier imbalance between top layer electron density and bottom layer hole density for graphene bilayers separated by 1 nm at 300 K with $e_r = 3$ and $\Delta = 0.5$ eV. (Adapted with permission from [22], copyright (2010) by the American Physical Society.)

for low-voltage switching in the BiSFET, the effect of single particle coupling on these two aspects of the condensate are minimal.

Finally, we point out that screening of the interlayer Coulomb interactions substantially affects the strength of the condensate and associated critical temperature T_c. As already seen, larger permittivities ε and larger interlayer separations d reduce the interlayer Coulombic interactions and, thus, the exchange interaction and T_c. The strength of Coulomb interactions can also decrease due to screening from other charge carriers, a consideration neglected above except through possible adjustments to the effective permittivities. Moreover, there is no perfect theory of either superfluidity or screening. Above room temperature condensation was obtained in [3] based on calculations with unscreened interactions, although the qualitative effects of screening were discussed in some detail, while calculations based on static screening predict extremely low transition temperatures [24]. Recent calculations accounting for the dynamic nature of the interactions and the self-consistent reduction in the screening with condensate bandgap formation still predict the possibility of a room temperature T_c, although requiring smaller dielectric permittivity than originally estimated [15]. Figure 8.4 (adapted from [15]) shows the ratio of estimated bandgap 2Δ to Fermi level E_F for different models of screening, as a function of effective fine structure constant for graphene, $\alpha = e^2/(4\pi\varepsilon\hbar v_D)$ in SI units. Here, v_D is the fixed group velocity magnitude for carriers near the Dirac point in graphene. When the self-consistent reduction in screening with formation of the condensate induced bandgap is considered, there is an abrupt rise in the strength of the condensate at $\alpha \approx 1.5$, beyond which the full dynamical and gapped screening results approach (but do not reach) the result obtained with only dielectric screening. Still, with the vacuum fine structure being approximately 2.2, a low dielectric constant environment is clearly required based on the results. However,

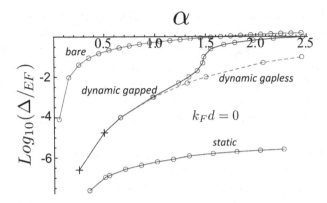

Fig. 8.4 Spontaneous gap as a function of effective fine structure constant $\alpha = e^2/(4\pi\varepsilon\hbar V_D)$ in SI units, where v_D is the fixed group velocity magnitude for carriers near the Dirac point in graphene, for (left to right) unscreened/bare interactions, and for static screening, dynamic gapless screening, and dynamic self-consistent screening, where 2Δ is the condensate induced bandgap. We require $E_F = 200$ mV and $\Delta = 50$ mV, such that required values of Δ/E_F are in the vicinity of 0.25 and $\log_{10}(\Delta/E_F)$ is in the vicinity of -0.6. If the dynamic self-consistent reduction in screening in the coherent state is neglected, the sudden rise at $\alpha \approx 1.5$ is absent. The full dynamical and gapped screening result approaches the result obtained with bare interactions for strong coupling, but differs from it by several orders of magnitude at weak coupling. (Parts (a) and (b) reprinted with permission from [15], copyright (2012) by the American Physical Society.)

theoretical numbers could vary somewhat either way from reality, making it either impossible or more readily possible to achieve condensation in dielectrically separated graphene layers. Ultimately, experiment will be required to resolve this issue.

Notably, with the fine structure constant inversely proportional to the carrier velocity, these results also suggest that use of an analogous material such as silicene or germanene (much like graphene but with carbon atoms replaced with silicon or germanium atoms respectively) would be better than graphene in this respect.

8.3.2 Low-voltage NDR

Part of the essential physics of interlayer current flow in the presence of the condensate is that while the associated exchange interaction alone can self-consistently support a spontaneous condensate, it cannot by itself support an interlayer current. In the employed tight-binding model, the electron charge current flow form from any point (orbital) in the top layer $\mathbf{R}_T$ to any point (orbital) in the bottom layer $\mathbf{R}_B$ (which has units of simply amps) is $J(\mathbf{R}_T, \mathbf{R}_B) = -qV_{\text{bare}}(\mathbf{R}_T, \mathbf{R}_B)2\hbar^{-1}\text{Im}\{[V_{\text{bare}}(\mathbf{R}_T, \mathbf{R}_B) + V_F(\mathbf{R}_T, \mathbf{R}_B)]\rho^\dagger(\mathbf{R}_T, \mathbf{R}_B)\}$, where $V_{\text{bare}}(\mathbf{R}_T, \mathbf{R}_B)$ is the single-particle/bare interlayer coupling potential. With the Fock interaction V_F being proportional to the density matrix ρ (Eq. (8.1)), the current J becomes simply

$$J(\mathbf{R}_T, \mathbf{R}_B) = -qV_{\text{bare}}(\mathbf{R}_T, \mathbf{R}_B)2\hbar^{-1}\text{Im}\rho^\dagger(\mathbf{R}_T, \mathbf{R}_B), \tag{8.2}$$

where V_{bare} has been taken to be real for simplicity (but not necessity). Eq. (8.2) does not explicitly include the Fock interaction V_F and holds independent of condensate formation, but the interlayer density matrix ρ is modified greatly by condensate formation via the Fock interaction. Because the magnitude of the pseudospin is greatly enhanced by condensate formation, the interlayer tunneling current also is greatly enhanced compared to what it would be in the absence of the condensate. Indeed, the layers are expected to be nearly shorted together independent of the bare coupling for small interlayer voltages, at least in the sense approaching the ballistic limit of performance in Landauer–Büttiker conductance theory (as will be illustrated through simulation results below).

The interlayer density/pseudospin matrix $\rho(\mathbf{R}_T, \mathbf{R}_B)$ has both magnitude and phase angle as a function of $\mathbf{R}_T$ and $\mathbf{R}_B$. However, through the feedback between the interlayer density matrix/pseudospin and the exchange interaction that primarily supports it, in practice the pseudospin phase becomes a roughly homogeneous property of the condensate as a function of $\mathbf{R}_T - \mathbf{R}_B$. We do not depend on this approximation for Eq. (8.2) or for the current flow simulations results to follow, but it does aid estimation of the critical current and explaining the more rigorously calculated behavior. Moreover, the magnitude of ρ (i.e., the strength of the condensate) remains essentially fixed for interlayer voltages on the scale of interest for the BiSFET even while the current flow increases essentially linearly with increases in interlayer voltage, leaving the increasing current to be accommodated through increases in the pseudospin phase angle and, thus, in the imaginary component of the pseudospin consistent with Eq. (8.2).

However, the steady-state (DC) current that can be supported through variations in the pseudospin phase angle is maximized when the angle reaches $\pm\pi/2$ and the pseudospin is purely imaginary. This maximum interlay current is referred to as the "critical current," and is analogous to the critical current in a superconducting Josephson junction [17]. Consistent with Eq. (8.2), however, this critical current varies linearly with the single-particle interlayer coupling V_{bare}.

If the interlayer voltage is increased beyond the critical current, the pseudospin phase cannot be stabilized and the steady-state current vanishes, although the condensate itself does *not* break down. Rather the pseudospin phase angle and, thus, the interlayer current is expected to oscillate with time, with an oscillation frequency of roughly of $2qV_{il}/h$ where V_{il} is the interlayer voltage and h is Plank's constant, analogous to the Josephson effect. An interlayer voltage of 25 mV, as commonly used for BiSFET circuit simulations, would produce oscillation frequencies greater than 12 THz! Averaged over one clock cycle of a 10 Ghz clock frequency and/or filtered out by load capacitances, this signal would essentially vanish. On the other hand, again analogous to a Josephson junction, a BiSFET or just a BiS junction could also be used as linear voltage-to-frequency THz generator above the interlayer DC NDR onset voltage V_{NDR} associated with the critical current.

Critically, so long as the condensate exists, there is nothing in this discussion of interlayer current flow that is sensitive to temperature. In principle, the condensate can have interlayer NDR onset voltages comparable to or smaller than the thermal voltage $k_B T/q$, allowing the proposed BiSFET to operate at or below thermal voltages in principle.

To illustrate the above transport physics, quantum transport simulations of interlayer current flow have been performed with the Fock exchange potential self-consistently calculated via an iterative method [6, 7]. We considered a voltage biasing arrangement much like that of Fig. 8.5(b), except that the right contacts were grounded rather than left floating in these initial simulations. All simulations shown here were performed at 300 K (as defined by the occupation probabilities of the incident propagating quantum mechanical energy eigenstates). The length of the condensate region was defined simply by setting the exchange interaction to external points to zero. For the simulations shown here, the length of the condensate regions was 15 nm. A resulting condensate bandgap of approximately 200 mV (even if not completely formed with some probability density remaining in the gap via evanescing state) eliminates most current flow between left and right contacts under this biasing scheme. As a result, the device is effectively contacted the same as in Fig. 8.5(b), i.e., the same as a BiS junction or BiSFET without the control gate(s). A single-particle interlayer A-sublattice atom to nearest-neighbor A-sublattice atom coupling (as would be expected to dominate A-to-B coupling through a dielectric [5]) of 0.5 mV was assumed in the example. The pseudospin phase as a function of iteration step after applying the interlayer voltage V_{il} across and equally divided between the left contacts is shown in Fig. 8.6(a) (adapted from [6]). A stable phase angle of

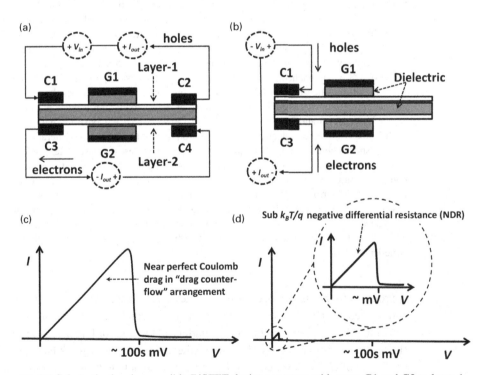

Fig. 8.5 Schematic showing possible BiSFET device geometry with gates G1 and G2 to layer-1 and layer-2 respectively and (a) four independent contacts C1 to C4, (b) two independent contacts C1 and C3, (c) illustration of current–voltage characteristics for the device in the drag-counterflow arrangement shown in (a), and (d) current–voltage characteristics for the device geometry shown in (b).

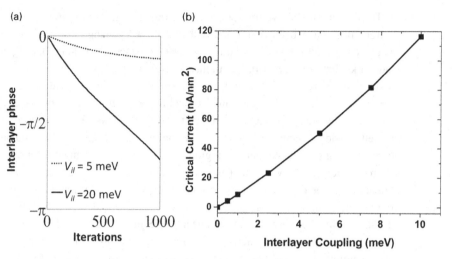

Fig. 8.6 Illustration of a sub-$k_\mathrm{B}T/q$ interlayer NDR onset voltage. (a) Variations in pseudospin phase difference between layers (it is roughly independent of position) during iterative calculations of the condensate and associated steady state current flow between layers, as obtained through quantum transport calculations at a simulation temperature of 300 K (figure adapted from [6]). The shown pseudospin phase angle was calculated between adjacent atoms at the center of the condensate region, the pseudospin also was confirmed to be almost spatially uniform throughout the condensate consistent with expectations as discussed in the text. An A-sublattice to A-sublattice hopping potential of 0.5 mV was assumed. The largest possible current flow between layers occurs when the interlayer phase angle is $\pm\pi/2$. For a 5 meV interlayer voltage, a stable phase of $\pi/8$ is approached. As the interlayer current flow is proportional to both the interlayer voltage and the sine of the phase angle, below the critical current, this result would suggest a critical current density and NDR onset voltage of, respectively, 6 nA/nm^2 × sin $(\pi/2)/$ sin $(\pi/8) \cong 16$ nA and 5 mV × sin $(\pi/2)/$ sin $(\pi/8) \cong 13$ mV. Consistent with this expectation, for a 20 meV interlayer voltage, no stable solution is found with increasing iteration count beyond what is shown here. Note that the rate of change of the phase increases after $-\pi/2$ is reached. These results from quantum transport calculations are perhaps two to three times the 0.5 meV bare coupling results shown on the critical current vs. hopping potential curve of (b) for the same type (A atom to A atom) of interlayer coupling, which was obtained earlier from more "back of the envelope" calculations [5]. However, the condensate here (as measured by the bandgap) is stronger than that for which the prior results were obtained as well, so that the agreement is as good as could be expected. (Part (b) adapted with permission from [5], copyright (2011) by the American Physical Society.)

approximately $\pi/8$ is approached for an interlayer voltage of 5 meV, with an interlayer current density per unit area in the center of the condensate region of approximately 6 nA/nm^2. This current, around three-quarters of the limit of Landauer–Büttiker theory, shows the layers to be nearly shorted together despite the very weak bare coupling. The prediction of an interlayer current flow proportional both to the interlayer voltage and to the sine of the pseudospin phase up to the critical current, suggests a critical current density and associated NDR onset voltage of, respectively, 6 nA/nm^2 × sin $(\pi/2)/$ sin $(\pi/8) \cong 16$ nA and 5 mV × sin $(\pi/2)/$ sin $(\pi/8) \cong 13$ mV. Consistent with this expectation, no stable interlayer phase can be achieved at an interlayer voltage of

20 mV; the phase angle magnitude exceeds $\pi/2$ and, in extended simulations not shown, continues to increase until it reaches $2\pi = 0$ and then repeats indefinitely in these iterative calculations. This latter behavior is consistent with expectations of oscillatory time-dependent behavior of the interlayer current. This calculated critical/switching voltage is one-half of $k_B T$ at the 300 K simulation temperature, again supporting the possibility of sub-$k_B T/q$ voltage switching in BiSFETs. Notably, these results from quantum transport calculations are reasonably consistent with the results shown in the critical current vs. hopping potential curve of Fig. 8.6(b) that was obtained from more "back of the envelope" calculations previously [5], if under somewhat different conditions. Among the additional findings was that the interlayer current below the NDR onset voltage fell by over three orders of magnitude with a 1 mV interlayer coupling if the Fock exchange interaction was turned off during simulation. Moreover, this ratio should increase with the square of reductions in bare hopping potential. More extensive quantum transport simulation results, including still lower NDR onset voltages at room temperature, and details of the simulation method can be found in [6, 7].

For comparison, the so-called "drag-counterflow" arrangement of Fig. 8.5(a) also has been simulated and the results presented in [7]. Consistent with experiments in III-V systems [14] and theory [16], there was nearly perfect interlayer Coulomb drag in which current *in* the unbiased layer closely approached in magnitude the external voltage-driven current *in* the biased layer flowing in the opposite direction, up to a relatively large voltage of approximately 100 mV here. Beyond the 100 mV bias, the condensate collapsed taking the current in the bottom layer away with it. The effect of the large voltage, i.e., of splitting of the Fermi level, is to fill states above the condensate bandgap and to empty ones below, leading to condensate collapse in much the same way that increasing the temperature or charge imbalance does. This behavior contrasts fundamentally with the source of NDR for DC interlayer current flow with the biasing condition of Fig. 8.5(b) considered above, where the condensate remains but its pseudospin phase becomes time-dependent. Moreover, while physically interesting and potentially useful, this latter biasing scheme does not appear to offer the same opportunity for low-voltage switching as that of Fig. 8.5(b), and is not that relied upon for the proposed BiSFET.

8.4   BiSFET design and compact modeling

The essential physics discussed above govern the design elements of the BiSFET, including the evolution of the latter with our improving understanding of the former.

8.4.1   BiSFET (and BiS junction) design

Four basic elements are required to create BiSFETs: graphene layers separated by a dielectric tunnel barrier (or perhaps dielectrically separated layers of other semi-metallic or semiconducting 2D semiconductors); a region of condensation created and localized by some means; some means of gating the critical current; and contacts to the individual graphene layers (although one can imagine all-graphene layouts where the contacts would just be graphene leads).

While varying the layer separation can serve to localize the condensate, with the condensate localized by other means (see below) there is little need to vary the dielectric thickness beyond the region of condensation. Outside of the region of condensation, the interlayer current is expected to fall by several orders of magnitude absent the condensate, as discussed in Section 8.3.2. Therefore, it becomes reasonable to think of the dielectrically separated graphene layers as a spatially uniform graphene–dielectric–graphene sandwich of naturally 2D materials, such as graphene–hBN–graphene, graphene–TMD–graphene, or graphene–intercalant–graphene.

Creating good contacts to graphene is a non-trivial business, but not one specific to BiSFETs, although doing so independently to two graphene layers becomes still more problematic. In this section, we will simply presume that it can be done and focus on the layout issues. Ideally, the contacts should be on the same side of the region of condensation either for a BiS junction or for a BiSFET. The current is injected to the condensate region within the condensate bandgap and, thus, is transferred between layers via quantum mechanical wave functions evanescing into the region of the condensate. However, it is also arguable that transport could still be achieved across a short condensate regions such as considered in [6, 7] where the nominal gap is never completely empty of density of states. Moreover, in the simulations of [6, 7] the interlayer current flow was more delocalized throughout the condensate region than the single-particle current. Although more work is required on this subject, for now we have assumed that the contacts are made on the same side of the condensate region.

As per initial estimates [1] and the quantum transport simulations discussed in Section 8.3.2 [6, 7], the length of the region of condensation may need to be only on the scale of tens to a few tens of nanometers. The regions of condensation can be defined by varying anything that affects the condensate, charge concentration, dielectric environment, or spatial separation. To date, we have presumed the required charge densities to be produced via gating, nominally above and below the graphene bilayer system, so that the region of condensation is intrinsically localized to the vicinity of these gates. Initially we considered the possibility of achieving the required gate fields via gate work function engineering alone for one or both of the gates. However, based on our improved understanding of screening [15] as discussed in Section 8.3.1, it now seems more likely that use of substantial but fixed voltages applied to gates located a few nm or more away from the condensate will be required. Notably, if an intercalant could be used with its associated heavy doping of the graphene layers [25], perhaps only one gate would be required to strongly invert one layer, if it also lowered the carrier density in the other.

With a region of condensation, and with contacts established to the individual graphene layers, one already has a BiS junction, as illustrated in Fig. 8.7(c). As previously noted, a BiS junction would be expected to behave much like a Josephson junction [17], producing a symmetric low-voltage-onset DC NDR characteristic and THz oscillations beyond the DC NDR onset voltage (although the latter would be expected to be filtered out via resistance–capacitance time constants in BiSFET logic circuits).

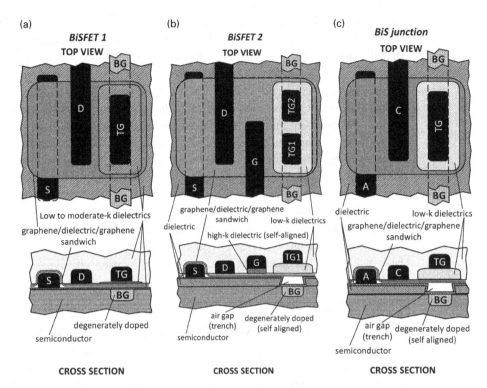

Fig. 8.7 Illustration (only) of (a) the initial BiSFET concept, "BiSFET 1," (b) the updated concept, "BiSFET 2," and (c) an un-gated BiS junction. The black regions are metallic contacts: "source" (S) and "drain" contact (D) contacts to the opposite layers; top gate(s) (TG) and bottom gate (BG) for creating the condensate, and also used for V_{NDR} control in BiSFET 1; and V_{NDR} control gate (G) for BiSFET 2. One may think of BiSFET 2 as two BiS junctions with an integrated conventional BiSFET.

However, for the BiSFET, a control gate is needed to alter the critical current, although only to a limited degree, to determine which device reaches its critical current first. However, the change in the approach to creating the region of condensation as discussed above has precipitated a basic shift in approach to gating the critical current for BiSFETs, which has led to shift in layout of the prototypical BiSFET. In the initial BiSFET design, the gate or gates for which work function engineering was used to create the condensate, could serve double duty as a control gate for the critical current. Switching would be achieved via small gate voltage-induced variations in the carrier densities, and resulting limited changes in the condensate strength and critical current, as described in Section 8.3. We refer to this device approach as "BiSFET 1," as schematically illustrated in Fig. 8.7(a). However, both because the gates used to create the condensate have been moved further away and because substantial voltages are applied to them to create the condensate, these gates are no longer available for low-voltage control of the critical current. Instead, a separate "control gate" must be provided to alter the critical current. Moreover, with the condensate hidden between the fixed-voltage gates, controlling the critical current via the strength of the condensate is no longer a readily available option. Therefore, in what we will call BiSFET 2, schematically illustrated in Fig. 8.7(b), a gate is used to control current crowding in parallel conduction paths via limited changes in the lead resistance. For example, to the extent that the current can be shunted to one side in two identical parallel paths (identical perhaps not being optimal) the combined critical current can be effectively halved. Once one condensate region falls into the NDR regime, current is shunted back, and the other is forced into the NDR regime as well. However, halving of the critical current is not nominally necessary; again, gate control need only overcome unintended variations in critical current within and among devices. Note that it is probably not sufficient to simply crowd the current into one side of a large condensate region, as the pseudospin phase angle tends to become a homogeneous property of the condensate as noted above; rather two separate regions of condensation are required. In the schematically illustrated BiSFET 2, there is also a greater effort to minimize the dielectric environment about the condensate, again not only between but, more importantly, above and below the condensate, including perhaps by using air gaps.

8.4.2   Compact circuit models

For circuit modeling purposes, we have created simple compact device models intended to capture essential device elements and behavior. As previously noted, we include a capacitance between each gate and the adjacent graphene layers, a capacitance between the two graphene layers, and an intra-layer quantum capacitance (associated with moving the Fermi level with respect to the band edge) for each layer. We include contact resistances. We include a model of interlayer current as a function of the interlayer voltage (Fermi level difference) and charge balance/imbalance. These models are not meant to be precise as we are not concerned with fractional changes in performance at this stage of development. For BiSFET 2, we also considered two parallel conduction paths – essentially two parallel BiS junctions – with gate-controlled resistance added along (at least) one of the paths. This gating has been modeled as an effective variation of the contact resistance of one (or more) leads, and has been modeled after a graphene FET with thermal and inhomogeneous smearing of the conductance minimum. However, the capacitive coupling to the gates used to create the condensate is significantly weakened. Moreover, while these gates have fixed large voltages applied, they should act as virtual grounds in the circuit. So for BiSFET 2, the gates used to create the condensate should have little effect on power consumption. However, charging and discharging the control gate, which is inherently strongly coupled, becomes a new source of power consumption. These compact models for BiSFET 1 and BiSFET 2 are shown in Fig. 8.8(a) and Fig. 8.8(b), respectively. The layout and compact model for a simple BiS junction is shown in Fig. 8.7(c).

For circuit purposes for BiSFET 2, we distinguish between "p-type" and "n-type" (vs. p-channel and n-channel) devices, only in that we assume that application of a negative gate voltage to an "n-type" device decreases its critical current, and application of a positive voltage to a "p-type" device decreases its critical current. With the gates separate from the region of condensation and degree of current balance/imbalance

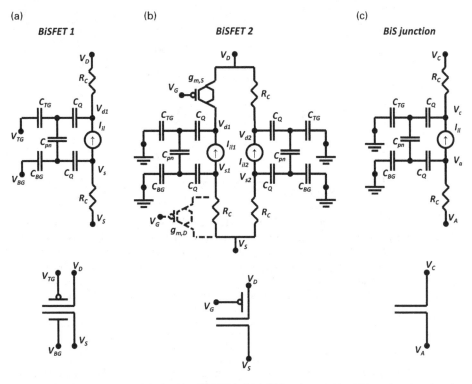

Fig. 8.8 Simple compact models for (a) BiSFET 1, (b) BiSFET 2, and (c) BiS junction showing the essential elements including gate capacitances, interlayer capacitances, quantum capacitances, contact resistances and, for BiSFET 2, integrated graphene FET gate control, where a p-type BiSFET is illustrated here. (Note that p-type is only meant to imply that a negative voltage increases the critical current through improved current balance, which could be the result of increasing hole current in a p-type layer, or decreasing electron current in an n-type layer. I–V characteristics for the interlayer current I_{pn} as a function of interlayer bias and charge balance, and (stronger than necessary) graphene FET characteristics are illustrated in Fig. 8.9(a) and Fig. 8.9(b), respectively.

between the two parallel regions of condensation being the basis of switching, there are many scenarios under which a device could be either "p-type" or "n-type." No assumption is made to whether or not the graphene layer adjacent to the gate, or both layers, are actually p-type or n-type, or whether the carrier type in the layers under the gate correspond to those in the region of condensation. (With Klein tunneling over a short region between the gate and condensate regions, the resistance associated with the transition between p-type to n-type should be less than the nominal contact resistances.)

For specificity, while an abrupt conductance drop beyond the NDR onset voltage, as illustrated in Fig. 8.5(d), with a large on/off ratio is expected for the interlayer condensate-enhanced DC current [6, 7, 14, 16], neither trait is actually necessary for the circuits we consider. We have been conservative with both to reinforce the latter point. A smooth characteristic has been used for most simulations of both BiSFET versions [1, 2], as shown in Fig. 8.9(a), and in the following equation:

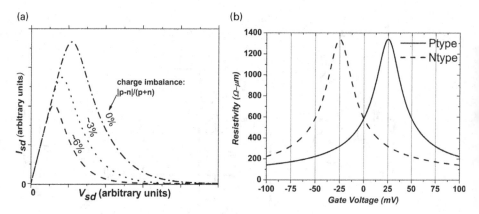

Fig. 8.9 (a) Model *I–V* characteristics for a condensate depending on the interlayer charge balance, and (b) gate modulated resistance of the graphene lead in BiSFET 2 as used in the simulations shown in this work with $E_{F\min} = 10$ meV. Circuits have also been shown to work for substantially weaker dependencies on charge imbalance, slower decay of the tail of the interlayer *I–V* characteristic (although more rapid is expected) and reduced on/off gate-controlled lead resistivity (at least without variances in device characteristics among devices.)

$$I_{SD} = G_o V_{SD} \left\{ 1 + \left[\frac{|V_{SD}|/V_{NDR}}{\exp(1 - |V_{SD}|/V_{NDR})} \right] \right\}, \qquad (8.3)$$

where I_{SD} is the source-to-drain interlayer current, G_o is the Landauer–Büttiker ballistic conductance for injection into these highly charged regions, V_{SD} is the source-to-drain voltage drop across the condensate $-(E_{F,S} - E_{F,D})/q$ (distinguished from the source to drain voltage by intermediate resistive losses), and V_{NDR} is the critical-current-associated single-particle interlayer coupling-dependent magnitude of V_{SD} at which NDR onset occurs. V_{NDR} has been approximated as

$$V_{NDR} = V_{NDR,n=p} \exp\left(-10 \frac{|p - n|}{p + n} \right), \qquad (8.4)$$

where p and n are the hole and electron concentration on the p-type and n-type layers in the region of condensation, $V_{NDR,n=p}$ is the magnitude of the NDR onset voltage for balanced charge distributions. In retrospect, Eq. (8.4) (introduced first) is somewhat overly sensitive to charge imbalance as compared to the (subsequently performed) simulations of Fig. 8.3. However, we have checked using the results of Fig. 8.3, and circuits based on BiSFET 1 continued to work with the associated more limited gate control. More importantly, for the updated BiSFET concept, BiSFET 2, which does not rely on charge imbalance, this approximation is nominally conservative in terms of desired device operation.

For the BiSFET 2 compact model, the resistance of the gated lead as a function of the control gate voltage is illustrated in Fig. 8.9(b). The peak resistance is assumed to be limited by thermal and inhomogeneous smearing about the Dirac point, as

$$R^{-1} = \sqrt{\sigma_g^2 + \sigma_{\min}^2}, \qquad (8.5)$$

where $\sigma_g = 8q^2 E_F W/(v_F h^2)$ is the quantum conductance due to the number of modes in a graphene layer of width W and Fermi level E_F, h is the Planck's constant, q is the electron charge, v_F is the essentially fixed Fermi velocity magnitude of electrons in graphene, and W is the channel width [1]. σ_{min} represents the minimum conductivity due to thermal and inhomogeneous smearing about the Dirac point. The Fermi level in graphene under the gate is calculated using an equivalent series capacitance network of gate oxide and graphene quantum capacitance. In practice, we only consider ± 50 mV range ($2V_{clock}$ in simulations to follow) about the maximum resistance. We also have considered considerably more smearing than shown in Fig. 8.9(b).

8.5   BiSFET logic circuits and simulation results

With a strong NDR characteristic with increasing interlayer voltage rather than current saturation with increasing source-to-drain voltage, and gate control that only alters the NDR onset voltage rather than turning the device on and off directly, BiSFETs are not simple low-voltage drop-in replacements for MOSFETs in CMOS-like circuits. Alternative circuit architectures must be found.

In this section, we illustrate one way in which Boolean logic gates could be implemented using BiSFETs, as well as how they cannot be for reference, and we provide SPICE-level circuit simulations. We emphasize that the approach discussed here is only a first attempt. How to best use the condensate for logic and other purposes needs to be explored further. With both BiSFET versions exhibiting a conductance peak intrinsically centered about zero interlayer voltage and a gateable critical current followed by NDR, essential circuit architectures and scale of power consumption are similar. Although most of the original circuit work was performed with the BiSFET 1 version [1, 2], we exhibit currently available proof of concept results obtained with the BiSFET 2 version, returning to those for BiSFET 1 for an earlier more complex example.

8.5.1   BiSFETs logic circuit basics: how to use and how not to use BiSFETs

If BiSFETs are inserted into a conventional CMOS inverter circuit, if anything, a memory element is produced [2]. As schematically illustrated in Fig. 8.10(a), there will be three separate operating points given fixed power supply voltages $\pm V_{supply}$ as shown: two operating points with one *or* the other BiSFET within its low conductance/off-state beyond the NDR onset voltage, and the other one still in its high conductance/on-state, with the output voltage either high ("1") or low ("0"); and one operating point with both BiSFETs in their low conductance/off-state beyond their NDR onset voltages, with the output in a perhaps not entirely stable intermediate state with little available current to drive the next device in series. Moreover, altering the input voltage to produce limited changes in the NDR onset voltages will have essentially no effect on the high- or low-output states, and no apparent useful effect on the intermediate state. This indifference to input voltage with a fixed supply voltage, however, has led to at least one memory element design [2] as noted, and others are no doubt possible.

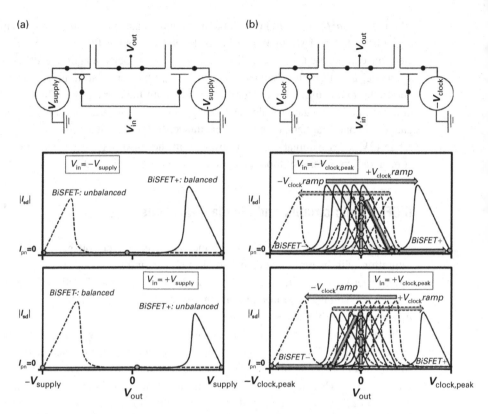

Fig. 8.10 Ilustration of how to and how not to make a BiSFET-based inverter. (a) CMOS-like inverter circuit with BiSFETs and fixed power supplies $\perp V_{\mathrm{supply}}$, and an associated output voltage V_{out} that is essentially independent of input voltage V_{in}. The solid and dashed curves illustrate the interlayer current magnitude vs. common source/output voltage V_{out}, for "BiSFET+" and "BiSFET−," respectively. The, intersection points of the paired BiSFET $I{-}V$ characteristics represent possible circuit operating points/values of V_{out}, of which there are three for each input voltage. These operating points remain largely fixed independent of V_{in}; there is no path between operating points with changes in V_{in}. (b) CMOS-like inverter circuit with the BiSFETs and power supply ramped to $\pm V_{\mathrm{clock,peak}}$ after the input voltage is set. Again, the solid and dashed curves illustrate the interlayer current magnitude vs. common source/output voltage V_{out}, for "BiSFET+" and "BiSFET−," respectively. However, these characteristics move with respect to V_{out} with increasing $\pm V_{\mathrm{clock}}$, and, therefore so does the intersection/circuit operating point/value of V_{out}. With $\pm V_{\mathrm{clock}} = 0$, there is only one possible operating point, $V_{\mathrm{out}} = 0$. As the supply voltages are ramped, initially both BiSFETs are in their high conductivity/on-states. The current increases, while the intersection/circuit operating point/value of V_{out} remains approximately zero. However, at some point with the increasing supply voltages, the BiSFET with the lower critical current can support no more current and falls into its low conductivity/off-state beyond its NDR onset voltage. With the same current funneled through both BiSFETs, the other BiSFET can never reach its critical current and, thus, remains in its high conductance/on-state. The intersection/circuit operating point/value of V_{out}, in turn, then follows the supply voltage connected to the on device until $\pm V_{\mathrm{clock,peak}}$ is reached, and the desired inversion operation is achieved, although delayed by the power supply ramp time.

However, if the inputs are set and the power supplies are then ramped up, the desired output logic state will be obtained, as schematically illustrated in Fig. 8.10(b). In this way, initially there is only one possible operating point at $V_{out} = 0$ with both BiSFETs in their high conductance/on-state. As the balanced supply voltages are ramped, initially both BiSFETs remain in their high conductivity/on-states, the current increases, while V_{out} remains approximately zero. However, at some point with the increasing supply voltages, the BiSFET with the lower critical current reaches its critical current and falls into its low conductivity/off-state beyond its NDR onset voltage. With the same current funneled through both BiSFETs, the other BiSFET can never reach its critical current and, thus, remains in its high conductance/on-state. As a result, V_{out} follows the supply voltage connected to the on device until $\pm V_{clock,peak}$ is reached, and the desired inversion operation is achieved, although delayed by the power supply ramp time. Moreover, once the clock voltage is raised and the gate output is set, the input signal can be turned off with no effect; each gate doubles as a latch.

This basic procedure can be used to realize more complex logic elements as well. With the BiSFET 1 version, we realized a full array of logic elements, including but not limited to those discussed in [2], and without the dual power supplies that make design easier but are not strictly necessary. With the more recent BiSFET 2 version, we have created proof of concept circuits at the time of writing.

In order to perform a series of computations with the same BiSFET-based gate, a clocked power supply is required; to perform calculations by a sequence of BiSFET gates, a multi-phase clocked power supply is required [2]. For a single gate, the clock voltage has to be raised after the input is set to perform the required logic function. The clock voltage and the associated gate output then must be held high long enough for the next gate to read that output. The clock has to be lowered again so that the gate can accept a new input. Now the clock should be held low while the state of the preceding gate – its next input – is being set so as to not interfere with the process. For multiple gates, there also must clearly be a time lag between gates. In a series of gates, the state of the first gate must be set before the state of the second gate can be set, which must be set before the state of the third gate can be set, and so forth. To this end, we have employed a four-phase clocking scheme as shown in Fig. 8.11. Circuit-level simulations [26] suggest that such a low-voltage clocked supply can be created even in CMOS technology, with 50% or less overhead. However, the shape of the signal and the number of phases have not been fully optimized.

A disadvantage of this clocked power supply scheme is that the activity factor for each logic element is effectively unity. An advantage of this scheme is that each logic element is free to consider new inputs once the next logic element in the series is set, no matter how many logic elements there are in series.

8.5.2   Circuit simulation results

For the BiSFET 2-based circuits, VerilogA model-based SPICE-level circuit simulations were performed using the device models described in Section 8.4 and shown in Fig. 8.10(b). Except for that of the gated lead of the compact model, the contact

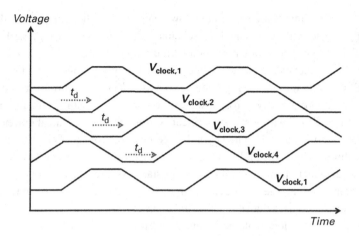

Fig. 8.11 Four-phase clocking scheme, $V_{clock,n}(t)$, with inter-clock time delay t_d.

resistances were taken to be the value of resistance 320 Ω-μm at zero gate voltage, corresponding to a Fermi level of $|E_F| = 50$ mV and a carrier concentration of about $2 \times 10^{11}/cm^2$ in the leads. The contact resistance of the gate lead was modeled consistent with Eq. (8.5), which behaves as illustrated in Fig. 8.9(b), for the simulation results shown here. However, much larger σ_{min} and accompanying smaller on/off ratio as 1.06 for the included graphene FETs also have been used successfully in the same circuits. The interlayer capacitance was taken to be ≈ 0.35 fF/μm corresponding to an SiO$_2$-*equivalent* oxide thickness (EOT) of 1 nm. The capacitance to control gate was taken to be ≈ 0.41 fF/μm corresponding to an EOT of 0.85 nm. The capacitance to each of the gates used to create the condensate was taken to be ≈ 0.03 fF/μm corresponding to an EOT of 10 nm, representing a low-k gate dielectric environment shown in Fig. 8.7(b) with closer gates. The interlayer current–voltage relation through the condensate is shown in Fig. 8.9(a). The width W of the condensate regions, by which all of the above was scaled, was taken to be 20 nm unless otherwise noted. The peak value of clocked power supply voltage $V_{clock,peak}$ was taken to be 25 mV in all cases for illustrative purposes, approximately room temperature $k_B T/q$.

Simulation of the inverter illustrates the potential for power savings possible by using peak power supplies on the scale of room temperature $k_B T/q$. Figure 8.12 shows the SPICE-level simulated response for the BiSFET 2-based inverter circuit of Fig. 8.10(b) with a four-inverter load, input signal similarly supplied from a preceding inverter, and a clock frequency of 10 GHz. The calculated average energy per BiSFET per operation is about 10 zJ (10^{-2} aJ), which includes all power consumed during the clock cycle. The drastic reduction in power consumption per switching event relative to the even end of the roadmap MOSFETs, which would be on the scale of a few aJ in principle, is primarily due to the voltage reduction.

More complex logic elements can provide similar power savings. Figure 8.13(a) shows the schematic NAND circuit based on BiSFET 2. Figure 8.13(b) shows the SPICE-level simulated response of the NAND gate with a 10 GHz clock, verifying its

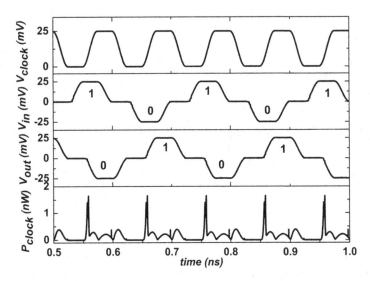

Fig. 8.12 SPICE-level simulated response of the BiSFET 2-based inverter shown in Fig. 8.10(b), with dual 10 GHz, ±25 mV peak clocked power supplies ±V_{clock}(t) driving a four-identical-inverter load (fan out of four). The energy per operation/clock cycle per BiSFET is about 10 zJ. The input signal was also taken from a preceding inverter in a similar fashion. The instantaneous power was conservatively calculated as the *magnitude* of the product of the current times voltage summed over both supplies. (Actually, there would be some power return during part of the clock cycle.)

(a)   (b)

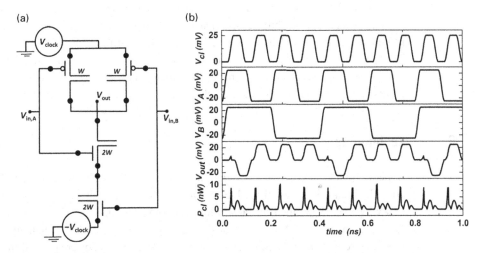

Fig. 8.13 (a) Schematic of BiSFET 2-based NAND circuit. The "2W" indicates device of twice the peak critical current, i.e, nominally twice the width. (b) SPICE-level simulated response of NAND with 10 GHz, 25 mV clock for BiSFET 2. The energy per operation for the NAND gate as a whole with no load is 140 zJ. Again, the switching power was obtained conservatively by integrating over the magnitude of the instantaneous power.

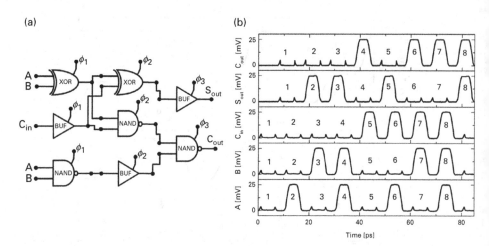

Fig. 8.14 (a) Circuit schematic of a 1-bit full adder incorporating XOR, NAND, and buffer gates, with clocked powers supplies φ_1, φ_2, and φ_3. The buffer is required for synchronizing signals. (b) Output of the 1-bit adder, including sum S_{out} and carry C_{out}. These older results were obtained with BiSFET 1-based logic elements and with 100 GHz clocked power supplies separated by a 2.5 ps delay. The higher speeds were allowed because parasitic contact resistances were not considered.

functionality and providing the instantaneous power consumption. The energy per operation for the NAND gate as a whole with no load is about 140 zJ.

We have considered more complex circuits with the older BiSFET 1 version as noted. Figure 8.14(a) shows a one-bit adder with sum and carry, incorporating smaller logic elements of XOR, NAND, and buffer gates (where the latter are used to maintain signal synchronization), and three clock phases. The parameter values for BiSFET 1 are provided and explained in [1, 2]. Figure 8.14(b) illustrates the adder's functionality. This one-bit adder also has been used to form a four-bit ripple carry adder (results not shown). We again note that no matter how many stages are considered in series, new inputs can be considered on each clock cycle.

8.6   Technology

If theory holds and creation of condensates and even sub-thermal switching voltages are possible, then realization of the BiSFET and its switching voltage and power consumption well may be defined by fabrication technology limitations, not theoretical limitations (much as for CMOS previously). The greatest technology challenges for the BiSFET are providing the dielectric environment necessary for creation of the condensate, control of the single-particle/bare interlayer tunneling via the interlayer dielectric, and providing only limited but reliable gate control.

The dielectric requirements for creating a condensate between dielectrically separated graphene layers, as discussed in Section 8.3.1 are fundamentally different from those required for optimal gate control in a MOSFET. Low-k dielectrics are desired, rather than high-k. The dielectric constants of the material above and below the dielectrically separated graphene layers generally will be more important than that of the dielectric between. It also is likely that the high-frequency dielectric constant is more important than the low-frequency one. Despite this positive note, air gaps in the vicinity of the condensate may be necessary for graphene-based BiSFETs. For this purpose, one may start with sacrificial oxides that are then etched away. While certainly technologically challenging, the use of air gaps between the gate and the source and the drain has already been proposed to reduce parasitic gate-to-source and gate-to-drain capacitances in conventional MOSFETs [19], and top and bottom gated suspended graphene bilayers have already been reported, although without the required interlayer dielectric at the time of writing [20]. The dielectric requirements may be somewhat less stringent for the analogous materials of silicene or germanene, or for TMDs as previously noted. However, achieving sufficient carrier concentrations to create the condensate may be more difficult, and challenges for producing these latter materials may be greater. Theory is imperfect, however, and we will not know just how stringent the requirements are for any of these systems until and unless a condensate can be realized experimentally.

In contrast, the dielectric requirements for the control gate region of BiSFET 2 are much the same as for "conventional" graphene FETs. The amount of gate control required will be significantly smaller, but so should be the voltages that are available for gating. The primary challenge may be reproducibility.

The interlayer dielectric, while also preferably low-k, must also be thin and provide both limited and reproducible interlayer single-particle/bare coupling. Remember that the critical current varies with the interlayer bare coupling. This requirement suggests naturally layered systems such as hexagonal boron–nitride (hBN), transition metal dichalcogenides (TMD), or graphene intercalants, all of which are being explored currently. The interlayer coupling must also be on the appropriate scale.

Rotational alignment of the opposing graphene, or silicene, or TMD layers, is also important. Preliminary work (not exhibited here) has suggested that the rotational alignment of the graphene layers has little impact on the spontaneous condensate. However, rotational alignment will certainly affect the single particle tunneling and, thus, the critical current. Indeed, simply rotating graphene layers may be enough to decouple them sufficiently to produce the condensate, and no interlayer dielectric may be required. However, with the corresponding increase in interlayer capacitance, gating the required charge densities to achieve room temperature condensation would become more problematic. Use of graphene intercalants as the interlayer dielectric would be advantageous in this regard, as one starts with crystallographically aligned bilayers into which the intercalant is then chemically inserted.

As intercalants also heavily dope the graphene layers [25], their use could lead to being able to produce the condensate with just one gate to create the condensate, which would be used to strongly invert one layer if while reducing the carrier concentration in the other in the process.

While daunting, having technology as the primary roadblock to progress to be chipped away over time as for CMOS previously, rather than having the intractable physics of thermionic emission and source-to-drain and channel-to-gate tunneling as roadblocks for continued long-term progress in CMOS, would be a win.

8.7   Conclusions

The goal of this work is to restart the "roadmap" beyond CMOS. Of course, that is the goal of all "beyond CMOS" work. However, with zeptojoule scale switching energies and novel functionality, the BiSFET could add a substantial amount of road!

Moreover, as previously noted, colleagues and we have only considered application to logic circuits in simulation, and by only one basic approach; there should be more opportunities if BiSFETs can be realized. Some of these opportunities are suggested by the above-discussed similarities of the $I-V$ characteristics of the proposed BiSFETs to those of gated resonant tunneling diodes but with lower voltages and larger on/off ratios, and to those of Josephson junctions but with gating and potentially room temperature operation. The possibility of multiple connections to the same condensate region, all cumulatively subject to the same critical current, also suggests the possibility of current controlled logic rather than voltage controlled gating, which we are just now beginning to explore. The possibility of multiple contacts to the same coherent region of condensation, or coupling or gate-controlled coupling and decoupling of coherent regions of condensation even if at higher voltages, also suggests the possibility of quantum computing applications.

Theory continues to suggest that room temperature superfluidity is possible in dielectrically separated graphene layers. If not successful in graphene-based systems, analogous systems of 2D materials such as silicene, germanene, and/or transition metal dichalcogenides are better in some ways according to theory, even if still more technologically challenging. Theory can explain and predict the types of behavior that are to be expected. In part, the experimental efforts of [14] in III-V systems were informed by our colleagues' theory [16] (which in turn was informed by past work of the authors of [10]), where that theory effort was supported as part of the BiSFET effort. Qualitatively, almost all of the essential physics of interest has been seen experimentally in analogous physical systems including III-V double quantum wells and Josephson junctions, except room temperature superfluidity at the time of writing, of course.

However, we are well aware of the limitations of theory. The BiSFET is based on exotic physics in an exotic material system based on imperfect theory. Ultimately, the issue must be resolved experimentally. The challenges to observe the condensate are substantial. If we can observe the condensate in the laboratory, much more work will be required to extend those results to commercial production. Moreover, even if successful, initial efforts may not run at voltages comparable to of $k_B T/q$ or lower due to technological limitations. However, technology would be expected to improve over time, much as it did for CMOS.

The final scenario is precisely the above stated goal of this work, to restart the "roadmap," i.e., to turn making substantial long-term progress into a technological challenge rather than the physical impossibility that it is for CMOS. Moreover, room temperature superfluidity in nanoscale devices could offer new application avenues, not just ongoing progress along existing avenues.

Acknowledgments

We acknowledge and thank all of our collaborators and colleagues whose work we have reviewed above along with our own. These include Allan MacDonald, Emanuel Tutuc, Luigi Colombo, Gary Carpenter, and Arjang Hassibi, and numerous students and postdoctoral scholars, former and present. This work was supported by the Nanoelectronics Research Initiative (NRI) through the Southwest Academy of Nanoelectronics (SWAN). Supercomputing resources were provided by the Texas Advanced Computing Center (TACC).

References

[1] S. K. Banerjee, L. F. Register, E. Tutuc, D. Reddy, & A. H. MacDonald, "Bilayer pseudospin field-effect transistor (BiSFET): a proposed new logic device." *IEEE Electron Device Letters*, **30**(2), 158–160 (2009).

[2] D. Reddy, L. F. Register, E. Tutuc, & S. K. Banerjee, "Bilayer pseudospin field-effect transistor: applications to Boolean logic." *IEEE Transactions on Electron Devices*, **57**(4), 755–764 (2010).

[3] H. Min, R. Bistritzer, J.-J. Su, & A. H. MacDonald, "Room-temperature superfluidity in graphene bilayers." *Physics Reviews B*, **78**(12), 121401 (2008).

[4] C. H. Zhang & Y. N. Joglekar, "Excitonic condensation of massless fermions in graphene bilayers." *Physics Reviews B*, **77**(23), 233405 (2008).

[5] D. Basu, L. Register, A. MacDonald, & S. Banerjee, "Effect of interlayer bare tunneling on electron-hole coherence in graphene bilayers." *Physics Reviews B*, **84**(3) (2011).

[6] X. Mou, L. F. Register, & S. K. Banerjee, "Quantum transport simulation of bilayer pseudospin field-effect transistor (BiSFET) on tightbinding Hartree-Fock model." In *Simulation of Semiconductor Processes and Devices, 2013 International Conference on* (2013).

[7] X. Mou, L. F. Register, & S. K. Banerjee, "Quantum transport simulations on the feasibility of the bilayer pseudospin field effect transistor (BiSFET)." In *Electron Device Meeting (IEDM), 2013 International, Technical Digest*, pp. 4.7.1–4.7.4 (2013).

[8] International Technology Roadmap for Semiconductors. Available online at www.itrs.net.

[9] E. E. Mendez, L. Esaki, & L. L. Chang, "Quantum Hall effect in a two-dimensional electron-hole gas." *Physics Review Letters*, **55**(20), 2216 (1985).

[10] I. B. Spielman, J. P. Eisenstein, L. N. Pfeiffer, & K. W. West, "Resonantly enhanced tunneling in a double layer quantum Hall ferromagnet." *Physics Review Letters*, **84**(25), 5808 (2000).

[11] M. Pohlt, M. Lynass, J. G. S. Lok *et al.*, "Closely spaced and separately contacted two-dimensional electron and hole gases by in situ focused-ion implantation." *Applied Physics Letters*, **80**(12), 2105–2107 (2002).

[12] J. A. Seamons, D. R. Tibbetts, J. L. Reno, & M. P. Lilly, "Undoped electron-hole bilayers in a GaAs/AlGaAs double quantum well." *Applied Physics Letters*, **90**(5), 052103–3 (2007).

[13] L. Tiemann *et al.*, "Critical tunneling currents in the regime of bilayer excitons." *New Journal of Physics*, **10**(4), 045018 (2008).

[14] D. Nandi, A. D. K. Finck, J. P. Eisenstein, L. N. Pfeiffer, & K. W. West, "Exciton condensation and perfect Coulomb drag." *Nature*, **488**(7412), 481–484 (2012).

[15] I. Sodemann, D. Pesin, & A. MacDonald, "Interaction-enhanced coherence between two-dimensional Dirac layers." *Physics Reviews B*, **85**(19) (2012).

[16] J.-J. Su & A. H. MacDonald, "How to make a bilayer exciton condensate flow." *Nature Physics*, **4**(10), 799–802 (2008).

[17] K. K. Ng, *Complete Guide to Semiconductor Devices* (New York: Wiley, 2002), pp. 569–574.

[18] P. Mazumder, S. Kulkarni, M. Bhattacharya, S. Jian Ping, and G. I. Haddad, "Digital circuit applications of resonant tunneling devices." *Proceedings of the IEEE*, **86**(4), 664–686 (1998).

[19] K. Wu, A. Sachid, F.-L. Yang, & C. Hu, "Toward 44% switching energy reduction for FinFETs with vacuum gate spacer." In *Simulation of Semiconductor Processes and Devices, 2012 International Conference on*, pp. 253–256 (2012).

[20] R. T. Weitz, M. T. Allen, B. E. Feldman, J. Martin, & A. Yacoby, "Broken-symmetry states in doubly gated suspended bilayer graphene." *Science*, **330**(6005), 812–816 (2010).

[21] Y. E. Lozovik & V. I. Yudson, "Feasibility of superfluidity of paired spatially separated electrons and holes; a new superconductivity mechanism." *Soviet Journal of Experimental and Theoretical Physics Letters*, **22**, 274–275 (1975).

[22] D. Basu, L. Register, D. Reddy, A. MacDonald, & S. Banerjee, "Tight-binding study of electron-hole pair condensation in graphene bilayers: Gate control and system-parameter dependence." *Physics Reviews B*, **82**(7), (2010).

[23] P. Avouris, Z. Chen, & V. Perebeinos, "Carbon-based electronics." *Nature Nanotechnology*, **2**(10), 605–615 (2007).

[24] Y. E. Lozovik & A. A. Sokolik, "Electron-hole pair condensation in a graphene bilayer." *JETP Letters*, **87**(1), 55–59 (2011).

[25] P. Jadaun, H. C. P. Movva, L. F. Register, & S. K. Banerjee, "Theory and synthesis of bilayer graphene intercalated with ICl and IBr for low power device applications." *Journal of Applied Physics*, **114**(6), 063702 (2013).

[26] L. F. Register, X. Mau, D. Reddy *et al.*, "Bilayer pseudo-spin field effect transistor (BiSFET): concepts and critical issues for realization." Invited presentation at the 221st Electrochemical Society Meeting, Seattle, Washington, May 7, 2012.

Section III

Alternative field effect devices

9 Computation and learning with metal–insulator transitions and emergent phases in correlated oxides

You Zhou, Sieu D. Ha, and Shriram Ramanathan

9.1 Overview

Electron devices with components that undergo phase transitions can add new functionality to classical devices such as field effect transistors and p-n junctions. In this chapter we examine recent research on utilizing phase transition materials, such as but not limited to vanadium dioxide (VO_2), for electronics and provide a perspective on how phase transition electronics may complement and add function to complementary metal–oxide–semiconductor devices in emerging computing paradigms. In parallel, there is continuous need to innovate in high-frequency communications, reconfigurable devices and sensors. These fields sometimes may not directly overlap with research directions in computing, however, when new materials are being explored, a variety of interesting properties are uncovered and there is cross-pollination of ideas. In correlated oxides too, studies motivated by fast switching properties have created broad interest such as in the microwave device arena and are considered here for completeness. It is finally pointed out that ionic conduction in oxides or ion-mediated electronic phase transitions induced for example in electric double layer transistors or their solid-state counterparts could play a significant role in future research and development of such correlated electron material systems. Although operationally slower than solid-state devices, liquid gates offer new directions to explore paradigms in reconfigurable fluidic devices that have seen substantial growth in the soft matter fields.

9.1.1 Introduction

Scaling of the metal–oxide–semiconductor field effect transistor (MOSFET) has sustained the growth of the microelectronics industry for numerous decades. As the gate length of these transistors approaches the sub-10 nm regime, it is becoming increasingly difficult to enhance device performance metrics such as energy efficiency and switching speed accordingly. The fundamental limit of complementary metal–oxide–semiconductor (CMOS) scaling that originates from the basic operation principle of MOSFETs has motivated researchers to look for alternative computation component/architecture to complement current CMOS technology. Currently, this is perhaps one of the most important problems in the condensed matter community and has great significance to continued growth of the hard sciences in academia.

One of the efforts is to use a set of materials referred to as electron correlated insulators as the channel in place of conventional Si semiconductors [1–3]. The electron–electron interaction is significantly strong in these materials that they are insulators at low temperatures even if one would expect them to be metallic from a classical electronic band theory [4]. When electrons or holes are doped into these materials, the change in the free carrier concentration can induce phase transitions, such as insulator-to-metal transitions and magnetic transitions [5]. These phenomena have prompted the idea of using them as the channel layer in a MOSFET structure: if there is no gate voltage applied, the channel is insulating and the device is off. When a gate voltage is applied, carriers are doped into the channel electrostatically triggering an insulator-to-metal transition and the device is switched on. These devices have a fundamentally different operating principle than MOSFETs based on conventional semiconductors such as Si, Ge, and III-V materials. Therefore it may have different scaling potential, as will be discussed in this chapter.

Among many of these correlated materials, vanadium dioxide (VO_2) has piqued particular interest [6]. It shows a metal–insulator transition near room temperature ($T_C \sim 340$ K) with a drastic change in its conductivity (3–5 orders of magnitude). The transition temperature is sufficiently close to current CMOS temperatures, providing possibilities of realizing room temperature application and integrating VO_2-based devices onto CMOS architectures. In addition, it has been recently shown that the transition can occur at an ultrafast time scale that could be potentially interesting for memory and logic, as shown in Fig. 9.1 [7–14]. In this chapter, we review the progress made on both theoretical understanding and experimental demonstration of 3-terminal devices based on VO_2. We also introduce VO_2-based 2-terminal devices and discuss the possibilities of utilizing these devices in alternative computation architectures.

9.1.2   Outline of chapter

We first give a brief introduction to the metal–insulator transition phenomena in VO_2. The possible transition mechanisms proposed are reviewed with a picture of the band structure. Then we briefly compare the field effect devices using correlated oxides as the channel with conventional MOSFETs to reveal different operation mechanisms and the scientific underpinnings of why such devices could be of technological interest. The next section reviews recent progress on fabricating FET devices using VO_2 as a channel layer. Both solid-state devices and ionic liquid-gated devices are discussed. The challenges and outlook of fabricating such devices are also presented. We then continue to discuss how an electrically triggered metal–insulator transition can occur in 2-terminal devices and how this phenomenon can be used for resistive switching, analog and microwave applications. Finally, we look beyond the idea of "transistor" and current computational paradigms, i.e., von Neumann architectures. Novel computational paradigms such as neural computation will be introduced with a special focus on how phase transition elements/correlated materials in general could be implemented in these architectures.

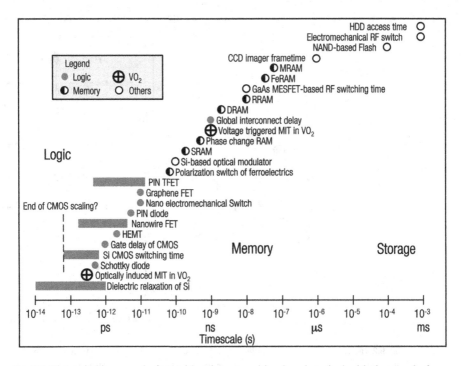

Fig. 9.1 The switching speed of metal-insulator transition benchmarked with the speed of some of the state of the art technologies. Data taken from [7–14] and references therein. A bar in the graph denotes a range of switching speed, where the right end represents the fastest demonstrated and the left end stands for the projected limit. A circle in the graph indicates the demonstrated switching speed for the technology. Logic, memory, and other technologies are represented by symbols with different fillings as indicated by the legend. Vanadium dioxide is impressively fast in the grand scheme of electronic switches.

9.2 Metal–insulator transition in vanadium dioxide

9.2.1 Electronic and structural transition

The metal–insulator transition in bulk vanadium dioxide was observed by Morin in the 1950s [15]. In single crystal form, VO_2 is semiconducting with an optical bandgap of about 0.6 eV below the transition temperature ($T_C \sim 340$ K) [16]. When heated above the transition temperature, the resistivity drops by about four orders of magnitude and subsequently the resistivity increases with temperature indicating metallic conductivity. The electronic phase transition is also accompanied by a structural transition, where the insulating state has a different crystal structure from the metallic phase [17].

In the metallic phase, vanadium dioxide crystallizes in a rutile structure (often referred to as R phase) as shown in Fig. 9.2(a). Its crystal system is body-centered tetragonal with lattice constant $a = b = 4.555$ Å, $c = 2.851$ Å, and c-axis perpendicular to a and b. The vanadium atoms occupy the vertices and the center of the tetragonal unit cell. Each of the vanadium atoms is surrounded by six oxygen atoms and sits in the

(a)

(b)

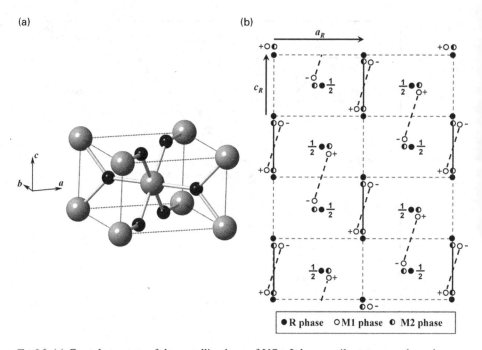

Fig. 9.2 (a) Crystal structure of the metallic phase of VO_2. It has a rutile structure where the black atoms are oxygen and gray atoms are vanadium. (b) The structural relation between the rutile R phase (black dot), the monoclinic M2 phase (black and white dot), and the monoclinic M1 phase (white circle), all viewed from the lattice b-direction. Oxygen atoms are omitted in the figure for clarity. From R phase to M1 phase, vanadium atoms become dimerized and tilted. In M2 phase, half of the vanadium atoms form dimers and the other half becomes tilted. (Part (b) reprinted with permission from M. Marezio, D. B. McWhan, J. P. Remeika, & P. D. Dernier, *Physical Review B*, **5**, 2541 (1972). Copyright 1972 by the American Physical Society.)

center of the octahedron made by the neighboring oxygen atoms. These octahedra share their edges with their neighboring octahedra along the c-axis and share vertices with the neighboring octahedra within the (001) plane. When going across the transition temperature into the insulating phase, the vanadium atoms dimerize along the original rutile c-axis and the dimers tilt with respect to the rutile c-axis. The oxygen atoms stay roughly at the same positions. Figure 9.2(b) shows the relation between the metallic and insulating phase structure viewed from the rutile b-axis (note that the oxygen atoms are omitted in the drawing for clarity) [17]. The small black dot shows the position of the vanadium atoms in the rutile phase and the rectangle with dashed line is the unit cell. The center of the unit cell is the vanadium atom at the body center with the c-coordinate equal to 1/2. The open circles in Fig. 9.2(b) indicate the vanadium positions in the insulating phase. Because of the dimerization of the vanadium atoms, the two originally equivalent vanadium atoms in the dimer become non-equivalent. Therefore, the new unit cell in the insulating phase must be twice as large as the original rutile primitive unit cell to include both vanadium atoms of the dimer. The tilt of the dimers makes the vanadium atoms displaced away from the rutile c-axis and also out of the rutile (010) plane as indicated in Fig. 9.2(b). The dimerization and tilting lowers the crystal symmetry and the crystal system transforms from tetragonal to monoclinic (often called M1 phase). The insulating phase unit cell is related to the metallic unit cell by the relations: $a_{M1} = 2c_R$, $b_{M1} = b_R$, $c_{M1} = a_R - c_R$. The choosing of the lattice vector is based on the convention for monoclinic crystals, but the essential point is the unit cell doubles along the rutile c-axis. As one may imagine, the change in crystal structure could possibly lead to change in the electronic band structure and act as a mechanism for the metal–insulator transition, which will be discussed in detail later. By applying stress or doping, it is possible to achieve another monoclinic insulating phase in VO_2 called M2 [17]. In the M2 phase, half of the vanadium atoms dimerize along the rutile c-axis without tilt while the other half tilts without dimerization as shown by the black and white dots in Fig. 9.2(b).

The above structural change leads single crystal bulk VO_2 to crack when phase transitions occur and it has limited the electronic application of the bulk materials [18]. Recent developments in thin-film technology have enabled the growth of high-quality VO_2 films on various substrates, such as, but not limited to, Si, Ge, sapphire, TiO_2, and GaN [19–22] by several different chemical and physical vapor deposition techniques. These thin films show stable thermally or electrically triggered metal–insulator transitions for thousands of cycles [13, 23]. The substrate clamping prevents the material from cracking and this renders the possibilities of reliably applying such materials in real electronic applications.

In general, the drastic change in conductivity across the metal-to-insulator transition could either be due to the vanishing free carrier density or the divergence of the effective mass of electrons, which corresponds to a change in the carrier density or carrier mobility, respectively [5]. In vanadium dioxide, it has been found that the conductivity change is primarily due to the change in the carrier density whereas the carrier mobility is almost invariant. Hall measurements on both single crystal and thin-film VO_2 have shown that the majority carriers are electrons in both phases and the carrier density increases from $\sim10^{18}\,cm^{-3}$ to $\sim10^{23}\,cm^{-3}$ from the insulating to the metallic phase as shown in Fig. 9.3 [18, 24]. On the contrary, the carrier mobility stays at a roughly constant value of $\sim0.1\,cm^2/Vs$ across the transition.

9.2.2   Proposed mechanisms of the phase transition

The success of Si CMOS technology owes much to the understanding of Si band structure and the ability to model its interface properties. Similarly, understanding the transition mechanism and thus the band structure in both phases is crucial to accurately model and predict any device functionality with such oxides. However, the metal–insulator transition mechanism in VO_2 has been debated over the last four decades and is still under active study. The central question is whether the transition is due to electron-lattice interaction (Peierls transition), electron–electron correlation (Mott transition), or a collaborated Peierls–Mott transition [5, 25–27]. A related question is whether one can trigger an electronic phase transition and reverse to insulating phase *before a structural change happens*. This is crucial for having fast on and off time

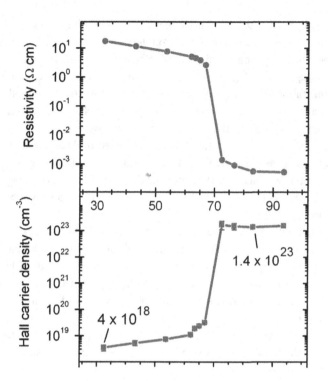

Fig. 9.3 The measured resistivity and carrier concentration as a function of temperature in VO_2 thin films on c-plane sapphire grown by magnetron sputtering. The increase of conductivity across the phase transition is primarily due to the change in carrier density but not in carrier mobility. (From D. Ruzmetov, D. Heiman, B. B. Claflin, V. Narayanamurti, & S. Ramanathan, *Physical Review B*, **79**, 153107 (2009). Copyright 2009 by the American Physical Society.)

constants for a switch. In other words, one may ask if it is possible to decouple the electronic phase transition from the structural transition. As a result, it may not be surprising that the band structure of VO_2 is also not well calibrated. Below we describe proposed transition mechanisms and corresponding electronic band structures.

In VO_2 compound, vanadium has a d^1 electron configuration and oxygen has fully filled $2p$ electron shells. According to band theory, the electronic band structure is derived from these molecular orbitals. This picture is especially useful in describing band structure of materials with narrow bandwidth such as transition metal oxides with d and f valence electrons. Therefore, the closed oxygen $2p$ shell does not contribute to electron conduction [28]. The d band derived from vanadium d orbitals, however, is partially filled with only one electron per molecule and becomes the conduction band. The above picture roughly explains why VO_2 is a metal but does not show why it can become an insulator. Based on this simple picture, we discuss below the metallic phase band structure and from there continue to describe the insulating phase.

In the rutile phase, the vanadium atoms are located at the centers of the oxygen octahedra. The d electron in the vanadium atom will thus experience Coulomb interaction from oxygen anions. In a perfectly spherical Coulomb potential, for example in

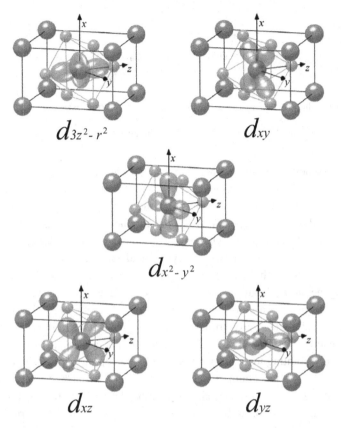

$d_{3z^2-r^2}$

d_{xy}

$d_{x^2-y^2}$

d_{xz}

d_{yz}

Fig. 9.4 The five originally degenerate d orbitals in the rutile structure. They exhibit different energy in the crystal structure due to the presence of a crystal field. The small spheres denote oxygen atoms while the large spheres denote vanadium atoms. (Adapted with permission from Eyert, *Annalen der Physik*, **11**, 650–704 (2002). © 2002 Wiley-VCH Verlag GmbH & Co. KGaA, Weinheim.)

an isolated vanadium atom, the d shell is five-fold degenerate without considering the spin degeneracy. When the atoms are positioned in the VO_2 crystal, the Coulomb interaction from oxygen anions will lift the five-fold degeneracy, which is often referred to as crystal field splitting. Figure 9.4 shows the crystal structure of VO_2 and how the five d orbitals of vanadium appear within the lattice. The x-coordinate is directed along the rutile c-axis as shown [29]. If we look at the electron distribution of the five different d orbitals with respect to the surrounding oxygen atoms, it is evident that the lobes of the d_{xy} and $d_{3z^2-r^2}$ orbitals are oriented to the oxygen atoms. On the other hand, the lobes of the other three d orbitals are not pointed to the neighboring oxygen atoms, but are in fact directed in a way that they are farthest from oxygen. Since each electron carries a negative charge, the d_{xy} and $d_{3z^2-r^2}$ orbitals will experience more Coulomb repulsion from the oxygen and thus have a higher energy than the other three d orbitals. (Note that the choice of the coordinates in Fig. 9.4 is different from what is usually used in crystal field theory for convenience.) As a result of this, the original d orbitals with

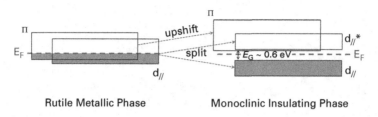

Rutile Metallic Phase   Monoclinic Insulating Phase

Fig. 9.5 The proposed band structure of VO$_2$ in metallic and insulating phases.

five-fold degeneracy split into higher two-fold degenerate e_g (from $d_{3z^2-r^2}$ and d_{xy} orbitals) and lower three-fold degenerate t_{2g} bands (from $d_{x^2-y^2}$, d_{xz}, and d_{yz} orbitals). Because there is only one d electron per molecule in VO$_2$, the higher-energy e_g band will be empty and the lower t_{2g} band will be the conducting band. Also note that the t_{2g} band will be six-fold degenerate when spin is taken into consideration. The above d band splitting is a classical example of crystal field theory, which happens in many transition metal oxides with oxygen octahedra in their crystals. In vanadium dioxide, the three-fold degenerate t_{2g} band is further lifted by the interaction between electrons in nearest neighbor vanadium atoms. As can be seen in Fig. 9.4, the lobe of the $d_{x^2-y^2}$ orbital is directed along the x-axis (rutile c-axis). Imagine that we draw the whole crystal lattice by repeating the unit cell along the x-direction; notice that the lobe of this $d_{x^2-y^2}$ orbital is actually pointing to a vanadium atom. On the contrary, the d_{xz} and d_{yz} orbitals are not pointing to any vanadium atoms as shown in Fig. 9.4. The consequence is that the $d_{x^2-y^2}$ orbital will have a lower energy than the d_{xz} and d_{yz} orbitals. Based on the above discussion, we can draw the formation of the energy bands for rutile VO$_2$ as shown in Fig. 9.5 [30]. Conventionally a different notation of the band for a crystal is used than for the molecular orbital. Because the $d_{x^2-y^2}$ orbital is directed along the rutile c-axis and almost one-dimensional, it is called $d_{//}$ band to reflect its specific geometry. On the other hand, the d_{xz} and d_{yz} orbitals are rather isotropic and referred to as π band. The conduction band of the rutile VO$_2$ is therefore the sum of the $d_{//}$ band with two-fold degeneracy and the π band with four-fold degeneracy as shown in Fig. 9.4. Since VO$_2$ only has one d electron, the conduction band is partially filled and rutile VO$_2$ is metallic.

The above band structure is generally accepted for the metallic phase, but exact band structures such as the band width, the Fermi level, and the offset between $d_{//}$ and π band are not well characterized. It has been estimated that the total bandwidth of $d_{//}$ and π bands is ~ 2 eV and the Fermi level is located ~ 0.6 eV above the conduction band minimum [25, 29, 31, 32]. It is also interesting to note that because of the quasi-one-dimensional character of the $d_{//}$ band, there is a conductivity anisotropy of VO$_2$. The conductivity parallel to the rutile c-axis is a few times larger than the conductivity perpendicular to the c-axis at room temperature both in single crystal [33] and epitaxial thin films [34]. The conductivity difference is not extremely large, however, probably due to contribution from the π band [33, 34].

Goodenough proposed a metal-to-insulator transition mechanism based on conventional band theory mainly considering structural effects [30]. In the insulating phase, the vanadium atoms become paired along the rutile c-axis. From the molecular orbital perspective, such pairing will lead to the splitting of $d_{//}$ into two bands: a bonding and an anti-bonding band, because $d_{//}$ is almost one-dimensional along the c-axis. In other words, (the Peierls transition mechanism) the doubling of the unit cell (due to pairing) will halve the unit cell in reciprocal space and the first Brillouin zone. This opens up a bandgap at the new Brillouin zone boundary, which leads to the splitting of the $d_{//}$ band into two bands as shown in Fig. 9.5. At the same time, the tilting of the vanadium pairs causes an upshift of the π band. The net result of the dimerization and tilting is that there is a bandgap opening as shown in Fig. 9.5. Note that the above discussions are solely based on how the structural change can modify the band structure and no electron–electron correlation is considered. Therefore the above transition mechanism is often considered as Peierls transition.

Mott *et al.*, on the contrary, explained the metal-to-insulator transition in terms of electron correlation [35]. It was argued that, in the insulating phase, the $d_{//}$ band splits into two Hubbard bands due to electron correlation. When transformed into the metallic phase, the π band is energetically lowered and becomes partially filled. Thus the electrons in the π band could screen the electron correlation in the $d_{//}$ band and merge the two Hubbard bands into a single band.

Although the discussion is still under active study, it has been proposed by many that the metal–insulator transition may not have a single origin, that is, it might be due to a collaborative Mott–Peierls transition where both mechanisms help to promote the transition [32].

9.3   Field effect devices using phase transitions

9.3.1   Mott FET: theory and challenges

The schematic of a Mott FET is drawn in Fig. 9.6(a). It is a 3-terminal device similar to a conventional FET. The channel is built from a correlated insulator with uniform dopant concentration. The source and drain are simply metal contacts providing good Ohmic contact to the channel. The gate stack is formed of a thin layer of gate dielectric and a metal contact. When $V_G = 0$ V, the channel material is in its insulating phase and the device is in its off-state. When a gate voltage is applied, the carriers induced can trigger an insulator-to-metal transition in the channel and turn the device on. Figure 9.6(b) shows how the band structure of the channel material changes across the threshold gate voltage.

There are several possible predicted advantages of these Mott FETs over conventional MOSFETs [3]. First, the insulator-to-metal transitions in Mott FETs could serve as a charge enhancement mechanism that may achieve sharper sub-threshold switching than conventional MOSFETs. In the insulating phase, electrons need to overcome the Coulomb repulsion to hop/transport through the lattice. For Mott insulators with d^n electron configuration, the conduction can be characterized as electrons hopping through the lattice:

$$d^n + d^n + U \rightarrow d^{n+1} + d^{n-1}. \tag{9.1}$$

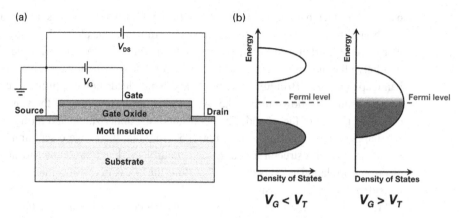

Fig. 9.6 (a) Schematic drawing of a Mott FET. It has a similar structure as conventional MOSFET using a Mott insulator as channel material. (b) When the gate voltage is below the threshold voltage, the channel is insulating due to electron correlations, i.e., electron–electron interactions. Above the threshold voltage, the electrostatically doped carriers can screen the interactions between electrons, which helps to induce a filling-controlled insulator-to-metal transition, resulting in the collapse of the bandgap.

In order for electron hopping to happen, one has to overcome the Coulomb repulsion energy given by

$$U = E(d^{n+1}) + E(d^{n-1}) - 2E(d^n), \tag{9.2}$$

where $E(d^n)$ is the total energy of a transition metal that has n d electrons and U is the Coulomb repulsion energy that effectively blocks electrons from hopping through the lattice. When electrons (or holes) are doped into the insulator by application of gate voltage, the electron configuration of some of the transition metal atoms becomes d^{n+1} (or d^{n-1}). As a result, electron hopping could happen without any barriers as indicated by the following mechanism:

$$d^{n+1} + d^n \rightarrow d^n + d^{n+1}. \tag{9.3}$$

Note that the initial state and final state have the same electron configurations and therefore the same energy. So there is no energy penalty for electron conduction to happen, driving an insulator-to-metal transition. Due to the phase transition, the free carrier density in the channel can be larger than the net carrier density induced electrostatically by the gate. This serves effectively as a charge-enhancement effect and helps to achieve a sharper transition (smaller sub-threshold swing) in a Mott FET [3]. It is also interesting to note that both electron and hole doping can in principle induce a phase transition, which means that either a positive or negative gate bias can turn on the Mott FET. The conductance modulation can be ambipolar from this simple picture. In reality, the phase diagram of a Mott insulator is usually asymmetric for electron and hole doping. This could lead to, for example, asymmetric conductance modulation for different gate voltage polarity.

Another important aspect of a switching device is the switching speed. For a conventional MOSFET, the intrinsic switching speed is the gate delay characterized by the time to charge the gate capacitance. For a certain charge accumulation Q, the switching speed could be enhanced by applying a larger drive voltage or using materials with higher mobility. In ideal Mott FETs, the switching speed would be limited by the charging of the gate dielectric as well as the material-limited carrier-induced phase transition speed. Because the free carrier density n is larger than the net charge accumulation density Q/e, the Mott FETs can have a larger driving current than conventional MOSFETs for the same amount of net charge Q, therefore having a faster switching speed [36]. On the other hand, in terms of material-limited phase transition speed, it has been found in many materials that the metal–insulator transition can be triggered at a very fast time scale by optical signals [12]. If these optically induced Mott transitions are due to the increase in carrier density by photo-doping as argued, it is reasonable to expect a similar switching time scale if these carriers are electrostatically doped.

In addition, because p-n junctions are not required in the channel Mott FETs, the depletion region along the channel could be eliminated [37]. Therefore, short channel effects caused by the slow scaling of the depletion layer in MOSFETs may not be significant in Mott FETs. The channel material is either highly doped or un-doped, which can help to reduce dopant fluctuation in the channel [37].

In spite of the aforementioned potential advantages, it is also important to point out that many of these correlated oxides have very low carrier mobility, which could be a significant limit for high-speed switches [3]. It is generally believed that the low carrier mobility is due to the hopping characteristics of the carriers. However, extrinsic effects such as scattering due to crystal defects are also not well studied. Whether one could break the bottleneck of the room temperature low mobility remains unclear.

The first few attempts to fabricate Mott FETs were mainly focused on using doped copper oxides [1, 37–39]. The cuprate composition was chosen so that it lies near the phase boundary between the metallic and insulating phase, so that a small number of carrier density change induced by the gate voltage can turn on/off the device. Other efforts involved fabricating such devices from other oxides and organic Mott insulators [40–43]. However, there are rather limited experimental reports in fabricating such Mott transistors that show both reproducible and comparable performance with conventional MOSFETs. Some of the major challenges impeding the demonstration of such FET devices include thin-film growth, source/drain contact resistance, and gate dielectric engineering. First, the short screening length in these correlated oxides requires the fabrication of a smooth interface between channel and gate dielectric. Because the carrier density in many of these correlated oxides is very large, especially at the phase boundary, the screening length estimated by the classical Debye length is on the order of a few nanometers. As a result, the correlated oxide thin films need to be only a few nanometers thick to maximize field penetration and achieve a large on/off ratio. For the same reason, it is also desirable for the channel/gate interface to be nearly atomically sharp. Another problem is that it is not easy to form Ohmic contacts with low resistance to some of the correlated materials [39]. The contact resistance could be similar to the channel resistance in the off-state, impeding any switching behavior. Finally, if the channel is not doped, the 2D carrier density required to trigger the metal-to-insulator transition is usually sufficiently large (10^{14}–$10^{15}\,cm^{-3}$) such that the corresponding electric field is approaching the dielectric breakdown field of gate oxides [2]. Thus the growth of high-quality gate oxide with high dielectric constants on these materials is required to maximize the breakdown voltage. Another approach may be using back gate FET design where the substrate or its native oxide serves as a gate dielectric. A common example is to use $SrTiO_3$, which serves as substrates for many perovskites, as back gate. The d.c. dielectric constant of $SrTiO_3$ is 10^2–10^4 near room temperature [44, 45]. For eventual application, however, the high-frequency dielectric constant becomes more important. There is usually a drop in the dielectric constant at such frequencies, because the polarization is not fast enough to respond to a.c. signals in this frequency range. New materials like $(La,Sr)_2NiO_4$ may be potentially interesting in this regard [46]; however, increased loss with increased dielectric constant is a problem that requires careful consideration. For instance, this can limit the maximum gate voltage that can be applied due to leakage currents. Nevertheless, to realize transistor devices that will be competitive with state-of-the-art silicon technologies, this will remain an area of interest.

9.3.2   Solid-state VO$_2$-based FETs

Ever since the proposal of Mott FETs, there have been intense efforts to build such devices with VO_2. One obvious advantage of VO_2 is that its transition temperature is slightly above room temperature. In addition, the metal-to-insulator transition is sharp enough to provide a reasonably large on/off ratio. And from the Mott criterion, it seems that the critical carrier density needed to trigger the transition is achievable by solid-state gate dielectrics, although the validity of the Mott criterion may not be accurate for this material system [47].

Chudnovskiy et al. proposed a 3-terminal VO_2 FET device: the channel material is VO_2 and other parts are exactly the same as in Fig. 9.6(a) [48]. The device proposed operates above the transition temperature and uses a gate voltage to turn it off. Due to the large carrier density in the metallic phase, this device has not been successfully demonstrated yet. Other works have mainly focused on how to induce the metallic phase from the insulating phase but within similar device geometry. Kim et al. have grown VO_2 on SiO_2/Si substrate and fabricated back gate 3-terminal VO_2 FET devices with SiO_2/Si as the gate [49]. There was no observation of gate modulation of the channel resistance. Instead, it was found that the gate voltage could tune the critical electric field (E_T) to trigger the electrically induced metal–insulator transitions (E-MIT). The phenomenon was explained in terms of carrier induced metal–insulator transitions caused by the change in carrier concentration. However, the critical electrical field, E_T, may not be a reliable parameter because there is usually rather a large variation in the E_T value when VO_2 is switched for multiple cycles by an external voltage. The data may not be easily reproducible, therefore, and remain inconclusive.

Ruzmetov et al. fabricated VO_2-based FETs with a similar device structure as shown in Fig. 9.7(a) [50]: they grew VO_2 on c-plane sapphire and deposited SiO_2 (100 nm thick) by magnetron sputtering as the top gate dielectric oxide. The sapphire substrate

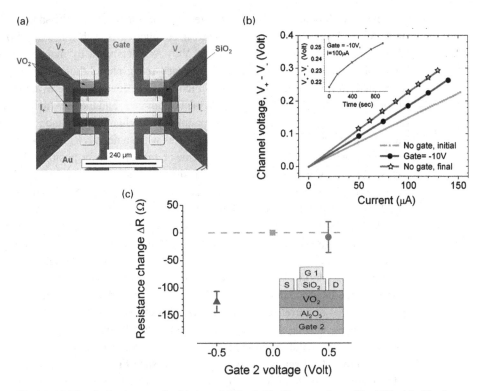

Fig. 9.7 (a) The device image of a fabricated VO_2 field effect transistor. The VO_2 thin film is patterned into a Hall bar geometry. Oxide dielectrics such as HfO_2 and SiO_2 have been used as the top gate. (b) Time-dependent conductance modulation in a top-gate device. This observation may be related to slow traps at the channel/dielectric interface. (c) The conductance modulation in a back-gated device. The inset shows the device schematics. (Reprinted with permission from D. Ruzmetov, G. Gopalakrishnan, C. Ko, V. Naryanamurti, & S. Ramanathan, *Journal of Applied Physics*, **107**, 114516 (2010). Copyright 2010, American Institute of Physics.)

enables the epitaxial growth of VO_2 and the VO_2 films as grown show a change in resistance of three to four orders of magnitude. But it was found that after the deposition of SiO_2 on top, the metal–insulator transition magnitude of VO_2 is degraded by two orders of magnitude. It is possible that there is diffusion of Si into VO_2 or the VO_2 surface is oxidized into V_2O_5 during the SiO_2 deposition. Although the exact reason is not clear, it can be seen that the proper choice of gate dielectric material and the engineering of the VO_2/gate dielectric interface is important to achieve a large on/off ratio. As shown in Fig. 9.7(b), the application of a negative gate bias (–10 V) causes an increase in the channel conductance. The response, however, shows a slow time-dependent phenomenon. The channel resistance even continued to decrease for ~10 min after the gate voltage was removed. In addition, a positive gate bias of up to + 10 V does not reverse the channel conductance back. The breakdown voltage (~15 V) is larger than the applied gate voltage, but the irreversible conductance modulation may be related to stoichiometric change of VO_2 at the channel/gate interface.

Devices with both top and bottom gates were also fabricated in the same study. First a thin Al_2O_3 layer (25 nm) was grown on conductive Si substrates by atomic layer deposition (ALD). Then a thin layer of VO_2 was deposited by magnetron sputtering onto Al_2O_3. The VO_2 thin film was patterned using photolithography and SiO_2 was deposited on top of the VO_2 channel. Finally metal contacts were evaporated. In this device, Al_2O_3 and Si substrate serve as the bottom gate, and the SiO_2 and metal serve as the top gate. Figure 9.7(c) shows the modulation of channel resistance as a function of the back gate voltage. The resistance decreases when a negative gate voltage of -0.5 V is applied, while there is no obvious resistance modulation for positive gate bias up to 0.5 V. At -0.5 V gate voltage, channel resistance changes by -0.26% ($\pm0.04\%$). The above effect was only observed in this device at 60 and 65°C, and there was no conductance modulation for either polarity of gate voltage at lower temperatures. In addition, unlike the top-gated devices, such devices have shown a reproducible electrical response to applied gate voltages with no time dependence upon the removal of the gate voltage. The elimination of the time dependence is thought to be related to the improved interface between the bottom ALD-grown gate insulator and VO_2 with respect to the interface in the top-gated devices. Such studies on time dependence are in general crucial in correlated oxides where traps at interfaces can lead to measurement artifacts. New techniques to determine the electrical quality of oxide interfaces are therefore of extraordinary importance for the field to move forward.

There have also been attempts to build solid-state FETs with VO_2 nanobeams [51]. VO_2 nanobeams with a width of 0.3–1 μm and thickness of 300–600 nm were gated by ALD-deposited top gate HfO_2 of 20 nm thick. The channel conductance increases on positive gate bias and decreases on negative bias, being different from previous studies. The change in channel conductance is also quite small (within a few percent). On the other hand, these devices also show a time-dependent conductance modulation, similar to the top-gated devices in early studies. When the gate voltage sweeps between -2.5 V and 2.5 V with a cycle time of 20 mins at 370 K, the conductance as a function of gate voltage shows a hysteretic loop with the largest change in resistance being -6%. And there is a phase lag between the conductance modulation and gate voltage. Such observations might be related to interfacial/bulk states, slow traps, or mechanical relaxation of VO_2.

9.3.3   Ionic liquid-gated VO_2 FETs

As can be seen from the above discussion, VO_2-based solid-state gate FETs show moderate channel conductance modulation primarily due to the large carrier density even in the insulating phase of VO_2. Recent developments of electrolyte gating using ionic liquids has enabled the induction of 2D carrier accumulation density that is more than one order of magnitude larger than what is achievable by conventional gate oxides such as SiO_2 and HfO_2 [52, 53]. Ionic liquids are molten salts at the relevant temperature range (room temperature for FET applications). They are electrolytes composed of solely cations (typically organic) and anions (typically inorganic) without any solvents. They are ideally ionic conductors but not electronic conductors. These materials are part

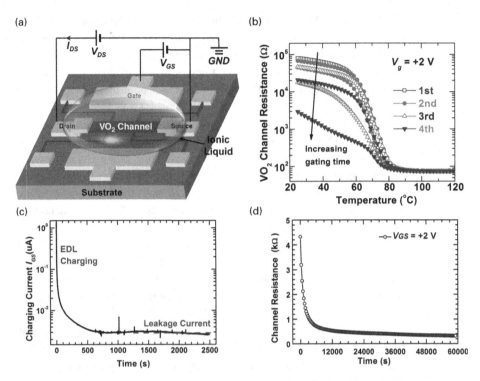

Fig. 9.8 (a) The schematic structure of an ionic liquid-gated VO$_2$ FET. (b) Resistance versus temperature curves measured consecutively under a fixed gate bias. (c) The charging current I_{GS} and (d) channel resistance as a function of time under a constant gate bias. Two different time constants indicate that the channel conductance modulation can be related to non-electrostatic effect. (Parts (a), (c), and (d) reprinted with permission from Y. Zhou & S. Ramanathan, *Journal of Applied Physics*, **111**, 084508 (2012). Copyright 2012, American Institute of Physics. Part (b) reprinted with permission from Z. Yang, Y. Zhou, & S. Ramanathan, *Journal of Applied Physics*, **111**, 014506 (2012). Copyright 2012, American Institute of Physics.)

of a well-plowed area of research in electrochemical energy conversion and storage in the condensed matter physics area, however, they are primarily considered as inert interfaces for building supercapacitors on exotic electronic materials. The schematic of an ionic liquid-gated transistor is shown in Fig. 9.8(a). It has a similar structure as conventional MOSFETs with the gate oxide replaced by an ionic liquid. When a gate voltage is applied across the ionic liquid and channel, an electric double-layer will form at the solid/liquid interface on the liquid side. To balance the charge on the liquid side, there would be some net charge accumulated in the solid channel. This electric double-layer therefore effectively acts as a gate capacitance to electrostatically control the carrier density within the channel. The effective capacitance per area is usually quite large, on the order of $10\,\mu\text{F/cm}^2$ due to the small charge separation distance in the electric double-layer. Typically a gate voltage of a few volts can be applied and this induces a 2D carrier density of $\sim 10^{14}\,\text{cm}^{-2}$ or even up to $\sim 10^{15}\,\text{cm}^{-2}$ [53–55]. At larger gate bias, electrochemical reactions can occur, and can modify the electronic properties of the channel. As a result, such a non-electrostatic effect can mimic a field effect and must be carefully examined/minimized. Transition metal oxide surfaces are notoriously sensitive to such interactions and hence need extra attention.

Figure 9.8(a) shows the device structure of a fabricated VO_2 electric double-layer transistor (EDLT) device. The VO_2 was grown on c-plane sapphire to achieve large metal–insulator transition ratio [56, 57]. As shown in Fig. 9.8(b), for the insulating phase, the channel resistance decreases with increasing positive gate bias (threshold voltage ~ 1.5 V), with conductance modulation of more than one order of magnitude. Negative gate bias does not change the insulating phase channel conductance. This agrees with Hall measurements that show that the majority carrier is n-type in insulating VO_2. For the metallic phase, the conductance stays the same under different gate bias, which is due to the large intrinsic carrier density. A similar phenomenon was also observed in a later study by Nakano *et al.* [58].

Interestingly, there are slow dynamics in the conductance modulation. It was found that the conductance modulation happens at a much slower speed than the charging of the gate capacitance as shown in Fig. 9.8(c) and (d) [57]. The slow dynamics raises questions on whether the channel conductance modulation is due to pure electrostatic effect or electrochemical reactions. Therefore to eliminate non-electrostatic effects, careful control of the chemical environment and selection of a proper gating time are required. Additionally, systematically studying how the dynamics scale with the device size in EDLTs would be an important step towards the understanding and improvement of such relaxations.

There have been several studies addressing the possibilities of electrochemical reactions in ionic liquid-gating experiments. For example, X-ray photoelectron spectroscopy reveals that there are strong electrochemical reactions and change of the vanadium valence state under –2 V gate bias [57]. It has also been shown there is essentially no field effect in electrolyte gated VO_2 nanobeams when using dehydrated ionic liquid DEME-TFSI [59]. On the other hand, if the ionic liquids were contaminated with water, there would be a rather large conductance modulation. This proposes another possible mechanism, hydrogen doping, which may induce large change in the resistance in these ionic liquid-gating experiments [59]. A schematic drawing of different mechanisms leading to channel conductance modulation in ionic liquid-gated transistors is shown in Fig. 9.9 for reference.

9.4   Two-terminal devices utilizing phase transitions

9.4.1   Threshold switches, resistive memory, and neural computing

A notable aspect of the MIT in VO_2 is that it can be triggered electrically (E-MIT) in 2-terminal devices [60]. The E-MIT is manifested in room temperature current–voltage measurements by an abrupt jump in current at some critical voltage, and is hysteretic, similar to the temperature-induced MIT. An example *I–V* curve showing the E-MIT switching behavior of VO_2 is shown in Fig. 9.10(a). Here, there are several small jumps

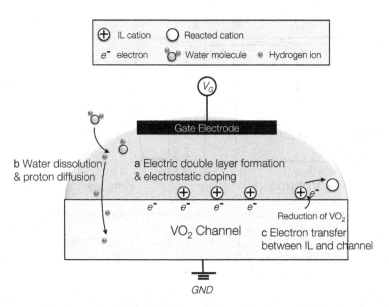

Fig. 9.9 A schematic drawing of the possibility of (a) electrostatic doping due to electric double layer, (b) hydrogen doping due to water dissolution, and (c) channel oxidizing/reduction due to electron transfer between ionic liquid (IL) and channel material. Careful control of the environment and selection of proper gating time are required to maximize the relative contribution of the electrostatic effect.

corresponding to partial switching of the film followed by a large jump at 3 V. The change in resistance from the low-voltage state to the high-voltage state is approximately three orders of magnitude, which is similar to the resistance change observed as a function of temperature in that device, indicating that the full VO_2 film volume can be switched using electrical excitation [61]. There is on-going discussion regarding the mechanism of the E-MIT, whether it is triggered by electric-field [62], Joule heating [63], carrier injection [64], or some other mechanism. Regardless, it has been shown that the current jump can occur as quickly as ~2 ns for a device with 400 nm spacing between electrodes, as shown in Fig. 9.10(b) [65]. The rise time being faster than estimated for the thermal time constant is an important result and points to evidence for Mott transition. The E-MIT can also be actuated using a conductive atomic force microscope tip [66], implying the possibility of nanoscale VO_2 switching devices. The abrupt current jump, ultrafast switching speeds, and nano-scalability have generated interest toward implementing 2-terminal VO_2 devices into novel electronic devices. In this section, we review applications of the VO_2 E-MIT that have been experimentally demonstrated.

The change in the resistance in VO_2 has been definitively shown to be caused by a metal–insulator transition rather than a defect-related conductive filament as in many other transition metal binary oxides. It was found that there is a clear correlation between the thermal transition magnitude and the E-MIT magnitude, i.e., stoichiometric VO_2 films with larger thermal MIT magnitude will show a larger on/off ratio in 2-terminal devices [65]. This is clearly distinct from defect-related resistive switching

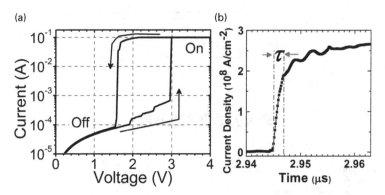

Fig. 9.10 (a) *I–V* curve of a 2-terminal VO$_2$ device showing electrically triggered metal–insulator transition (E-MIT). VO$_2$ thin films of 120 nm thick were grown on *c*-plane sapphire and fabricated into 50 μm (width) by 25 μm (length) 2-terminal devices by photolithography (b) Transient measurement of voltage driven MIT showing turn-on time of ~2 ns. (Part (a) reprinted with permission from S. D. Ha, Y. Zhou, C. J. Fisher, S. Ramanathan, & J. P. Treadway, *Journal of Applied Physics*, **113**, 184501 (2013). Copyright 2013, American Institute of Physics. Part (b), © 2013 IEEE. Reproduced, with permission, from Y. Zhou, X. Chen, Z. Yang, C. Mouli, & S. Ramanathan, *IEEE Electron Device Letters*, **34**, 220 (2013).)

phenomena and bi-polarity in many other transition metal oxides [67]. The RF properties of these 2-terminal devices also indicate that the switching mechanism is a metal–insulator transition. It was demonstrated that a thermally driven MIT can significantly switch the transmission of an RF signal through a 2-terminal VO$_2$ co-planar waveguide (CPW) [68]. The authors showed that the transmission parameter S_{21} (insertion loss) changes by an average of ~25 dB up to 35 GHz from the insulating phase at room temperature to the metallic phase at 127°C, with less than 3 dB insertion loss in the metallic phase for a series two-port configuration. It was later shown that similar RF switching can be achieved with a coupled DC bias to trigger the E-MIT as opposed to the thermally driven MIT [61, 69]. Insertion loss and isolation (S_{11}) for a VO$_2$ CPW in the on- and off-states of the E-MIT is shown in Fig. 9.11(a) and (b). Insertion loss ($|S_{21}|$) can be tuned by varying the maximum compliance current during *I–V* measurements, as shown in Fig. 9.11(c). The flat, broadband response of the VO$_2$ CPW can be attributed to the highly conductive nature of the metallic phase. In the insulating phase, transmission is low, but it monotonically increases as frequency increases because of the small capacitive coupling between electrodes, which behaves as a short at high frequency. Lumped circuit element modeling of the frequency response shows that the VO$_2$ CPW device can be considered as a parallel combination of a capacitor and variable resistor, with its resistance state determined by the DC bias level [61]. It was additionally shown that at low DC bias, ramping up the input power of the RF signal can have a similar effect on S_{21} as observed for high DC bias [61]. This suggests the possibility of triggering the VO$_2$ MIT solely with RF power. It is also inferred that the E-MIT is actually due to a bulk phase transition and not conductive filament formation as in many other oxides.

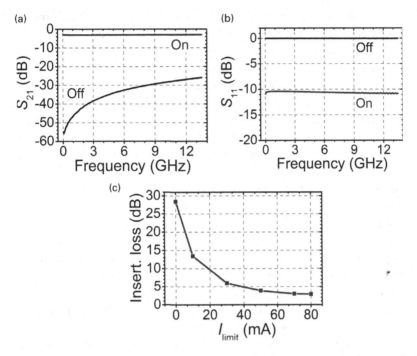

Fig. 9.11 (a) Transmission parameter S_{21} and (b) reflection parameter S_{11} in on- and off-states for a 2-terminal VO$_2$ co-planar waveguide. S-parameters measured with 70 mA compliance current. (c) Insertion loss ($|S_{21}|$) at 10 GHz as a function of maximum compliance current in on-state. (Reprinted with permission from S. D. Ha, Y. Zhou, C. J. Fisher, S. Ramanathan, & J. P. Treadway, *Journal of Applied Physics*, **113**, 184501 (2013). Copyright 2013, American Institute of Physics.)

In current-controlled E-MIT measurements, VO$_2$ exhibits negative differential resistance, which has been used to fabricate oscillator circuits [70–72]. For circuits composed of a 2-terminal VO$_2$ device and a series resistor, an applied voltage pulse near V_{th} across the circuit generates voltage oscillations across the VO$_2$ film [70]. At voltages significantly below or above V_{th}, the VO$_2$ output is flat and oscillations are not produced. The oscillation frequency is a function of the applied voltage pulse magnitude and the external resistor, and it is generally in the range 0.3–1 MHz [71]. The oscillations are due to discharging/charging of the internal VO$_2$ capacitance in accordance with rapid transitions from insulating to metallic phases and vice versa. Simulations suggest that electric field triggering is predominantly responsible for the oscillations, and that temperature transient variation controls the envelope function of the oscillatory output [72].

VO$_2$ is being investigated as a selector element in resistive switching crossbar arrays. There is intensive research focus on resistive switching for high-density, non-volatile memory technology due to advantages such as $4F^2$ scalability (F = minimum chip feature size), <10 ns write times, >10^{10} cycling endurance, and >1000 hour retention [73]. In resistive switching memory, the resistance state of the device (low or high) is used as the bit storage medium. The crossbar array, in which memory cells are sandwiched between perpendicularly crossing word and bit electrode lines, is a primary architecture for high-density memory. A significant issue that must be overcome for commercialization of resistive switching crossbar arrays is the current sneak-path problem. If a high-resistance state (HRS) cell in a crossbar is adjacent to several low-resistance state (LRS) cells, then when applying a voltage to read the resistance state of the HRS cell, some current may flow through the nearby LRS cells instead, resulting in an erroneous memory read event. Two-terminal VO_2 devices have been implemented in series with resistive switches to mitigate the sneak-path problem [74–77]. Using the volatile threshold switching of the VO_2 E-MIT, these selector devices serve to limit current for cells with applied voltage below the E-MIT threshold voltage (V_{th}). Thus, when attempting to read a HRS cell, less current will flow through adjacent LRS cells, and a read error will not occur. VO_2 was integrated in memory devices with Pt/NiO/Pt/VO_2/Pt heterostructures, where NiO was the resistive switching material [74]. Suppression of sneak-path current was demonstrated in VO_2-TiO_2 heterostructure crossbars and VO_2-ZrO_x/HfO_x series devices [75, 76]. The requirements for proper operation of a VO_2 selector element are (1) $V_{th} < |V_{reset}|$ and $V_{th} < |V_{set}|$, the voltages to switch the memory cell to/from LRS/HRS; (2) $R(VO_2, \text{off}) > R(\text{LRS})$; and (3) $R(\text{HRS}) + R(VO_2, \text{on}) < 2(R(\text{LRS}) + R(VO_2, \text{off}))$ [75].

The abrupt current jump in the VO_2 E-MIT has been implemented as a varistor for electrostatic discharge (ESD) surge protection [78]. The VO_2 varistor can be placed in parallel to circuitry that requires protection. If the voltage across the varistor reaches some critical value from a surge in circuit voltage, the VO_2 film will transition to the metallic phase due to the E-MIT, shunting current away from the circuit. The authors have tested the varistor properties of VO_2 films and have found comparable transient response to commercially available ZnO varistors with respect to peak varistor voltage drop (~230 V) and response time (~13 ns). The VO_2 varistors were determined to be robust for ESD voltages up to 3.5 kV.

In addition to VO_2, NbO_2 also exhibits a thermally driven MIT ($T_{MIT} \sim 800°C$) and an abrupt E-MIT in I–V measurements [79, 80]. The E-MIT of NbO_2 has been implemented in neuristor devices, which mimic action potential generation in biological neurons according to the Hodgkin–Huxley model [81]. Above some critical input voltage, a neuristor delivers a spike output ("all-or-nothing spiking") similar to that observed in neuronal ion channels. Two 2-terminal NbO_2 devices (labeled as devices $M_{1,2}$) connected as shown in Fig. 9.12(a) form a neuristor device. An input current signal charges the capacitors $C_{1,2}$ until the voltages across $M_{1,2}$ are above the threshold voltages for the E-MIT, after which discharge of the capacitors through the metallic-phase NbO_2 films causes the output signal to spike. An offset in charge times between channels and opposite polarity DC biasing of the channels results in an action potential spike. The spiking pattern of the neuristor can be controlled by modifying the capacitors $C_{1,2}$, as shown in Fig. 9.12(b). Most neuronal simulation is implemented in software or with complex integrated circuits. The ability to use phase transition materials to mimic neuron behavior may lead to novel electronics platforms capable of adaptive, non-Boolean computing.

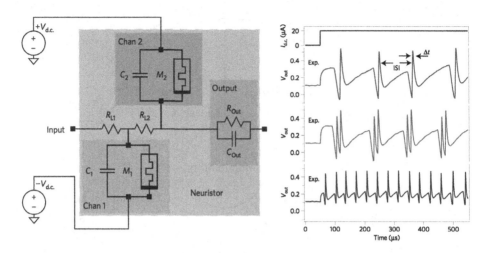

Fig. 9.12 (a) Circuit model of neuristor fabricated using two E-MIT devices ($M_{1,2}$). (b) Different spiking behaviors of the neuristor circuit that can be achieved by modifying capacitances $C_{1,2}$. (Reprinted by permission from Macmillan Publishers Ltd: *Natural Materials*, **12**, 114 (2013), copyright 2013.)

9.5   Neural circuits

While there are applications of metal–insulator phase transitions towards well-established device concepts in electronics, as discussed above, VO$_2$ and other MIT materials may have application in advanced computational electronics such as hardware artificial neural networks. Artificial neural networks perform computation in an interconnected, parallel manner and can be contrasted with von Neumann architectures used in modern computers and other computational models (see Fig. 9.13) [82]. Neural networks are intended to emulate brain function, which has significant advantages over sequential processing in computers, such as pattern recognition, adaptivity, and contextual processing [83]. Such networks are typically implemented in software, but hardware implementations have significant advantages with respect to speed, power consumption, and volume [84]. In the most basic model of brain function, biological neurons are composed of dendrites and an axon, which receive and propagate signals, a cell body, which generates signals, and synapses, which store information and modulate the strength of signals that are received. There has been substantial effort recently in utilizing the non-volatile memory behavior of transition metal oxide resistive switches to mimic synapse function [11]. The demonstration of neuronal signal spiking behavior using the E-MIT of correlated oxides may be another component for next generation hardware neural networks. It may be possible to cooperatively integrate an E-MIT device as the cell body and resistive switching devices as the synapse into nanoscale electronics that inherently mimic neuron behavior. Such devices offer natural solid-state analogs to biological systems and are of growing interest for low-power computing of relevance to mobile devices.

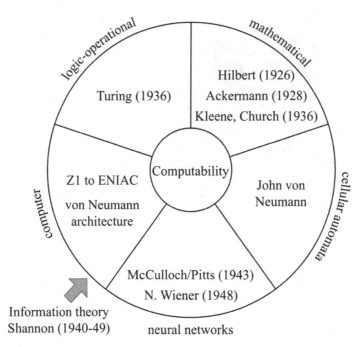

Fig. 9.13 Five models of computation, including the computer model (von Neumann architecture), Turing machines, and neural networks. (Reprinted from R. Rojas, *Neural Networks: A Systematic Introduction* (New York: Springer-Verlag, 1996), with kind permission from Springer Science+Business Media.)

Electronic phase transitions may have additional applications in hardware artificial neural networks besides action potential generation. It has been simulated that the interaction between coupled oscillators can emulate cellular neural networks [85, 86], which are artificial neural networks in which there is only communication between nearest neighbor neuronal units [87]. These simulations are based on experimental demonstration of phase-locking between two spin torque oscillators in close proximity [88]. When the oscillators are independently biased such that the output frequencies begin to approach each other, AC magnetic dipole or spinwave excitations between the oscillators interact to cause phase-locking between the two signals. In addition to artificial neural networks, simulations using coupled spin torque oscillators have been used to demonstrate associative processing and polychronous wavefront computation [89, 90]. Polychronous wavefront computation is motivated by brain function but is a non-neural network model of computation in which transponder nodes transmit and receive radial pulses [91]. Computation is encoded in the time delays between received pulses, and logic and memory applications have been simulated. The E-MIT of VO_2 may have direct utility in hardware implementations in the above computational models that use oscillators or spike generation. An important step would be to show that nearby E-MIT oscillators can phase-lock, which is common for interacting non-linear oscillators [88]. Such demonstrations may potentially lead to new paradigms of beyond-CMOS computational electronics and this is an area of anticipated growth.

9.6   Conclusions

Correlated oxides such as but not limited to vanadium dioxide show promise as an active element in computing devices. Mott FETs may show fast and energy-efficient switching utilizing electronic phase transitions. Although the phase transition has been known in some of these materials for several decades, studies on thin-film systems and gated devices are still in nascent stages. It is only recently conclusively demonstrated that vanadium dioxide can be switched over tens of millions of cycles reproducibly in thin-film devices, for example. Such results spur tremendous interest for practical devices and one can anticipate growth in the field. Many of the fundamental material properties, however, are still largely unknown for correlated materials. Demonstrating Mott FETs with performance comparable to current CMOS devices requires significant engineering efforts on material growth, gate oxide fabrication, and interface optimization, as well as theoretical developments to quantify the band structure, especially at interfaces. Further developments may lead to applications of correlated oxides in computation paradigms alternative to von Neumann architectures. The ability to switch states in 2-terminal Mott devices is a unique aspect of phase transition materials and further it serves to probe fundamental properties that we need to understand to gain a deeper understanding of the underlying physics. The tuning of electronic properties of correlated oxides reversibly using ionic interactions from solid or liquid electrolytes is yet another direction that could lead to interesting device/circuit concepts. A substantial amount of knowledge on ion transfer at interfaces exists in the electrochemistry community and can well be adapted to the study of oxides in the coming years.

The authors acknowledge NSF DMR-0952794 and ARO MURI W911-NF-09-1-0398 for financial support.

References

[1] D. M. Newns *et al.*, "The Mott transition field effect transistor: a nanodevice?" *Journal of Electroceramics*, **4**, 339–344 (2000).

[2] C. H. Ahn *et al.*, "Electrostatic modification of novel materials." *Reviews of Modern Physics*, **78**, 1185–1212 (2006).

[3] Y. Zhou & S. Ramanathan, "Correlated electron materials and field effect transistors for logic: a review." *Critical Reviews in Solid State and Materials Sciences*, **38**(4), 286–317 (2013).

[4] P. P. Edwards *et al.*, *Metal–Insulator Transitions Revisited* (London: Taylor & Francis, 1995).

[5] M. Imada *et al.*, "Metal-insulator transitions." *Reviews of Modern Physics*, **70**, 1039–1263 (1998).

[6] Z. Yang *et al.*, "Oxide electronics utilizing ultrafast metal-insulator transitions." *Annual Review of Materials Research*, **41**, 337 (2011).

[7] International Technology Roadmap for Semiconductors. Available at: www.itrs.net.

[8] W. R. Deal *et al.*, "Demonstration of a 0.48 THz amplifier module using InP HEMT transistors." *IEEE Microwave and Wireless Components Letters*, **20**, 289–291 (2010).

[9] K. Hei *et al.*, "A new nano-electro-mechanical field effect transistor (NEMFET) design for low-power electronics." In Electron Devices Meeting, 2005. IEDM Technical Digest. IEEE International, pp. 463–466 (2005).

[10] D. Loke *et al.*, "Breaking the speed limits of phase-change memory." *Science*, **336**, 1566–1569 (2012).

[11] S. D. Ha and S. Ramanathan, "Adaptive oxide electronics: a review." *Journal of Applied Physics*, **110**, 071101 (2011).

[12] A. Cavalleri *et al.*, "Picosecond soft x-ray absorption measurement of the photoinduced insulator-to-metal transition in VO_2." *Physical Review B*, **69**, 153106 (2004).

[13] Y. Zhou *et al.*, "Voltage-triggered ultrafast phase transition in vanadium dioxide switches." *IEEE Electron Device Letters*, **34**, 220–222 (2013).

[14] S. Hormoz & S. Ramanathan, "Limits on vanadium oxide Mott metal-insulator transition field-effect transistors." *Solid-State Electronics*, **54**, 654–659 (2010).

[15] F. J. Morin, "Oxides which show a metal-to-insulator transition at the Neel temperature." *Physical Review Letters*, **3**, 34–36 (1959).

[16] H. W. Verleur *et al.*, "Optical properties of VO_2 between 0.25 and 5 eV." *Physical Review*, **172**, 788–798 (1968).

[17] M. Marezio *et al.*, "Structural aspects of the metal-insulator transitions in Cr-doped VO_2." *Physical Review B*, **5**, 2541–2551 (1972).

[18] W. H. Rosevear & W. Paul, "Hall effect in VO_2 near the semiconductor-to-metal transition." *Physical Review B*, **7**, 2109–2111 (1973).

[19] Z. Yang *et al.*, "Metal-insulator transition characteristics of VO_2 thin films grown on Ge (100) single crystals." *Journal of Applied Physics*, **108**, 073708 (2010).

[20] T.-H. Yang *et al.*, "Semiconductor-metal transition characteristics of VO_2 thin films grown on c- and r-sapphire substrates." *Journal of Applied Physics*, **107**, 053514 (2010).

[21] Y. Muraoka & Z. Hiroi, "Metal–insulator transition of VO_2 thin films grown on TiO_2 (001) and (110) substrates." *Applied Physics Letters*, **80**, 583–585 (2002).

[22] Y. Zhou & S. Ramanathan, "Heteroepitaxial VO_2 thin films on GaN: Structure and metal-insulator transition characteristics." *Journal of Applied Physics*, **112**, 074114 (2012).

[23] C. Ko & S. Ramanathan, "Stability of electrical switching properties in vanadium dioxide thin films under multiple thermal cycles across the phase transition boundary." *Journal of Applied Physics*, **104**, 086105 (2008).

[24] D. Ruzmetov *et al.*, "Hall carrier density and magnetoresistance measurements in thin-film vanadium dioxide across the metal-insulator transition." *Physical Review B*, **79**, 153107 (2009).

[25] R. M. Wentzcovitch *et al.*, "VO_2: Peierls or Mott–Hubbard? A view from band theory." *Physical Review Letters*, **72**, 3389–3392 (1994).

[26] T. M. Rice *et al.*, "Comment on 'VO_2: Peierls or Mott–Hubbard? A view from band theory'." *Physical Review Letters*, **73**, 3042–3042 (1994).

[27] M. M. Qazilbash *et al.*, "Mott transition in VO_2 revealed by infrared spectroscopy and nano-imaging." *Science*, **318**, 1750–1753 (2007).

[28] R. J. Powell *et al.*, "Photoemission from VO_2." *Physical Review*, **178**, 1410–1415 (1969).

[29] V. Eyert, "The metal-insulator transitions of VO_2: a band theoretical approach." *Annalen der Physik*, **11**, 650–704 (2002).

[30] J. B. Goodenough, "The two components of the crystallographic transition in VO_2." *Journal of Solid State Chemistry*, **3**, 490–500 (1971).

[31] M. Abbate *et al.*, "Soft-x-ray-absorption studies of the electronic-structure changes through the VO_2 phase transition." *Physical Review B*, **43**, 7263–7266 (1991).

[32] S. Biermann et al., "Dynamical singlets and correlation-assisted Peierls transition in VO_2." Physical Review Letters, **94**, 026404 (2005).

[33] P. F. Bongers, "Anisotropy of the electrical conductivity of VO_2 single crystals." Solid State Communications, **3**, 275–277 (1965).

[34] J. Lu et al., "Very large anisotropy in the dc conductivity of epitaxial VO_2 thin films grown on (011) rutile TiO_2 substrates." Applied Physics Letters, **93**, 262107–3 (2008).

[35] A. Zylbersztejn and N. F. Mott, "Metal-insulator transition in vanadium dioxide." Physical Review B, **11**, 4383–4395 (1975).

[36] J. Son et al., "A heterojunction modulation-doped Mott transistor." Journal of Applied Physics, **110**, 084503–4 (2011).

[37] D. M. Newns et al., "Mott transition field effect transistor." Applied Physics Letters, **73**, 780–782 (1998).

[38] C. Zhou et al., "A field effect transistor based on the Mott transition in a molecular layer." Applied Physics Letters, **70**, 598–600 (1997).

[39] A. G. Schrott et al., "Mott transition field effect transistor: experimental results." In MRS Proceedings, p. 243 (1999).

[40] J. A. Misewich & A. G. Schrott, "Room-temperature oxide field-effect transistor with buried channel." Applied Physics Letters, **76**, 3632–3634 (2000).

[41] M. Sakai et al., "Ambipolar field-effect transistor characteristics of (BEDT-TTF)(TCNQ) crystals and metal-like conduction induced by a gate electric field." Physical Review B, **76**, 045111 (2007).

[42] K. Shibuya et al., "Metal-insulator transition in $SrTiO_3$ induced by field effect." Journal of Applied Physics, **102**, 083713 (2007).

[43] A. Yoshikawa et al., "Electric-field modulation of thermopower for the $KTaO_3$ field-effect transistors." Applied Physics Express, **2**, 121103 (2009).

[44] T. Sakudo & H. Unoki, "Dielectric properties of $SrTiO_3$ at low temperatures." Physical Review Letters, **26**, 851–853 (1971).

[45] H. M. Christen et al., "Dielectric properties of sputtered $SrTiO_3$ films." Physical Review B, **49**, 12095–12104 (1994).

[46] A. Podpirka et al., "Synthesis and frequency-dependent dielectric properties of epitaxial $La_{1.875}Sr_{0.125}NiO_4$ thin films." Journal of Physics D: Applied Physics, **45**, 305302 (2012).

[47] H. T. Kim et al., "Mechanism and observation of Mott transition in VO_2-based two- and three-terminal devices." New Journal of Physics, **6**, 52 (2004).

[48] F. Chudnovskiy et al., "Switching device based on first-order metal-insulator transition induced by external electric field" In Future Trends in Microelectronics: The Nano Millennium, pp. 148–155 (2002).

[49] H.-T. Kim et al., "Mechanism and observation of Mott transition in VO_2-based two- and three-terminal devices." New Journal of Physics, **6**, 52 (2004).

[50] D. Ruzmetov et al., "Three-terminal field effect devices utilizing thin film vanadium oxide as the channel layer." Journal of Applied Physics, **107**, 114516–8 (2010).

[51] S. Sengupta et al., "Field-effect modulation of conductance in VO_2 nanobeam transistors with HfO_2 as the gate dielectric." Applied Physics Letters, **99**, 062114 (2011).

[52] R. Misra et al., "Electric field gating with ionic liquids." Applied Physics Letters, **90**, 052905–3 (2007).

[53] K. Ueno et al., "Electric-field-induced superconductivity in an insulator." Nature Materials, **7**, 855–858 (2008).

[54] H. Y. Hwang *et al.*, "Emergent phenomena at oxide interfaces." *Natural Materials*, **11**, 103–113 (2012).

[55] K. Ueno *et al.*, "Discovery of superconductivity in $KTaO_3$ by electrostatic carrier doping." *Nature Nanotechnology*, **6**, 408–412 (2011).

[56] Z. Yang *et al.*, "Studies on room-temperature electric-field effect in ionic-liquid gated VO_2 three-terminal devices." *Journal of Applied Physics*, **111**, 014506–5 (2012).

[57] Y. Zhou and S. Ramanathan, "Relaxation dynamics of ionic liquid–VO_2 interfaces and influence in electric double-layer transistors." *Journal of Applied Physics*, **111**, 084508–7 (2012).

[58] M. Nakano *et al.*, "Collective bulk carrier delocalization driven by electrostatic surface charge accumulation." *Nature*, **487**, 459–462 (2012).

[59] H. Ji *et al.*, "Modulation of the electrical properties of VO_2 nanobeams using an ionic liquid as a gating medium." *Nano Letters*, **12**, 2988–2992 (2012).

[60] G. Stefanovich *et al.*, "Electrical switching and Mott transition in VO_2." *Journal of Physics: Condensed Matter*, **12**, 8837 (2000).

[61] S. D. Ha *et al.*, "Electrical switching dynamics and broadband microwave characteristics of VO_2 RF devices." *Journal of Applied Physics*, **113**, 184501–7 (2013).

[62] B. Wu *et al.*, "Electric-field-driven phase transition in vanadium dioxide." *Physical Review B*, **84**, 241410 (2011).

[63] A. Zimmers *et al.*, "Role of thermal heating on the voltage induced insulator-metal transition in VO_2." *Physical Review Letters*, **110**, 056601 (2013).

[64] X. Zhong *et al.*, "Avalanche breakdown in microscale VO_2 structures." *Journal of Applied Physics*, **110**, 084516–5 (2011).

[65] Y. Zhou *et al.*, "Voltage-triggered ultrafast phase transition in vanadium dioxide switches." *IEEE Electron Device Letters*, **34**, 220–222 (2013).

[66] J. Kim *et al.*, "Nanoscale imaging and control of resistance switching in VO_2 at room temperature." *Applied Physics Letters*, **96**, 213106–3 (2010).

[67] R. Waser and M. Aono, "Nanoionics-based resistive switching memories." *Nature Materials*, **6**, 833–840 (2007).

[68] F. Dumas-Bouchiat *et al.*, "RF-microwave switches based on reversible semiconductor-metal transition of VO_2 thin films synthesized by pulsed-laser deposition." *Applied Physics Letters*, **91**, 223505–3 (2007).

[69] A. Crunteanu *et al.*, "Voltage- and current-activated metal–insulator transition in VO_2-based electrical switches: a lifetime operation analysis." *Science and Technology of Advanced Materials*, **11**, 065002 (2010).

[70] Y. W. Lee *et al.*, "Metal-insulator transition-induced electrical oscillation in vanadium dioxide thin film." *Applied Physics Letters*, **92**, 162903–3 (2008).

[71] H.-T. Kim *et al.*, "Electrical oscillations induced by the metal-insulator transition in VO_2." *Journal of Applied Physics*, **107**, 023702–10 (2010).

[72] T. Driscoll *et al.*, "Current oscillations in vanadium dioxide: evidence for electrically triggered percolation avalanches." *Physical Review B*, **86**, 094203 (2012).

[73] H. S. P. Wong *et al.*, "Metal-oxide RRAM." *Proceedings of the IEEE*, **100**, 1951–1970 (2012).

[74] M. J. Lee *et al.*, "Two series oxide resistors applicable to high speed and high density nonvolatile memory." *Advanced Materials*, **19**, 3919–3923 (2007).

[75] S. H. Chang *et al.*, "Oxide double-layer nanocrossbar for ultrahigh-density bipolar resistive memory." *Advanced Materials*, **23**, 4063–4067 (2011).

[76] S. Myungwoo *et al.*, "Excellent selector characteristics of nanoscale VO_2 for high-density bipolar ReRAM applications." *IEEE Electron Device Letters*, **32**, 1579–1581 (2011).

[77] S. Myungwoo *et al.*, "Self-selective characteristics of nanoscale VO_x devices for high-density ReRAM applications." *IEEE Electron Device Letters*, **33**, 718–720 (2012).

[78] B.-J. Kim *et al.*, "VO_2 thin-film varistor based on metal-insulator transition." *IEEE Electron Device Letters*, **31**, 14–16 (2010).

[79] R. F. Janninck and D. H. Whitmore, "Electrical conductivity and thermoelectric power of niobium dioxide." *Journal of Physics and Chemistry of Solids*, **27**, 1183–1187 (1966).

[80] F. A. Chudnovskii *et al.*, "Electroforming and switching in oxides of transition metals: the role of metal–insulator transition in the switching mechanism." *Journal of Solid State Chemistry*, **122**, 95–99 (1996).

[81] M. D. Pickett *et al.*, "A scalable neuristor built with Mott memristors." *Nature Materials*, **12**, 114–117 (2013).

[82] R. Rojas, *Neural Networks: A Systematic Introduction* (New York: Springer-Verlag, 1996).

[83] A. K. Jain *et al.*, "Artificial neural networks: a tutorial." *Computer*, **29**, 31–44 (1996).

[84] J. Misra & I. Saha, "Artificial neural networks in hardware: A survey of two decades of progress." *Neurocomputing*, **74**, 239–255 (2010).

[85] T. Roska *et al.*, "An associative memory with oscillatory CNN arrays using spin torque oscillator cells and spin-wave interactions architecture and end-to-end simulator." In *Cellular Nanoscale Networks and Their Applications (CNNA), 2012 13th International Workshop on*, pp. 1–3 (2012).

[86] G. Csaba *et al.*, "Spin torque oscillator models for applications in associative memories." In *Cellular Nanoscale Networks and Their Applications (CNNA), 2012 13th International Workshop on*, pp. 1–2 (2012).

[87] L. O. Chua & L. Yang, "Cellular neural networks: applications." *IEEE Transactions on Circuits and Systems*, **35**, 1273–1290 (1988).

[88] S. Kaka *et al.*, "Mutual phase-locking of microwave spin torque nano-oscillators." *Nature*, **437**, 389–392 (2005).

[89] S. P. Levitan *et al.*, "Non-Boolean associative architectures based on nano-oscillators." In *Cellular Nanoscale Networks and Their Applications (CNNA), 2012 13th International Workshop on*, pp. 1–6 (2012).

[90] F. Macià *et al.*, "Spin-wave interference patterns created by spin-torque nano-oscillators for memory and computation." *Nanotechnology*, **22**, 095301 (2011).

[91] E. M. Izhikevich & F. C. Hoppensteadt, "Polychronous wavefront computations." *International Journal of Bifurcation and Chaos*, **19**, 1733–1739 (2009).

10 The piezoelectronic transistor

Paul M. Solomon, Bruce G. Elmegreen, Matt Copel, Marcelo A. Kuroda,
Susan Trolier-McKinstry, Glenn J. Martyna, and Dennis M. Newns

10.1 Introduction

In this chapter we introduce a new device, the piezoelectronic transistor, or PET. For the past decade or so there has been an extensive push to find a device with capabilities beyond the extant complementary metal–oxide–semiconductor (CMOS) technology. This is because CMOS has run up against fundamental voltage limitation, which in turn has led to increased power dissipation with increasing densities and speeds [1–4], or has had a severe impact on density and speed when the allowable power dissipation is restricted. The quest for the low-voltage switch has not yet produced a clear candidate technology that can replace CMOS. Even though theoretical expectations have been promising for devices such as the tunnel FET [5], experimental results have so far fallen short.

There are two sources of intrinsic voltage limitation for the electronic switches employed in digital circuits: limits to non-linearity (as manifested in the voltage required to switch from off- to on-state) caused by the Boltzmann statistics of electrons being emitted over a barrier (see Fig. 10.1), and thermal noise in resistors. The former has the value $k_B T/e$, the unit of voltage corresponding to the electron's thermal kinetic energy, where k_B is Boltzmann's constant, T is the absolute temperature, and e is the electronic charge. The electronic charge has the value $\sqrt{k_B T/C}$, since the Johnson (thermal) current noise of the circuit resistors is integrated by the capacitance, C, of the circuit node.

The $\exp(eV_G/k_B T)$ non-linearity gives rise to the famous 60 mV/dec log (I_D) vs. V_G (drain current vs. gate voltage) transistor characteristic in the *sub-threshold* regime, which, when accounting for sufficient on/off ratio, overdrive, and tolerances, leads to a practical power-supply voltage (V) limit of 0.8–1 V for CMOS. The $\sqrt{k_B T/C}$ noise, accounting for an ~8× factor to achieve a low enough error rate, gives a voltage limit of ~10 mV [1]. Thus we have a gap between 10 mV and 800 mV that can be exploited by new device types not operating on the principle of a gate-controlled electron barrier, leading to possible energy savings ($\propto V^2$) of up to four orders of magnitude.

Electromechanical devices [6–8], discussed elsewhere in the volume, are not limited by this $k_B T/e$ barrier. They are examples of *transductive* devices [1, 2], i.e., devices where an electrical input is transduced to some alternative force, e.g. mechanical, and then back to electrical, e.g. by closing a switch. This chain of action is obviously more complex than direct electrostatic modulation, as in an FET, and may

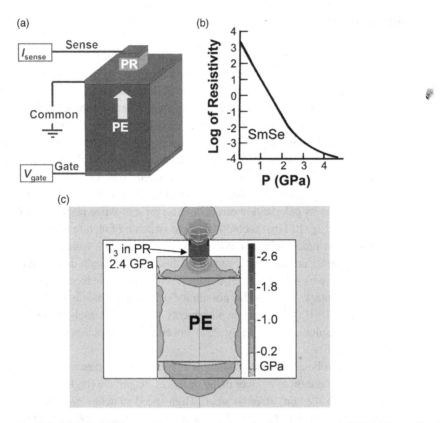

Fig. 10.1 (a) PET. (b) Piezoresistor characteristics after Jayaraman *et al.* [12]. (c) Stress distribution in a PET from ANSYS simulations. Stress is compressive except in small regions with dashed contour lines indicating tensile stress. (Parts (a) and (b) reproduced with permission from Newns *et al.* [9].)

involve trade-offs in density and performance; however, it comes with valuable properties: lower voltages and isolation.

When the intermediate force is derived from an input electric field, and when the force modulates an intensive property of the output transducer, voltages may be reduced (scaled) simply by reducing the dimensions of the input and output elements. The isolation property arises when the output transduction does not depend on electric field but only on the intermediate force. These properties will be put in context when we discuss the PET operating principles in Section 10.2. The value of the isolation property will become apparent when we discuss circuits in Section 10.4.

The PET is a simple yet revolutionary device. It is the ultimate scale-down of an electromechanical relay replacing a low-energy-density air capacitor with a high-energy-density piezoelectric actuator, a slow action due to a long and flexible mechanical path with a short, direct, transmission path, and an unreliable make-break vacuum contact with an all solid-state contact where the only motions are atomic level strains and movements of electron wave functions.

10.2   How it works

This section will give an overview of PET static operation using a typical piezoelectric (PE) and piezoresistive (PR) materials combination. More details on choices and physics of the PE and PR elements are given in Sections 10.3.1 and 10.3.2 respectively.

The PET (see Fig. 10.1) is a solid-state relay where a piezoelectric element (PE) provides the mechanical switch and a piezoresistive element (PR) provides the mechanical to electrical switching. Unlike the nanoelectromechanical systems (NEMS) relay, where a cantilever arm completes the mechanical circuit, in the PET this function is provided by a rigid yoke, called high-yield strength material (HYM) by Newns *et al.* [9–11]. Switching is caused by compression of the PR due to extension of the PE under a bias parallel to its polarization direction. The piezoresistive response of PR materials such as SmSe (Fig. 10.1(b)) accomplishes the switching function where strains of a few percent generate orders of magnitude resistivity change. In a scaled device the mechanical displacements are extremely small, of the order of hundreds of picometers (pm). Many perovskite-based piezoelectric materials have coefficients of several hundred to a few thousand pm/V creating the possibility of a direct tandem PR/PE stack such as shown in Fig. 10.1(a). Because displacements are so small, high switching speeds can be attained despite approximately $25\times$ lower sound velocities compared to typical electron velocities.

For the PR to be an effective switch in a high-speed logic environment it is essential for current modulation to be over many orders of magnitude (high on/off ratio) and that its resistivity in the "on" state be small. High-speed switches must have an "on" state resistance of less than 10 KΩ, which dictates resistivities of <100 KΩ-nm (<0.01 Ω-cm) since the size of the PR is anticipated to be a few nm. This requires pressures of a few GPa for the materials under consideration.

The tandem arrangement (Fig. 10.1(a)) means that both PE and PR share the same force, but the pressure required to switch the PR is considerably larger than the maximum pressure available from the PE ($\sim\frac{1}{2}$ GPa). Therefore the PE has to have a much larger area than the PR ("hammer and nail" effect) with a rigid plate on top of the PE to distribute the force and prevent the PE from being indented by the PR. Another important consideration is that the PE has to be close to 1:1 in aspect ratio. This is because most high strain piezoelectrics also undergo lateral contraction during the electrically driven vertical expansion (e.g., the piezoelectric d_{31} coefficient is finite and negative). Either the underlying substrate in piezoelectric films with large lateral dimensions or constricting sidewalls inhibit this free movement, reducing the piezoelectric expansion or attainable pressure.

The PET operating point is derived with the aid of Fig. 10.2. In analyzing the forces, we first note that the effect of the air gap is to eliminate any lateral body forces on either of the PR or PE elements. We shall also assume that there are no lateral surface forces acting on the PE/HYM, PE/PR, or PR/HYM interfaces. For the PE, this ignores *clamping* effects on the piezoelectric response. Practically, lateral surface forces can be minimized by making the aspect ratio of the device tall and thin. Secondly, we shall assume that the HYM and electrodes are incompressible, so that the total device height, $L+l$, is

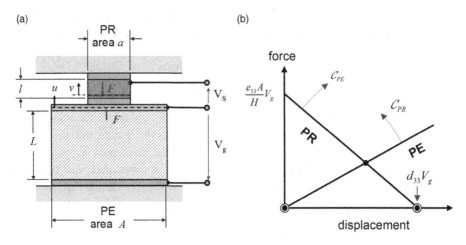

Fig. 10.2 (a) PE/PR stack with nomenclature. (b) PE/PR load lines showing limits of force and displacement and PET operating point given by their intersection.

conserved under application of voltage to the PE. Thirdly, we shall assume that the common contact is thick enough and made of a hard metal such as iridium (Young's modulus = 528 GPa) to ensure minimal bending of the PE/PR interface. Since the z-forces on the PE and PR at the PE/PR interface must be equal and opposite, a rigid common contact imposes the condition that the ratio (PR z-stress)/(PE z-stress) is in the ratio A/a of the cross-sectional areas of the two materials. Due to the very high dielectric constant, typically over 1000, of the PE, there will be hardly any leakage of the field (associated with applying a voltage V_G between the gate and common contacts) into the air gap. Thus electric field will be given by just a z-component equal to V_G/L.

The appropriate electromechanical equations to analyze this situation are

$$S = \mathfrak{s}^E T + d^T E; \quad S' = \mathfrak{s}' T', \tag{10.1}$$

where S, S' are the strain tensors in the PE and PR respectively, and T, T' the corresponding stress tensors, $\mathfrak{s}^E$, $\mathfrak{s}'$ are the mechanical compliances in the PE and PR respectively, the former being defined at constant electric field E, with d^T the transpose of the piezo coefficient tensor, in the PE. We write these relations in compressed Voigt [13] notation:

$$\begin{aligned} S_i &= \mathfrak{s}^E_{ij} T_j + d_{\alpha i} E_\alpha; \\ S'_i &= \mathfrak{s}'_{ij} T'_j, \end{aligned} \tag{10.2}$$

where i, j are defined to run from $1, 2, \ldots, 6$ and α from 1 to 3, the Einstein summation convention being assumed. Now the assumptions about no lateral forces ensure that T_3, T'_3, the z-components of stress, will be the only non-zero ones. The z-components of strain, from Eq. (10.2), will then be

$$\begin{aligned} S_3 &= \mathfrak{s}^E_{33} T_3 + d_{33} E_3; \\ S'_3 &= \mathfrak{s}'_{33} T'_3. \end{aligned} \tag{10.3}$$

At this point the four unknowns, T_3, T'_3, S_3, S'_3, can be reduced to two, enabling solution of the simultaneous equations (Eq. (10.3)), by using the boundary conditions

$$\Delta l = -\Delta L; \text{ hence } S'_3 = -(L/l)S_3$$
$$T'_3 = (A/a)T_3. \tag{10.4}$$

The first condition in Eq. (10.4) requires that the upward displacement ΔL of the PE top surface be equal to the displacement Δl of the PR bottom surface, hence relating the strains in these materials. The second condition uses equal action and reaction at the PE/PR interface plus the assumed stiffness of the common contact to relate the stresses in the two materials. Substituting these two boundary conditions Eq. (10.4) into Eq. (10.3) and solving yields

$$T'_3 = \frac{-V_g d_{33}}{l s'_{33} + (a/A)L s^E_{33}}. \tag{10.5}$$

This expression for the z-stress in the PR (*negative implies compression*) is proportional to the applied gate voltage V_G, and depends on the thicknesses L and l of the PE and PR, the material properties, the piezoelectric response d_{33}, and the compliances s^{33}_E, s'_{33} of the PE and PR. In order for the PET to operate at low voltage, there are two obvious requirements: (1) (on the PE) the piezoelectric coefficient needs to be large, and (2) (on the PR) the change in resistivity with pressure needs to be large.

We can alternatively express Eq. (10.5) in terms of the compliances $C_{PE} = s^E_{33}L/A$ of the PE and $C_{PR} = s'_{33}l/a$ of the PR as

$$F = \frac{-V_g d_{33}}{C_{PE} + C_{PR}}. \tag{10.6}$$

This is illustrated graphically in Fig. 10.2(b) where the limits of displacement, for small a/A, and force, for large a/A, are shown.

10.3   Physics of PET materials

10.3.1   Polarization rotation in relaxor-based piezoelectric single crystals

The new class of relaxor-based piezoelectric single crystals, such as PMN–PT $(PbMg_{1/3}Nb_{2/3}O_3–PbTiO_3)$ and PZN–PT $(PbZn_{1/3}Nb_{2/3}O_3–PbTiO_3)$ [14] has exceptionally large d_{33} coefficients and hence is well-suited for use as the actuator in the PET. In Fig. 10.3(a) we illustrate the mechanism believed to be primarily responsible for their high piezoelectric response [15]. In the rhombohedral phase poled along [001], the single-crystal material consists of nanodomains [16] with global polarization along [001] but local polarizations oriented along the various pseudocubic <111> directions. When a z-directed electric field E_z (with sign parallel to the poling direction) is applied to the crystal, the tendency is to rotate the polarization towards the [001]. This dilates the crystal along the z-axis via the piezoelectric effect.

Table 10.1 Selected properties of PMN–0.33PT single crystals [14]

ε_{33}^{T}	ε_{33}^{S}	d_{33} (pm/V)	s_{33}^{E} (GPa^{-1})	$(c_{33}^{E})^{-1}$ (GPa^{-1})
8200	680	2820	0.120	0.0097

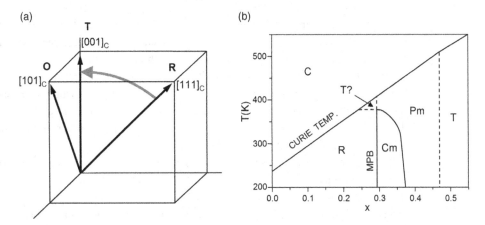

Fig. 10.3 Relaxor piezoelectrics. (a) Rotational mechanism for piezoelectricity in relaxor piezoelectrics. Polarization along [111] is rotated toward the z-axis by field E_z, dilating crystal along [001]. (b) PMN–PT phase diagram in temperature–composition plane, showing the cubic (C) phase, Curie temperature (lower bound of C phase), the rhomobohedral (R), and tetragonal (T) phases, and the morphotropic phase boundary (MPB – right-hand boundary of R phase). (After Shuvaeva *et al.* [17].)

To see why this can be a big effect we turn to the phase diagram [17] of PMN–PT in Fig. 10.3(b). The rhombohedral phase just discussed transitions to the tetragonal phase over a narrow composition region at which reduced symmetry is reported, producing a morphotropic phase boundary (MPB). The free energy surface for the system shows shallow barriers between polarizations with different orientations. Hence the direction of the polarization is extremely labile. In compositions like $(0.67\text{PbMg}_{1/3}\text{Nb}_{2/3}\text{O}_3$–$0.33\text{PbTiO}_3$, or PMN–0.33PT)) – this produces a large sensitivity for rotational displacement of the polarization by the E_z field. This sensitivity leads to large values of characteristic susceptibilities (see Table 10.1).

A closer look at Table 10.1 reveals the effects of extremely strong electromechanical coupling in this material. The dielectric constant is much less at constant strain S, where the coupling cannot contribute, than at constant stress T. The elastic properties are highly anomalous since electroelastic coupling plays a large role; we see from Table 10.1 that the large value of the elastic compliance, s_{33}^{E}, at constant electric field is not the same order as the inverse of the elastic stiffness, c_{33}^{E}, as would be the case for a conventional material. In Eq. (10.5) it is noteworthy that the large PE compliance s_{33}^{E} term can dominate the denominator, leading to an approximate dependence of the PR strain on the ratio d_{33}/s_{33}^{E}.

10.3.2   4f→5d electronic promotion under pressure in rare earth chalcogenide piezoresistors

The piezoresistor materials suitable for the PET application need to satisfy several criteria. Primarily, the resistivity needs to change by at least four orders of magnitude under a reasonable pressure (typically a few GPa), and the insulator–metal transition needs to be reversible. The second criterion typically rules out systems where a pressure-induced structural change occurs. Two classes of materials, Mott insulators such as Cr-doped V_2O_3 [18] and the rare earth chalcogenides, have been investigated in some detail. Here we shall focus on the rare earth chalcogenides.

The pressure dependence of resistivity in monochalcogenides has been extensively studied by Jayaraman *et al.* [12, 19–21]. The mechanism underlying piezoresistivity in the rare earth chalcogenides can be thought of as a form of pressure-dependent doping. The materials of interest, including SmSe, have a rocksalt-structure with a simplified band shown in Fig. 10.4(a) (for SmSe).

The electronic structure of SmSe consists of a filled Se 4p band (not shown), an empty, light mass 5d band, and a "half-filled" 4f shell. In the first half of the 4f series, the 4f states are $j = 5/2$, where the Sm^{2+} ion is occupied by 6 out of a possible 14 electrons. Note that even when not fully occupied there is still no 4f conduction since the holes are strongly localized in these materials [21]. Under pressure (or strain), the approximately 0.5eV 4f–5d energy gap (Fig. 10.4(a)) gradually closes (Fig. 10.4(b) and (c), [22], enabling electrons to be promoted into the 5d band where they cause n-type conduction. For low pressures the promotion is thermal, but under a pressure of several GPa the energy gap closes and the material becomes a metal with ~1 conduction electron per Sm.

Some properties of the SmSe material are given in Table 10.2.

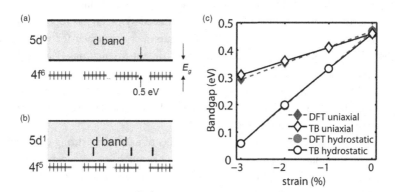

Fig. 10.4 Rare earth chalcogenide piezoresistors. (a) Electronic structure at ordinary pressure showing an empty light-mass *5d* band and fully occupied *4f* half-shell approximately 0.5 eV below the *5d* band lower edge. (b) Under pressure the *4f* band is pushed up, enabling *4f* electrons to be promoted into the *5d* band, leading to conduction. (c) *Ab initio* calculations showing closing of *4f*–*5d* energy gap E_G under uniaxial and hydrostatic strain. (Reproduced with permission from Jiang *et al.* [22].) DFT and TB refer to density functional theory and tight binding method, respectively.

Table 10.2 Selected properties of SmSe [23]

Structure: rocksalt	Piezoresistive gauge $d_{\log 10}\, \rho/dp\,(\text{GPa}^{-1})$	Metallic threshold p_{M} (GPa)	Bulk modulus K (GPa)
$a = 0.622$ nm	2.3	~3.5	43

10.3.3   Equivalent sub-threshold slope

The exponential pressure dependence of the SmSe resistance with pressure, i.e., with gate voltage of the PET, invites comparison with the exponential sub-threshold characteristics of the FET. As discussed above, the cause is similar in both cases. For the FET the Fermi level in the channel is controlled directly with gate voltage while in the PET it is controlled indirectly with the intermediate agency of pressure.

To compare these two situations we introduce a deformation potential, Ξ_{33}, describing the bandgap shift with uniaxial strain:

$$\Xi_{33} = \frac{\partial E_{\text{G}}}{\partial S_3'}. \tag{10.7}$$

This is computed as $\Xi_{33} \approx 6$ eV from the DFT uniaxial calculations in Fig. 10.4(b) [23]. Assuming now that carriers in the SmSe 5d band are thermally excited across the gap, that the Fermi level is pinned in the 4f band, and that the mobility in the 5d band is constant, we obtain a sub-threshold like exponential dependence of resistivity on voltage with an inverse slope of

$$\left(\frac{d\ln\rho}{dV_{\text{G}}}\right)^{-1} = \frac{k_{\text{B}}T}{e}\frac{l}{d_{33}\,\Xi_{33}}\left(1 + \frac{C_{\text{PE}}}{C_{\text{PR}}}\right). \tag{10.8}$$

The figure of merit compared to FETs is how much smaller the right-hand side of Eq. (10.8) is compared to $k_{\text{B}}T/e$. Thus the second term $l/d_{33}\Xi_{33}$ needs to be much less than unity. Using $d_{33} = 2.8$ nm/V (from Table 10.1) we obtain a requirement that $l \ll 16.7$ nm. This is a fairly easy target to achieve since the tunneling limit on the thickness is predicted to be about 3 nm [22]. Some of the above factor will be lost in the compliances of the PE and other elements in the system (right-hand bracket of Eq. (10.8)) yet enough should remain to retain a significant advantage over FETs.

10.4   PET dynamics

10.4.1   Electromechanical (Tiersten) equations

The dynamics of the PET are determined from the equations of motion for the PE and PR combined with the time-dependent charge on the PE and the relation between resistance and pressure in the PR. Figure 10.5(a) shows an idealized case with PE and PR displacements u and v, respectively, source voltage V_0 and circuit impedance R. Considering one-dimensional motions in the z direction, the equations for the PE that

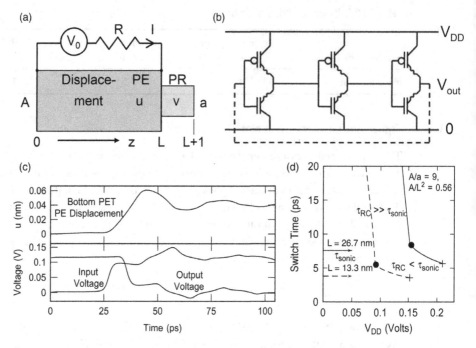

Fig. 10.5 (a) Essential diagram of the PET illustrating the nomenclature. (b) A complementary PET 3-inverter stage ring oscillator. The opposite type PETs (with an "o" attached to the gate) are poled in the opposite direction to the normal PETs. (c) Waveforms showing displacement and output voltage (periodic) for one-stage of a ring oscillator. (d) Time delay for switching as a function of supply voltage for two cases with $L = 13.3$ nm and 26.7 nm, $l = 2$ nm, and all other dimensions scaled.

relate the stress T and strain S to the electric field E and dielectric displacement D are, following Tiersten's textbook [24] (henceforth we will just refer to them as Tiersten's equations),

$$T = S/s_{33}^E - eE, \quad D = eS + \varepsilon^S E, \quad \text{with}$$
$$S = \partial u/\partial z \text{ and } \partial D/\partial z = 0, \tag{10.9}$$

where s_{33}^E is the normal coefficient of the compliance tensor for a constant electric field, comparable to the inverse of Young's modulus, e is the piezoelectric coefficient related to the piezoelectric expansion coefficient d_{33} as $e = d_{33}/s_{33}^E$, and ε^S is the dielectric constant measured at constant strain. The variables T, u, and E are functions of z and t. The fourth equation states that charge responsible for D is entirely on the PE surface, and may be written as a time-dependent surface density, $D = \sigma(t)$. The time dependence for the PET motion enters from Newton's law of motion,

$$\partial T/dz = \rho \partial^2 u/\partial t^2, \tag{10.10}$$

where ρ is the mass density in the PE.

As in Section 10.2 the equations for the PR are the same but without the piezoelectric effect. Again we use primes to denote the PR:

$$T' = S'/s'_{33}, \quad S' = \partial v/\partial z, \text{ and } \partial T'/\partial z = \rho' \partial^2 v/\partial t^2. \tag{10.11}$$

For PE and PR lengths L and l and areas A and a, respectively (see Fig. 10.5(a)), the boundary conditions are a fixed displacement at both ends, $u(0, t) = 0$, $v(L + l, t) = 0$, a co-moving common surface, $u(L, t) = v(L, t)$, $du(L, t)/dt = dv(L, t)/dt$, and an equal total force, which corresponds to a pressure ratio proportional to the area ratio, $T(L, t) = (a/A)T'(L, t)$.

What remains is a dependence of the PET dynamics on the circuit, which introduces a time-dependent impedance $R(t)$ that includes what other PETs are doing. For current $I(t)$, the impedance enters as

$$V_0 - R(t)I(t) = \int_0^L E(z, t)dz, \tag{10.12}$$

and the charge density on the PE surface is

$$A\sigma(t) = \int_0^t I(t)dt. \tag{10.13}$$

The resistance from the PR is the integral over the local PR resistivity, $\rho_{PR}(z, t)$, combined with some associated external wire resistance R_w:

$$R_{PR} = \frac{1}{a} \int_L^{l+L} \rho_{PR}(z, t)dz + R_w$$

where

$$\rho_{PR} = \rho_{PR,0} \exp(BT') + \eta_{sat} \tag{10.14}$$

includes the exponential dependence of the resistivity of the PR on the local tension T' characterized by the uniaxial gauge factor B, which is related but not identical to the piezoresistive gauge defined in Table 10.2 for hydrostatic pressure, down to some saturated resistivity η_{sat} at which the declining exponential flattens out. Note that T' is negative during PR compression because dv/dz is negative.

These equations can be written in dimensionless form with all of the material constants absorbed into one dimensionless parameter, the electromechanical coupling constant, $g = c^E d_{33}{}^2/\varepsilon^S$. In the following examples, $g = 10.9$. Simulations with g in the range from 1 to 10 all show good switching characteristics for the PET, but with higher g tending to be better.

The equations were solved numerically by direct integration with time steps short enough to resolve the changes in PR resistance, which is a sensitive function of pressure. Thevenin equivalence is used to simplify the representation of the circuit. One case is a chain of inverters making a ring oscillator (Fig. 10.5(b)). For each inverter, one can consider the voltage across the top (V_{top}) and bottom (V_{bot}) of the PEs, where

$V_{top} + V_{bot}$ is the applied voltage, V_{dd}. These voltages are the input drivers for the motions and resulting resistances in the PRs. If R_{top} and R_{bot} are the PR resistances in the top and bottom PETs of the previous inverter stage then the Thevenin voltage for the driving circuit is $V_{th} = V_{dd} R_{bot}/(R_{bot} + R_{top})$ and the Thevenin resistance is the parallel sum of R_{bot} and R_{top}. The current entering the inverter is related to the rate of change of the surface charges in the PEs as $I = A\frac{d}{dt}(\sigma_{bot} - \sigma_{bot})$ for PE area A. The rate of change of the surface charge is equal to the expansion rate of the PE plus the rate of change of voltage,

$$\frac{d\sigma_{bot}}{dt} = \frac{d\Delta u_{bot}}{dt} + \frac{\varepsilon^S}{e}\frac{dV_{bot}}{dt}, \qquad (10.15)$$

where Δu_{bot} is the total expansion of the bottom PE. Then from Ohm's law, $V_{th} - V_{bot} = IR_{th}$, from which the rate of change of both V_{bot} and V_{top} can be determined. In particular,

$$\frac{dV_{bot}}{dt} = \frac{e}{2\varepsilon^S}\left[\frac{V_{th} - V_{bot}}{R_{th}A} + \frac{d}{dt}(\Delta u_{top} - \Delta u_{bot})\right]. \qquad (10.16)$$

At each integration step, all terms on the right-hand side are known from the current PET configurations and motions in each inverter, and this allows the time rate of change of the input voltages for those inverters to be determined. The input voltage at the next time step is thus the current input voltage, V_{bot}, plus $(dV_{bot}/dt)dt$ for integration time step dt.

Figure 10.5(c) shows the response of one inverter in a nine-stage ring oscillator. The model assumes an area ratio $A/a = 25$, a length ratio $L/l = 13.3$, a PE aspect ratio $A^{0.5}/L = 0.75$, and dimensionless resistivities $\rho_{PR0} = 740$ and $\eta_{sat} = 0.15$. Conversion to dimensional resistivity involves multiplication by $s_{33}^E/(d_{33}^2 v_s) = 1.49 \times 10^4 \Omega$ times the PE length $L = 26.6$ nm, for $d_{33} = 2.8$ nm/V, $s_{33}^E = (8.4 \text{ GPa})^{-1}$, and sound speed $v_s = 10^{20}$ m/s. The resistivity formula also assumes $B = 5.4$ and $s_{33}' = (110 \text{ GPa})^{-1}$. Figure 10.5(c) shows that when the input voltage increases, the displacement of the bottom PE increases (shown in the top panel) causing the resistance in the bottom PR to decrease, and the displacement of the top PE decreases causing the resistance in the top PR to increase. The output voltage decreases as a result (bottom panel). The oscillations come mostly from the inertia in the PEs, with damping from resistance elsewhere in the circuit.

The operating range for source voltage lies between the lower limit required to compress the PR to the saturated regime, and an upper limit depending on the circuit. For an inverter, the upper limit is approximately twice the lower limit, at which point both the top and bottom PETs can be driven to their saturated resistances by a constant input voltage, which is equal to the output voltage for identical PETs and half of the source voltage. Between these limits, the frequency of the PET operation increases slightly with voltage because the PE expansion speed does. Figure 10.5(d) shows two curves for the switching time versus the source voltage in a nine-stage ring oscillator. One curve is for a thicker PE than the other, as indicated in the figure; both have a PR thickness of 2 nm. The thicker PE has a longer sonic propagation time and a correspondingly longer switching time in the operating range. The sharp increases in switching time for both curves occur at the lower limits to the operating voltage. For Fig. 10.5(d), the area ratio $A/a = 9$; the other parameters are the same as in Fig. 10.5(c).

10.4.2   Simulation of logic circuits

Logic gates can be constructed with PETs, as demonstrated above for the inverter chain. Because the sign of the voltage across the PE depends on the poling direction when the PE is made, PETs with opposite voltage drops have to be poled differently. This is the case in the inverter, for example (Fig. 10.5(b)), where the voltage across the top PET is lower on the gate terminal than the common terminal, while the voltage across the bottom PET is lower on the common terminal. The circuit symbol for the first type has an "o" at the gate terminal. By poling the PE in opposite directions, PETs with complementary transfer characteristics can be obtained. A more complex NAND gate, shown in Fig. 10.6(a), has one of the PETs (input A, bottom) isolated and requires a 4-terminal PET (4-PET) in which the two terminals on the PR are electrically isolated from the two terminals on the PE. The 4-PET is structurally more complex but electrically much more flexible; both normal and complementary versions can be made with the PE poled in the same direction, simply by reversing the two PE terminals.

Figure 10.6 illustrates the application of a PET to a 2-bit ripple-carry adder. Figure 10.6(a) shows the circuit diagrams for a PET NAND gate and a PET inverter. Figure 10.6(b) has the logic for a 1-bit full adder, which is one-quarter of the full circuit when 2 bits are added to 2 bits, including the carry bits. The 2-bit adder consists of four XOR logic gates that each contain four NAND gates (that each contain four PETs), plus three AND gates. The AND gates are inverters (two PETs) connected to the first NANDs in each XOR. There are 70 PETs in total. The PET parameter values are the same as in Fig. 10.5(c) but in this case the wire resistance, R_w, is set to zero.

The panels in Fig. 10.6(c) show the PET voltages at key places in the logic representing bit values as a function of time. The input numbers are designated A and B in the panels, with the low-order bit for A designated A0 and the high-order bit for A designated A1, etc. Thus the sum starts with A = 0 and B = 0 for time <0.05 ns, and then transitions to A = 11 and B = 10, in binary, and so on. These are the input values of the voltages shown in the two panels on the right, second from the top. The lower panel on the right is the sum of the two low-order bits, and the top panel in the middle is the carry bit (CB) from these, called C0. The fourth panel down in the middle column is the sum of the high-order bits A1 and B1, and the bottom panel in the middle column is the sum of this and the carry bit. The left-hand column consists of two of the carry bits, as indicated, plus the sum of them. Although there are slight oscillations at each value switching, the oscillations are short-lived and the transitions are generally sharp for the sum. All of this happens at a source voltage of about 0.08 V, much lower than in today's CMOS processors. The speed of a single transition is essentially a multiple of order unity times the sonic time for a pressure pulse to move through the PE. For a PE thickness of 26.6 nm, the sonic time is $\tau_s = L/(v_s\sqrt{1+g}) = 7.6$ ps.

10.4.3   Electromechanical lumped element circuit model for PET

The dynamic Tiersten equations are expressed in the form of coupled electrical and mechanical transmission lines (TL) with the mechanical TL being expressed in

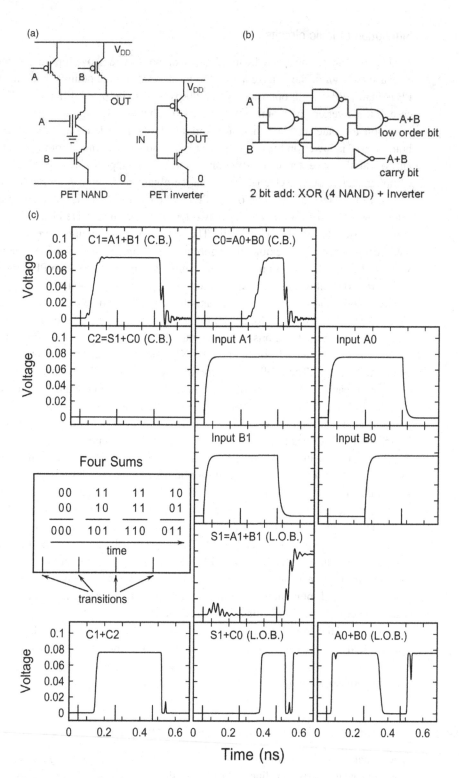

Fig. 10.6 (a) NAND and inverter circuits used in a 2-bit ripple-carry adder. A 4-terminal PET (4-PET) is used for the bottom A input of the NAND circuit. (b) Diagram of a 1-bit full

electrically equivalent form. This allows one to represent the piezoelectric transducer with conventional circuit elements and dependent sources, which opens the way to transient modeling of Piezotronic circuits using standard circuit simulators.

The mechanical-to-electrical transformations employed are given in Fig. 10.7(a). By choosing force to represent current [25], Kirchhoff's law is obeyed: mechanically, the sum of forces at a node is zero while electrically the sum of currents is zero. In this representation mass plays the role of capacitance with the force absorbed in acceleration represented as current being shunted to ground through the capacitor; compliance takes the role of inductance, while differential velocity is proportional to the rate of change of force. Since mass really has just one "terminal" and its acceleration is absolute, one side of the mechanical equivalent capacitor is always connected to ground. Note that velocity rather than displacement is used as a primary variable to avoid second order time derivatives (Newton's law). Displacements can be derived by integration after setting initial conditions.

The electrical equivalent elements (using standard electrical sign convention) become:

$$I_m = N\mathcal{F}, \quad V_m = -\mathfrak{v}/N, \quad C_m = N^2 m, \quad L_m = \mathcal{C}/N^2, \tag{10.17}$$

With the notation of Fig. 10.7(a), where N is an arbitrary transformation ratio with units of current/force and the subscript m is used for electrical equivalents of mechanical quantities. N scales the electrical equivalents much like a transformer ratio without changing the underlying mechanical quantities.

To proceed we convert the Tiersten equations, Eqs. (10.9) and (10.10), to their extensive form in terms of forces, voltage, and current,

$$\mathcal{F} = \frac{\mathcal{E}A}{s}\frac{d\mathfrak{v}}{dx} + eA\frac{dV}{dx}, \quad I = eA\frac{d\mathfrak{v}}{dx} - s\varepsilon A\frac{dV}{dx}, \quad d\mathcal{F} = \rho A dx s \mathfrak{v}, \tag{10.18}$$

where $s = \partial/\partial t$ is the Laplace variable, e is the density and $\varepsilon = 1/S_{33}^E$ is Young's modulus. We have used the following substitutions: $E = -dV/dx$, $S = d(\mathfrak{v}/s)/dx$, and $I = s\sigma A$, where A is the area, and have introduced the symbol for Young's modulus. To represent a thin slab of a piezoelectric transducer of length Δx and area A, these equations can be cast in terms of a series electrical capacitance, C_e, a series electro-mechanical inductor, L_m, and a parallel electromechanical capacitor, C_m, where

$$C_m = N^2 \rho A \Delta x, \quad L_m = \frac{\Delta x s_{33}^E}{N^2 A}, \quad C_e = \frac{\varepsilon A}{\Delta x}. \tag{10.19}$$

Since N is arbitrary, we choose N such that under clamped conditions $(\mathfrak{v} = 0 \rightarrow I_m = \varepsilon A N V_e/L)$ the output current is equal to the input voltage $\times$ the

---

Caption for Fig. 10.6 (continued) adder using conventional notation for NAND gates and inverters; this is one-quarter of the circuitry for the 2-bit ripple-carry adder. (c) Simulation of the sum of two 2-bit integers with PET technology. Four sums are shown, with voltage vs. time in each panel and transitions at the vertical markers. The sum result is in the lower three panels. The low-order bit in the sum is in the lower right, the next higher order bit is in the lower middle, and the high-order bit is in the lower left. High voltage represents a 1 bit and low voltage a 0 bit.

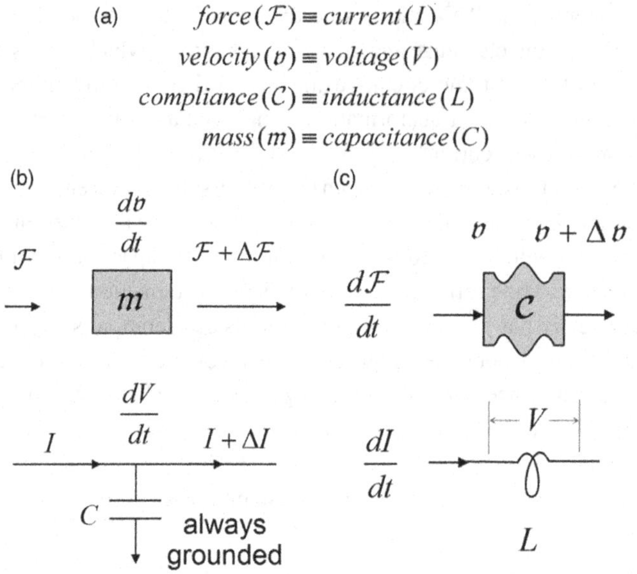

(a) $force\,(\mathcal{F}) \equiv current\,(I)$
$velocity\,(v) \equiv voltage\,(V)$
$compliance\,(C) \equiv inductance\,(L)$
$mass\,(m) \equiv capacitance\,(C)$

(b)   (c)

Fig. 10.7 (a) Mechanical to electrical equivalents. (b) Dynamic response of mass to force cf. capacitance to current. (c) Dynamic response of compliance to force cf. inductance to current.

characteristic admittance of a transmission line formed from the L_m and C_m elements, i.e. $I_m = V_e\sqrt{C_m/L_m}$. This ensures that V_m and V_e are roughly on the same scale. Thus $N = e/\mathcal{E}\tau_m$, where $\tau_m = L\sqrt{\rho/\mathcal{E}}$, which is the undressed sonic crossing time of the PE.

With these conditions the Tiersten equations can be solved, in the finite element limit, yielding the following representation:

$$\Delta I_m = -sC_mV_m, \quad \Delta V_m = -sL'_mI_m - \frac{\tau_m}{C'_e}I_e, \quad \Delta V_e = \frac{\tau_m}{C'_e}I_m - \frac{I_e}{sC'_e},$$

$$\text{where } C'_e = (1+g)C_e, \quad L'_m = \frac{L_m}{1+g} \quad \text{and} \quad g = \frac{e^2}{\varepsilon\mathcal{E}}. \tag{10.20}$$

Here g is the scale-independent coupling factor, as above. These equations can be translated into an equivalent circuit for the slab, containing both electrical and mechanical components, as illustrated in Fig. 10.8. Note that the mutual coupling terms are equal, as they must be to satisfy reciprocity. The factor $\sqrt{g/(1+g)}$ is the same as the electromechanical coupling factor k_{33} [13] and is equivalent to the coupling factor for electro-magnetic transformers [26]. For high-performance piezoelectrics, $g \rightarrow \infty$ and the electromechanical coupling approaches unity. A manifestation of the coupling is the well-known increase in the input capacitance by a factor of $1 + g$ when the output is open circuited (unclamped) compared to the clamped value. This ratio is used to measure k_{33} [13]. Strong coupling is essential for the PET since insufficient coupling will not permit an electrical resistor to be reflected into the mechanical circuit to provide the required damping. Strong coupling also can reduce sonic delay times, depending on the electrical load, due to the reduction of L_m by $1 + g$.

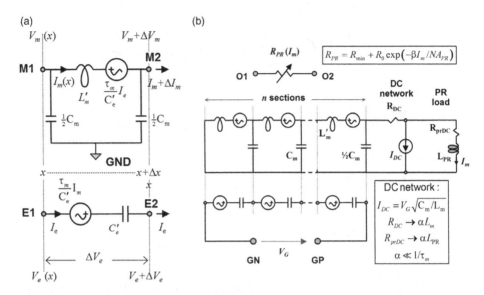

Fig. 10.8 (a) Electrical model of slab of a piezoelectric material of thickness Δx. The voltages and currents at the terminals M1 and M2 represent the *mechanical* velocities and forces applied to the slab boundaries with *electrical* voltages and currents at terminals E1 and E2. (b) Multi-segment electromechanical equivalent circuit of the PET.

The PET equivalent circuit is shown in Fig. 10.8(b). At the heart is a pair of coupled electromechanical transmission lines consisting of multiple PE segments. The mechanical circuit is loaded with the compliance of the PR represented by L_{PR}, and the PR itself, modulated by I_m, is part of the separate electrical output circuit.

The circuit is thus far incomplete because it does not specify initial conditions. This is done automatically by including the DC network consisting of R_{DC} and a current source, I_{DC}, as shown. I_{DC} gives the clamped (short circuit) force produced by V_e and the resistors R_{DC} and R_{prDC} allocate this force according to the relative compliances. The resistance values (but not the *ratios*) of the various R_{DC} are arbitrary as long as they are small enough so as not to load the AC circuit.

The model was implemented in IBM's AS/X circuit simulation tool and compared against the finite-difference solutions discussed in Section 10.4.1. Since the multi-segment PET employs the same equations, the results should be identical for a large enough number of segments. This is indeed shown to be the case in Fig. 10.9 where a circuit consisted of an unloaded PE (one end free) with 100 segments, and an input resistor to provide damping: the two solutions overlap precisely.

10.4.4   Mechanical and electrical parasitics

To complete the PET equivalent circuit, one needs to add both electrical and mechanical parasitics. The electrical parasitics are conventional coupling capacitors between the different electrodes as well as series resistors for the various electrodes and resistance internal to the PR itself. They can simply be incorporated into the electrical parts of the

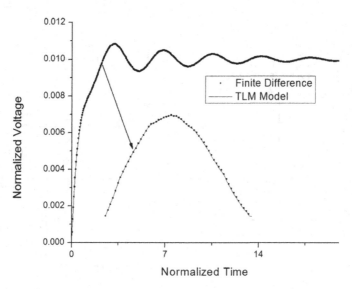

Fig. 10.9 Electrical input voltage to an open/fixed end piezoelectric element comparing the direct finite element solution. A section of curve has been magnified to show degree of agreement. Parameters were: $L = 100$ nm, $A = 10^4$ nm^2, $\varepsilon = 20$ GPa, $\rho = 2500$ kg/m^3, $d_{33} = 1$ nm/V, $g = 10.94$, $R_{\text{in}} = 1.77$ kΩ.

PET equivalent circuit. The mechanical parasitic elements are modeled as compliances and masses represented by inductors and capacitances connected to the mechanical terminals of the PET equivalent circuit. For instance (see Fig. 10.10(a)), the cap materials on top of the PE may be modeled by a series inductor and parallel capacitor and a constraining sidewall by a parallel inductor and capacitor. This is shown in Fig. 10.10(b) where a simplified one-segment PET circuit is used. For clarity the DC network has been left out, although, to account for the steady state force distribution, resistors would have to be added in series to the inductors as in Fig. 10.10(b).

The equivalent circuit of Fig. 10.10 shows how the mechanical parasitics affect the frequency response of the system. It is interesting that the dominant component is a product $C_{\text{cap}} \sqrt{L_m''/C_m}$, where the $''$ indicates that the magnitude of the inductor changes with the electrical load, i.e., due to the electromechanical coupling, the mechanical parasitics interact with the electrical input resistance.

10.5   Materials and device fabrication

10.5.1   SmSe

Making a functional piezoresistive thin film involves both fabrication and verification of properties. We have pursued this by co-sputtering a rare earth monochalcogenide thin film on a patterned substrate, with an in situ deposited metal capping layer. The structure was tested with a microindenter, monitoring the resistivity during compression.

(a)

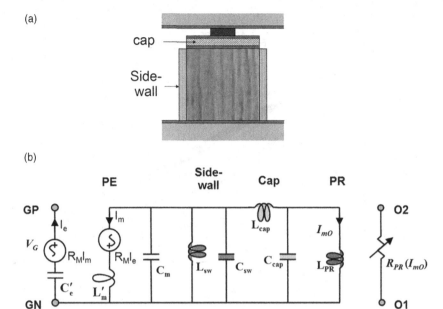

(b)

Fig. 10.10 (a) PET including PE sidewall and cap layer mechanical parasitics. (b) PET equivalent circuit consisting of a single transmission-line segment, including mechanical parasitics due to cap and sidewall layers. The DC network has been omitted for clarity.

Thin films of SmSe were made by sputtering from independent Sm and Se sources. The sources were mounted 17 degrees off-axis in an ultra-high vacuum deposition system capable of handling 8-inch wafers, creating a composition gradient across the sample. Although this caused non-uniform samples, it guaranteed that regions of the sample were monochalcogenide in composition. During deposition, the sample was held at 300°C. Sample temperature proved to be a highly significant factor in determining piezoresistivity; samples grown at room temperature exhibited only a 20× change in resistance under pressure, compared to 1100× resistance modulation for samples grown at 300°C. Modest growth rates were maintained (1–2 nm per minute), to avoid excessive heating of the selenium source. Initial results were obtained by careful calibration of sputtering power. While this was adequate for early demonstrations, process drift between runs created difficulties. Accordingly, a more sophisticated procedure was adopted using quartz crystal monitors in a feedback loop to control both Sm and Se deposition rates. An added complication came from the high vapor pressure of Se. Radiant heating of the Se source by the sample was found to accelerate the deposition rate during growth runs. Augmented source cooling as well as increased throw distance greatly improved process control.

Verification of film composition was obtained by several methods. Rutherford backscattering (RBS) was the principal technique used, with selected samples cross-checked using X-ray diffraction. Patterned devices were also examined with transmission electron microscopy energy dispersive X-ray spectroscopy (TEM-EDS) using RBS as a calibration standard. With these techniques, composition can be obtained with approximately 1% accuracy.

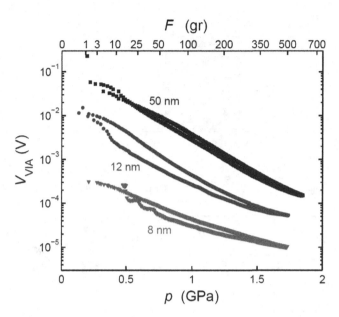

Fig. 10.11 SmSe pressure dependence of resistivity. (Reproduced with permission from Copel *et al.* [27].)

In order to quantitatively measure piezoresistance, a well-defined sample geometry is essential. By using a patterned sample with a narrow conductance aperture, we were able to reduce geometrical uncertainties inherent in the use of a microindenter. Substrates were made with a blanket tungsten ground-plane. A silicon nitride layer was deposited on the tungsten, and apertures were etched with diameters ranging from $0.5\,\mu m$ to $50\,\mu m$. A blanket SmSe layer was followed by an in situ TiN capping layer. Above the TiN, a $1\,\mu m$ thick aluminum layer was deposited. The aluminum serves to create a uniform pressure from the microindenter, minimizing the effect of irregularities of the indenter tip.

Piezoresistance was measured using a microindenter to apply pressure while monitoring resistance voltage. A constant current was fed through the indenter tip, while slowly increasing the applied force. Finite element analysis of the experimental configuration was crucial to converting the applied force into pressure, allowing quantitative interpretation of the results. Using the microindenter, we were able to apply pressures as high as 1.85 GPa, which was sufficient to cause an 1100× decrease in resistance in a 50 nm thick layer (Fig. 10.11 [27]). Additional measurements on thinner SmSe layers revealed piezoresistance extending down to 8 nm thickness. However, the thinnest films showed a drop-off in the piezoresistive response. We observe that the thinner films have smaller crystallite grain size, so perhaps a further increase in temperature will especially improve the response of the thinner films.

Future work will be motivated by the need to scale both lateral and vertical dimensions of the piezoresistor, as well as the need to increase piezoresistive response. Increasing the growth temperature is an option although process stability may become more difficult at higher temperatures. Challenges in scaling involve two different aspects; lateral scaling is primarily an integration problem, while vertical scaling will depend on improved growth techniques either to enhance microstructure or optimize interface characteristics. Obtaining higher performance piezoresistors is also an enormous opportunity for materials development to make a contribution to novel device research.

10.5.2   PMT-PT

High strain piezoelectric thin films such as $PbMg_{1/3}Nb_{2/3}O_3$–$PbTiO_3$ can be grown by a variety of techniques. Here, chemical solution deposition was employed, as it enables rapid changes in composition, as well as ready scaling to large wafer sizes (200 mm) without significant changes in the deposition process. Ultimately, transition to either a sputtering or MOCVD process should be straightforward.

To grow $0.7\ PbMg_{1/3}Nb_{2/3}O_3$–$0.3\ PbTiO_3$ films, a 2-methoxyethanol-based solution was utilized, following the work of Park *et al.* [28]. Strong {001} orientation could be achieved either using a $PbO/\{111\}Pt/\ Ti/SiO_2/Si$ substrate, or a $LaNiO_3/SiO_2/Si$ substrate, and crystallization temperatures of 700–740°C. Figure 10.12(a) shows a typical example of a blanket film. Dense, fiber-textured films can be achieved by controlling nucleation from the bottom electrode. When clamped to the underlying substrate, the permittivity of films between 300 and 350 nm thick is typically 1400–1600 with ~1% loss when measured at 1 kHz with 30 mV_{ac}. At an electric field of 1.5 MV/cm, piezoelectric strains exceeding 1% can be achieved in the clamped film [29].

On laterally patterning these films using reactive ion etching (RIE), the film can be laterally declamped from the thick substrate, significantly increasing the dielectric permittivity, polarization, and piezoelectric coefficients. Towards this end, an electron beam lithography process has been developed to allow piezoelectric films to be patterned down to feature sizes of 100 nm. For this purpose, PMN–$PT/Pt/TiO_x/SiO_2/Si$ films were coated 50 nm thick Pt top electrodes. Then, ZEP 520 resist was spun onto a wafer, and patterned using a Vistec 5200 electron beam system. Following development, and

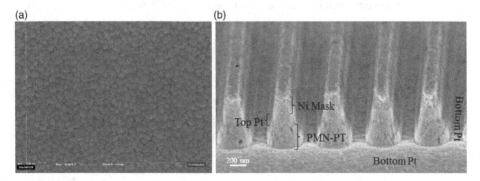

Fig. 10.12 (a) Scanning electron microscope image of a fiber-textured PMN–PT film surface. (b) Cross-sectional scanning field emission scanning electron microscope image of a patterned PMN–PT film.

plating of a Ni hard mask into the resist mold, the resist was removed and the piezoelectric film was patterned by reactive ion etching. The etch conditions were 10 sccm Cl_2/10 sccm CF_4, MHz bias power of 700 W, kHz power of 100 W, and a 2 mTorr chamber pressure. This etched the stack at a rate of ~1.7 nm/s. Sidewall angles of >80° were achieved. An example of a series of arms with feature sizes of ~100 nm is shown in Fig. 10.12(b). Measurements of finely patterned features shows that the remanent polarization can be doubled (to ~14 $\mu C/cm^2$), and that the permittivity and piezoelectric coefficient also rise.

10.5.3   PET fabrication – progress and issues

An experimental program is underway at IBM and Pennsylvania State University aimed at fabricating an integrated PET using current silicon microelectronics techniques.

The task of making a fully encapsulated PET is challenging and there are many different possible approaches. An approach using sacrificial material to form an isolated structure is shown in Fig. 10.13. The PE pillar and top metal are formed, followed by side pillars for the bottom half of the HYM. A sacrificial filler material is added between the PE and the HYM, followed by planarization, and creation of the top half of the structure. After the PR and its cladding layers are deposited and patterned, additional filler material is used to create a planar surface for the buried PR sense metal and the HYM cap. The structure is completed by removing the filler material.

This generically straightforward scheme is made more challenging by the incompatibilities of the key materials, PR and PE, to each other and to the standard processes employed. For instance, the perovskite type PE materials do not withstand many of the standard chemical etchants such as HF, and RIE selectivity requires the use of exotic hard mask materials such as nickel. The PE materials also do not tolerate a hydrogen ambient which is common in plasma processing, although their properties can be recovered with an oxygen anneal. The PR material by contrast, is readily oxidized so needs to be protected from the ambient by suitable encapsulation (or is perhaps protected by its native oxide [27]). The encapsulation, on the other hand, is detrimental to the PET's mechanical properties since it constrains the motion of the PR pillar.

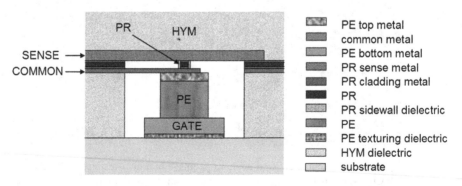

Fig. 10.13 Possible approach to fabricate a PET using VLSI and MEMs technology.

An issue is thermal compatibility. We have perfected a low-temperature process for synthesizing the PR [27] so that the PE can maintain high-quality properties throughout the process.

Although integration challenges remain, none of the issues is fundamental, and in a way these challenges are to be expected in a technology using a completely different material set from silicon. At IBM and PSU we have learned to cope with some of these challenges and completed many of the steps necessary for making a fully integrated device under contract to DARPA (DARPA MESO N66001–11-C-4109).

10.6   Performance estimation

The performance of the PET relative to other competing technologies was estimated based on the switching speed of a logic NAND gate as a function of fan-in/out. The comparison is affected by the assumptions regarding layout and wiring capacitance, etc., so it is important to keep the same set of assumptions across technologies. While a broad range of technologies have been benchmarked by others [30], here we concentrate on technologies capable of switching at CMOS-like speeds but at lower power, comparing the PET with Si FinFets, carbon nanotube FETs (CNTFETs), and tunnel FETs (TFETs). The cases were compared at a ground rule for a circuit layout of 7 nm.

This work makes use of simulations of PET inverters as described in Section 10.4.2 and then adds in RC delays to represent fan-in/out. Fan-in/out could have been included explicitly, but here we wanted to use the same methodology across technologies where possible in order to get the closest comparison. At this point we have not included the mechanical parasitics into the calculations, but their effects will be mainly to increase the intrinsic delay rather that the fan-in/out terms.

The switching speed is determined, for the case of the FET circuits, by the formula

$$\tau_{\text{ckt}} = \frac{C_{\text{node}} V_{\text{DD}}}{I_{\text{eff}}} + (FI - 1) R_{\text{on}} C_{\text{node}}, \tag{10.21}$$

where I_{eff} is the effective drive current [3] given by

$$I_{\text{eff}} = \frac{1}{2} \left(I_d|_{V_D = V_{DD}, V_G = V_{DD}/2} + I_d|_{V_D = V_{DD}/2, V_G = V_{DD}} \right) \tag{10.22}$$

and C_{node} is the total capacitance on the output node given by the sum of active device capacitance (C_a), parasitic edge capacitance (C_p), and wiring capacitance (C_w),

$$C_{\text{node}} = FO(C_{a\_\text{pfet}} + C_{a\_\text{pfet}}) + C_{p0} \left[w_p \left(FI + 2\tfrac{1}{2}FO \right) + w_n \left(1 + 2\tfrac{1}{2}FO \right) \right]$$
$$+ C_{w0} 5 L_{\text{cell}} FO \tag{10.23}$$

where C_{p0} and C_{w0} are per unit length. The factor of 2 is due to the Miller effect and we assume half the stages are switching, and L_{cell} is the cell pitch,

$$L_{\text{cell}} = \sqrt{2 A_{\text{device}} FI}. \tag{10.24}$$

The multiplier of 5 in the wire length equation is an application-dependent assumption. Assuming 20 circuit delays per clock period, the clock frequency $f_{clk} = \frac{1}{20}\tau_{ckt}$. The switching energy is

$$U_{ckt} = \frac{1}{2}C_{node}V_{DD}^2 + \frac{I_{off}V_{DD}}{\alpha f_{clk}}. \tag{10.25}$$

where the duty factor $\alpha = 1/20$.

For the PET an additional intrinsic τ_i delay is added to account for the sonic transit time, and an RC delay is added to this:

$$\tau_{ckt} = \tau_i + \frac{1}{2}(1 + FI)R_{on}C_{node}, \tag{10.26}$$

where the factor of ½ accounts for the fact that only half the circuit has a series fan-in. The values of τ_i are from inverter simulations. The capacitance of the PET is not a simple electric capacitance but it includes the electromechanical component (see section on PET model) and this depends on the mechanical load. Capacitance was extracted from the simulations by measuring the input charge as a function of voltage.

The PET parameters used were aggressive, meant to represent the technology at the ~11 nm node, thus the PE and PR were 24/8 nm in size, respectively. The key enabler of the PET technology is the low resistivity achievable from the PR material at high pressures. Here we assumed

$$\rho_{PR}(\Omega\text{-cm}) = 1.65 \, e^{-5.4P(GPa)} + 3.3 \times 10^{-4}, \tag{10.27}$$

based on Jayaraman's data for SmSe [12]. Thus low-output resistances may be achieved in very small area structures resulting in small fan-out delay components. The relevant parameters were: PE height, $L = 32$ nm, PR height, $l = 2$ nm, PE area, $A = 24 \times 24$ nm², PR area, $a = 8 \times 8$ nm², PE Young's modulus, $Y_{PE} = 8.4$ GPa, PR Young's modulus, $Y_{PR} = 110$ GPa, PE piezoelectric coefficient, $d_{33} = 2.8$ nm/V, PE electromechanical coupling coefficient, $g = 10.9$, and PE density $\rho = 8060$ kg/m³. At these dimensions, inverters were operable over a 0.19–0.25V range with delays of 8.2 to 6.0 ps. A dynamic PET input capacitance, $C_{eff} = 0.16$ fF was calculated. It is notable that this capacitance is in the same range as the CMOS active gate capacitance in spite of the much higher (~50×) effective dielectric constant of the PE. This is because the PE thickness is larger (16×) and the high-performance FETs have larger gate area.

For the CNT active capacitances based on 1-DEG electron gas were used with currents based on best experiments to date [31, 32]. Gate capacitance as well as parasitic capacitances to drain and source were included. For the TFET, currents and capacitances based on a TFET model [33] assuming InAs /GaSb materials was used.

Results of the comparison are shown in Fig. 10.14. It is seen that the PET has a solid advantage in low power at high speed in spite of the slow sound velocity. The high speed is attributable to the low on-resistance of the PET, enabling small area and small parasitic capacitance structures to be made with good current carrying capability. The tunnel FET is a potential competitor in the low-power regime; however, low-voltage

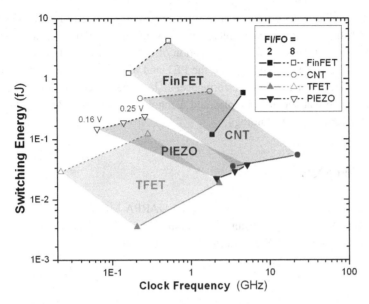

Fig. 10.14 Comparison of FinFET, CNTFET, and TFET with the PET for FI/FO = 2/2 and 8/8.

tunnel FETs have yet to be demonstrated [5] even though intense experimental effort is underway. The PET would need a scaled up experimental program as well in order to confirm these exciting predictions.

10.7   Discussion

The piezoelectronics technology investigation has to be evaluated against other efforts worldwide to explore alternatives to CMOS, capable of working at lower voltages than CMOS. Many of these technologies such as carbon nanotubes [31, 32], graphene, and two-dimensional dichalcogenides [34] are new embodiments of the FET, and will inherit the FET's voltage scaling difficulties; others such as the bilayer correlated tunneling transistor (BiSFET) [35] rely on unproved physical principles. Indeed even the low-voltage capability of the more conventional tunnel FET [5] is largely unproven. In contrast to this the PET relies on physical properties that are well-established in both theory and experiment. The main challenge is an engineering one of integrating the disparate PE and PR components into a single, strongly coupled, electromechanical system. The reliance on brute-force mechanical pressure permits extreme modulation of the resistivity of the PR to a degree not achievable in FETs, thus permitting extreme scaling of the PET size while maintaining high-current drive capability.

The technological challenges in implementing the PET are not to be discounted, yet theoretical issues may determine the eventual scope of the PET technology since only in the extreme scaled-down regime does the PET offer a large advantage over CMOS. Thus it is crucial to understand the physics of the PR at sizes of a few nm, where interface effects become dominant. It would be exciting if the onset of the pressure-induced metallic state produced a true low-resistance metal-to-metal interface. The physics of the PE is even more interesting, raising the question of how small can one make the PE without significant degradation of the piezoelectric properties.

In this chapter we have introduced the PET and offered a design guide and methodology for determining circuit performance. It is to be hoped that this will stimulate research into fundamental principles and technological solutions leading to the emergence of a viable PET technology.

Acknowledgments

This research was partially supported by the DARPA MESO Program (Mesodynamic Architectures) under contract number N66001–11-C-4109. We also want to acknowledge Thomas Shaw and Xiao Liu, for useful discussions.

References

[1] T. N. Theis & P. M. Solomon, "In quest of the 'next switch': prospects for greatly reduced power dissipation in a successor to the silicon field effect transistor." *Proceedings of the IEEE*, **87**(12), 2005–2014 (2010).

[2] T. N. Theis, "In quest of a fast, low-voltage digital switch." *ECS Transactions*, **45**(6), 3–11 (2012).

[3] W. Haensch, F. J. Nowak, R. H. Dennard, & P. M. Solomon *et al.*, "Silicon CMOS devices beyond scaling." *IBM Journal of Research and Development*, **50**(4/5), 339–358 (2006).

[4] H. Iwai, "Roadmap for 22 nm and beyond." *Microelectronic Engineering*, **86**, 1520–1528 (2009).

[5] A. C. Seabaugh, "Low-voltage tunnel transistors for beyond CMOS logic." *Proceedings of the IEEE*, **98**(12), 2095–2110 (2010).

[6] X. L. Feng, M. H. Matheny, C. A. Zorman, M. Mehregany, & M. L. Roukes, "Low voltage nanoelectromechanical switches based on silicon carbide nanowires." *Nano Letters*, **10**(8), 2891–2896 (2010).

[7] K. Akarvardar & H.-S. Wong, "Nanoelectromechanical logic and memory devices." *ECS Transactions* **19**(1), 49–59 (2009).

[8] T.-J. K. Liu, E. Alon, V. Stojanovic, & D. Markovic, "The relay reborn." *IEEE Spectrum*, April (2012).

[9] D. Newns, B. Elmegreen, X.-H. Liu, & G. Martyna, "A low-voltage high-speed electronic switch based on piezoelectric transduction." *Journal of Applied Physics*, **111**, 084509, (2012).

[10] D. M. Newns, B. G. Elmegreen, X.-H. Liu, & G. J. Martyna, "The piezoelectronic transistor: a nanoactuator-based post-CMOS digital switch with high speed and low power." *MRS Bulletin*, **37**, 1071–1076 (2012).

[11] D. M. Newns, B. G. Elmegreen, X.-H. Liu, & G. J. Martyna, "High response piezoelectric and piezoresistive materials for fast, low voltage switching: simulation and theory of transduction physics at the nanometer-scale." *Advanced Materials*, **24**(27), 3672–3677 (2012).

[12] A. Jayaraman, V. Narayanamurti, E. Bucher, & R. G. Maines, "Continuous and discontinuous semiconductor-metal transition in samarium monochalcogenides under pressure." *Physics Review Letters*, **25**(20), 1430–1433 (1970).

[13] P. Helnwein, "Some remarks on the compressed matrix representation of symmetric second-order and fourth-order tensors." *Computer Methods in Applied Mechanics and Engineering*, **190**(22–23), 2753–2770 (2001).

[14] S. Zhang & F. Li, "High performance ferroelectric relaxor-PbTiO₃ single crystals: Status and perspective." *Journal of Applied Physics*, **111**, 031301 (2012).

[15] H. Fu & R. E. Cohen, "Polarization rotation mechanism for ultrahigh electromechanical response in single-crystal piezoelectrics." *Nature*, **403**, 281–283 (2000).

[16] F. Bai, J. Li, & D. Viehland, "Domain engineered states over various length scales in (001)-oriented Pb(Mg$_{1/3}$Nb$_{2/3}$)O 3 -x%PbTiO₃ crystals: electrical history dependence of hierarchal domains." *Journal of Applied Physics*, **97**(5), 054103 (2005).

[17] V. A. Shuvaeva, A. M. Glazer, & D. Zekria, "The macroscopic symmetry of Pb(Mg$_{1/3}$Nb$_{2/3}$)$_{1-x}$Ti$_x$O₃ in the morphotropic phase boundary region ($x = 0.25$–0.5)." *Journal of Physics: Condensed Matter*, **17**, 5709–5723 (2005).

[18] D. B. Mc Whan & J. B. Remelka, "Metal-insulator transition in metaloxides." *Physics Reviews B*, **2**(9), 3734–3750 (1970).

[19] A. Jayaraman, V. Narayanamurti, E. Bucher, & R. G. Maines, "Pressure-induced metal-semiconductor transition and 4f electron delocalization in SmTe." *Physics Review Letters*, **25**(6), 368–370 (1970).

[20] A. Jayaraman & R. G. Maines, "Study of valence transitions in Eu-, Yb-, and Ca-substituted SmS under high pressure and some comments on other substitutions." *Physics Reviews B*, **19**(8), 4154–4161 (1979).

[21] A. Jayaraman, A. K. Singh, A. Chatterjee, & S. U. Devi, "Pressure-volume relationship and pressure-induced electronic and structural transformations in Ku and Yb monochalcogenides." *Physics Reviews B*, **9**(6), 2513–2520 (1974).

[22] Z. Jiang, M. A. Kuroda, Y. Tan *et al.*, "Electron transport in nano-scaled piezoelectronic devices." *Applied Physics Letters*, **102**(19), 193501 (2013).

[23] D. C. Gupta & S. Kulshrestha, "Pressure induced magnetic, electronic and mechanical properties of SmX (X = Se, Te)." *Journal of Physics: Condensed Matter*, **21**, 436011 (2009).

[24] H. F. Tiersten, *Linear Piezoelectric Plate Vibrations; Elements of the Linear Theory of Piezoelectricity and the Vibrations of Piezoelectric Plates* (New York: Springer, 1995).

[25] C. M. Close, D. K. Frederick, & J. C. Newell, *Modeling and Analysis of Dynamic Systems*, 3rd edn. (Chichester: Wiley, 2001).

[26] F. E. Terman, *Radio Engineers Handbook* (New York: McGraw Hill, 1943).

[27] M. Copel, M. A. Kuroda, M. S. Gordon, & X.-H. Liu *et al.* "Giant piezoresistive on/off ratios in rare-earth chalcogenide thin films enabling nanomechanical switching." *Nano Letters*, **13**(10), 4650–4653 (2013).

[28] J. H. Park, F. Xu, & S. Trolier-McKinstry, "Dielectric and piezoelectric properties of sol–gel derived lead magnesium niobium titanate films with different textures." *Journal of Applied Physics*, **89**(1), 568–574 (2001).

[29] R. Keech, S. Shetty, & M. A. Kuroda, "Lateral scaling of Pb(Mg$_{1/3}$Nb$_{2/3}$)O₃−PbTiO₃ thin films for piezoelectric logic applications." *Journal of Applied Physics*, **115**, 234106 (2014).

[30] D. E. Nikonov & I. A. Young, "Uniform methodology for benchmarking beyond-CMOS logic devices." In *Electron Devices Meeting (IEDM), 2012 IEEE International)*, pp. 10–13 (2012).

[31] A. D. Franklin, M. Luisier, S-J. Han *et al.*, "Sub-10 nm carbon nanotube transistor." *Nano Letters*, **12**(2), 758–762 (2012).

[32] A. D. Franklin, S. O. Koswatta, D. B. Farmer *et al.*, "Carbon nanotube complementary wrap-gate transistors." *Nano Letters*, **13**(6), 2490–2495 (2013).

[33] P. M. Solomon, D. J. Frank, & S. O. Koswatta, "Compact model and performance estimation for tunneling nanowire FET." In *Proceedings of the 69th Annual Device Research Conference*, pp. 197–198 (2011).

[34] V. Podzorov, M. E. Gershenson, Ch. Kloc, R. Zeis, & E. Bucher "High mobility field-effect transistors based on transition metal dichalcogenides." *Applied Physics Letters*, **84**(17), 3301–3303 (2004).

[35] S. K. Banerjee, L. F. Register, E. Tutuc, D. Reddy, & A. H. MacDonald, "Bilayer pseudo-spin field-effect transistor (BiSFET): a proposed new logic device." *IEEE Electron. Devices Letters*, **30**(2), 158–60 (2009).

11 Mechanical switches

Rhesa Nathanael and Tsu-Jae King Liu

11.1 Introduction

Mechanical switches have recently emerged as a promising alternative to CMOS for ultra-low power digital integrated circuits. The use of mechanical switches is far from a new concept. Early computing machines were built utilizing mechanical mechanisms (involving actuation and physical contact). The electromechanical relay was first built by Joseph Henry in 1835 using an electromagnet (induction coil) for actuation, and achieved superior performance than purely mechanical switches. In 1936, Konrad Zuse, a German engineer, built the first floating point binary mechanical calculator, the Z1, using electromechanical relays. Zuse then completed the first programmable, fully automatic digital computer, the Z3, in 1941 using 2000 relays. Separately, George Stibitz at Bell Labs used relays to built the "Model K" binary calculator in 1937 and a complex number calculator in 1940. Zuse and Stibitz are often credited as the inventors of computers. Through the twentieth century, computing machines were built using electromechanical devices. However, they were huge in size, slow, and very expensive to build. The era of electronic computing began with vacuum tubes, and took off in the 1950s after the inventions of transistors and integrated circuits. Today, relays have been virtually abandoned and digital computing devices utilize solid-state electronic switches (i.e., CMOS transistors). The ability to miniaturize and integrate billions of transistors on a single chip of silicon the size of a coin has enabled dramatic increases in functionality and performance, as well as reductions in cost [1]. As transistor scaling continues to the nanometer scale, however, CMOS technology faces a power crisis as it approaches a fundamental energy efficiency limit. Recall that the total energy dissipated in a CMOS digital circuit consists of two components: dynamic energy (E_{dynamic}) from charging and discharging capacitors and leakage energy (E_{leakage}) caused by transistor off-state leakage current (I_{off}),

$$E_{\text{total}} = E_{\text{dynamic}} + E_{\text{leakage}}. \tag{11.1}$$

Due to non-scalability of the thermal voltage (i.e., 60 mV/dec sub-threshold swing limit at room temperature), off-state leakage imposes a minimum energy limit. Overcoming this limit requires a device that has a more ideal switching behavior (i.e., abrupt on/off transition and zero off-state leakage). A relay is, in fact, such a device because switching is based on making and breaking physical contact. By leveraging advancements in planar processing and micro-machining technology over the past 40 or more years,

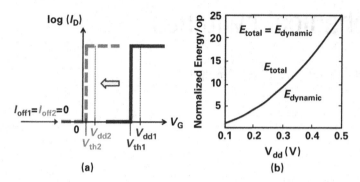

Fig. 11.1 Energy efficiency illustration for a mechanical switch. (a) I_D–V_G plot for a mechanical switch, showing the effect of lowering V_{dd} and V_{th} and its (b) energy per operation implications.

electromechanical relays can now be miniaturized to reap the functionality and cost benefits of monolithic integration as well, thus overcoming their historical disadvantages. A shift back to the computing device of old, the relay, suddenly looks compelling!

Figure 11.1(a) shows typical I_D–V_G behavior for mechanical switches, which operate based on making and breaking physical contact. There is no modulation of a potential barrier. Thus, the I–V characteristic is extremely abrupt (a step-like function). In the off-state, an air gap separates the electrodes, so that off-state leakage is effectively zero ($I_{off} = 0$). Therefore, the leakage energy term drops out completely:

$$E_{leakage} = 0. \tag{11.2}$$

As a result, the total energy dissipated in a mechanical digital circuit consists only of dynamic energy for charging and discharging capacitances (Fig. 11.1(b)):

$$E_{total} = E_{dynamic} = \alpha L_D f C V_{dd}^2. \tag{11.3}$$

Without a lower bound imposed by leakage, the energy per operation for a mechanical logic technology could be reduced to be less than the minimum energy per operation for CMOS technology, by simply scaling V_{dd}. This is because, in theory, V_{dd} and V_{th} can be lowered to be close to 0 V for an ideal switch, as shown in Fig. 11.1(a).

11.2   Relay structure and operation

11.2.1   Electrostatically actuated relays

Microelectromechanical relays that utilize electrostatic actuation are attractive for IC applications because they are relatively easy to manufacture using conventional planar processing techniques and materials, do not consume substantial active power, and are more scalable compared to those actuated by magnetic, piezoelectric, and thermal means. In an electrostatic switch, a movable electrode and a fixed electrode form a capacitor. When a voltage difference is applied to this capacitor, electrostatic force (due to attraction between oppositely charged electrodes) accelerates the movable electrode

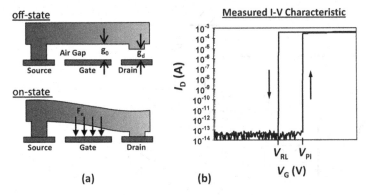

Fig. 11.2 (a) Schematic illustrations of a generic 3-terminal relay structure in the off-state and the on-state. (b) A typical I_D–V_G characteristic, showing the pull-in (V_{PI}) and release (V_{RL}) voltages.

toward the fixed electrode. This force is always attractive regardless of the polarity of the applied voltage (i.e., it is ambipolar), and its strength depends on the area of the electrodes and the separation between them (the actuation gap, g_0). When the voltage is removed, the spring restoring force returns the movable electrode to its original position.

A schematic of a basic 3-terminal (3T) electrostatic relay and a corresponding typical I_D–V_G characteristic are shown in Fig. 11.2. In the off-state, the source and drain are physically separated by an air gap, so that no current can flow. When a voltage (i.e., electrostatic force) is applied between the gate and source, the structure is actuated downward. When the source and drain are in contact, current can flow from drain to source and hence the relay is on. A typical I_D–V_G curve of an electrostatic relay is shown in Fig. 11.2(b). A relay turns on when the gate voltage exceeds the pull-in voltage (V_{PI}) and turns off when the gate voltage is lowered below the release voltage (V_{RL}). The switching behavior is abrupt and the off-state leakage is zero.

An electrostatically actuated beam can be modeled as a parallel plate capacitor on a mass–spring–damper system (Fig. 11.3). The motion dynamics of the movable electrode follows Newton's second law of motion,

$$m\ddot{z} + b\dot{z} + kz = F_{elec}(z),\qquad(11.4)$$

where z is the displacement of the movable electrode, m is the mass of the movable electrode, b is the damping factor, and k is the spring constant. F_{elec} is the electrostatic force when a voltage, V, is applied between the two electrodes, and is a function of the displacement, z. A full solution of the non-linear differential equation needs to be solved numerically.

For the purpose of gaining a fundamental understanding of the operation of an electrostatic actuator, we can model the structure as a simple parallel plate capacitor [2]. The electrostatic force is

$$F_{elec}(z) = \frac{\varepsilon_0(WL)V^2}{2(g_0 - z)^2} = \frac{\varepsilon_0(WL)V^2}{2g^2},\qquad(11.5)$$

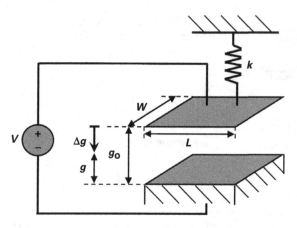

Fig. 11.3 Model of an electrostatically actuated beam as a dynamic parallel-plate capacitor and mass–spring–damper system.

where g is the actuation gap thickness, g_0 is the as-fabricated actuation gap thickness, W is the width of the actuation area, and L is the length of the actuation area.

The opposing force is the spring restoring force,

$$F_{\text{spring}}(z) = kz = k(g_0 - g).$$ (11.6)

At equilibrium, these two forces must balance,

$$\frac{\varepsilon_0(WL)V^2}{2g^2} = k(g_0 - g).$$ (11.7)

Note that F_{elec} increases superlinearly with displacement, while F_{spring} increases linearly with displacement. Therefore, there exists a point beyond which F_{elec} is always greater than F_{spring}, and the system becomes unstable. By stability analysis of Eq. (11.7), we can show that this critical displacement occurs when the electrode has moved by $\frac{1}{3} g_0$ (i.e., $g = \frac{2}{3} g_0$). At this critical point, the gap closes abruptly as the applied voltage is increased. This phenomenon is referred to as "pull-in" and the voltage at which it occurs is the pull-in voltage, V_{PI}, given by

$$V_{\text{PI}} = \sqrt{\frac{8kg_0^3}{27\varepsilon_0(WL)}}.$$ (11.8)

Note that the relay design in Fig. 11.2(a) employs a dimple at the contacting region to precisely define the apparent contact area. As a result, a smaller air gap exists at the dimple region (g_d) than the actuation region (g_0). Note that if $g_d < \frac{1}{3} g_0$, the pull-in phenomenon will not occur. The thicknesses of these two gaps can be defined independently. Thus having a dimple allows the relay's mode of operation (pull-in or non-pull-in) to be defined by adjusting the ratio of the two gap thicknesses.

For a more general (arbitrary) relay design, the turn-on voltage in pull-in mode can be expressed as

$$V_{PI} = \sqrt{\frac{8k_{eff}\,g_0{}^3}{27\varepsilon_0 A_{eff}}}, \qquad \text{when} \quad g_d \geq \frac{1}{3}g_0, \qquad (11.9)$$

where k_{eff} is the effective spring constant of the suspension beam(s) and A_{eff} is the effective overlap area between the movable and fixed electrodes.

For a relay designed to operate in non-pull-in mode, contact occurs when $g = g_0 - g_d$. The turn-on voltage can be expressed as

$$V_{PI} = \sqrt{\frac{2k_{eff}\,g_d(g_0 - g_d)^2}{\varepsilon_0 A_{eff}}}, \qquad \text{when} \quad g_d < \frac{1}{3}g_0. \qquad (11.10)$$

Recall that the release (turn-off) voltage (V_{RL}) is lower than V_{PI}. This hysteretic switching behavior is due to pull-in mode operation and surface adhesive force. A relay designed to operate in pull-in mode will snap down once the displacement reaches $\frac{1}{3}g_0$. Once pulled-in, the gap is effectively reduced to $g_0 - g_d$, so the electrostatic force is larger at the same voltage (V_{PI}). Thus, turning off the device requires the voltage to be lowered below V_{PI}. The spring restoring force must overcome both the electrostatic and surface adhesive forces that exist when contact is made, in order to turn off the relay. In non-pull-in mode, only the surface adhesive force causes the hysteretic switching behavior, so the hysteresis voltage is expected to be smaller.

Once contact is made, the force balance equation is as follows:

$$\frac{\varepsilon_0(WL)V^2}{2g^2} + F_A = k(g_0 - g), \qquad (11.11)$$

where F_A is the surface adhesive force.

At the release operating point, $g = g_0 - g_d$. The release voltage is found to be

$$V_{RL} = \sqrt{\frac{2(k_{eff}g_d - F_A)(g_0 - g_d)^2}{\varepsilon_0 A_{eff}}}. \qquad (11.12)$$

It should be noted that the hysteresis voltage sets the lower limit for relay voltage scaling. Because of its abrupt switching behavior, a relay can in principle be made to operate with V_{RL} close to zero, so that V_{dd} can be reduced to be as low as the hysteresis voltage.

11.2.2 Relay designs

The most common electrostatic relay design reported in literature has been the 3-terminal (3T) relay design (Fig. 11.4(a) and (b)) [3–11], but this is not necessarily the best design for digital logic applications. Similar to a MOSFET, the 3T relay is actuated by the voltage difference between the gate and source terminals (V_{GS}). When V_{GS} is increased to be greater than or equal to the pull-in voltage (V_{PI}), the relay turns on and current flows from drain to source terminals (I_{DS}). Control (V_{GS}) and signal (V_{DS}) are partially decoupled, so that despite V_{PI} being typically large, I_{DS} can remain small for high reliability. From a processing standpoint, this allows the electrode and structural materials to be optimized separately (the electrode material for good contact

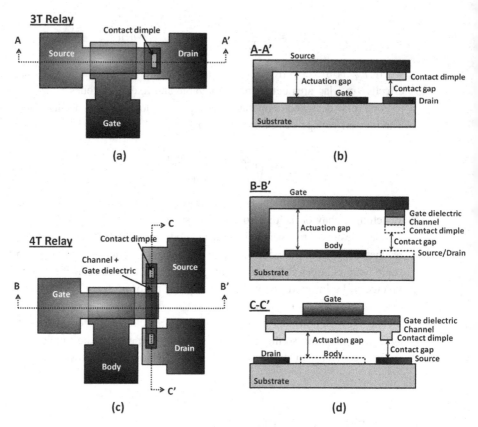

Fig. 11.4 (a) 3-terminal (3T) relay structure top view and (b) cross-sectional view along A–A′. (c) 4-terminal relay (4T) top view and (d) cross-sectional view along B–B′ and C–C′. When shown in dotted lines, structures are out-of-plane from the cross-sectional cuts.

reliability and sufficiently low on-resistance (R_{on}); the structural material for low-voltage operation). The main disadvantage of the 3T relay design is seen from a circuit perspective. When 3T relays are connected in series (examples include a conventional NAND or NOR gate), their source voltages (V_S) are not fixed. Therefore, their gate switching voltages (V_G) can vary, resulting in unreliable circuit behavior.

The 4T relay design (Fig. 11.4(c) and (d)) addresses the shortcoming of the 3T relay design with the addition of the body terminal and gate insulating layer. In this design, the actuated structure is the gate, whose position is controlled by the voltage applied between the gate and the body (V_{GB}), as opposed to the gate and source (V_{GS}) in the 3T design. Thus, the gate switching voltage can be fixed and independent of the source voltage (control and signal are fully decoupled). A metallic channel attached to the underside of the gate serves as a bridge connecting the source and drain when the relay is in the on-state. A layer of gate dielectric insulates the gate and the channel so that current flows only from the drain to the source during the on-state, not to the gate. Apart from fixing the gate switching voltages, the 4T relay allows the gate switching voltage to be tuned (post-process) by applying a body bias voltage, which can be leveraged to achieve low-voltage operation. Since electrostatic force is ambipolar, the same 4T relay structure can be operated either as a pull-down device or as a pull-up device, mimicking n-MOSFET or p-MOSFET operation with the body biased at 0V or V_{dd} respectively. Thus, 4T relays are attractive to implement low-voltage complementary logic circuits.

Another relay design consideration is the direction of actuation. While this chapter mainly focuses on relays actuated vertically (out-of-plane) from the substrate, relays actuated laterally (in-plane) with respect to the substrate have also been demonstrated [6–11]. In vertically actuated relays, tuning V_{PI} requires modifications in process (sacrificial layer thickness or structural thickness) or device footprint (actuation area). In laterally actuated relays, the actuation gap is defined by lithography, instead of the deposited sacrificial layer thickness. As a result, it is easy to adjust V_{PI} via layout without changing the device footprint by varying the width of the actuation gap. Laterally actuated relays have the benefit of simpler processing with fewer lithographic steps. The production of nanoscale gaps, however, will be limited by lithography. While it is simple to fabricate lateral 3T relays, the fabrication of laterally actuated 4T relays is not straightforward [10].

Relays with cantilever [3, 4, 6–11] and clamped–clamped beam [5, 7] designs have been demonstrated in the literature. Poor reliability due to stiction and welding at the contacts remains the key issue. Cantilever designs tend to be prone to stiction, not having enough spring restoring force to turn off reliably. While clamped–clamped beam design has the necessary restoring force, it is susceptible to buckling upon release due to compressive residual stress. Figure 11.5 shows a robust 4T relay design proposed in [12], with a representative I_D–V_G characteristic. It comprises a movable plate, the gate electrode, suspended at its corners by four folded-flexure suspension beams. Having four symmetric flexures gives extra stability and prevents rotation of the plate due to residual stress. The folded-flexure design relieves residual stress to prevent buckling, while still taking advantage of the higher spring restoring force of a clamped–clamped design for reliable turn-off. The folded-flexure also helps to mitigate any effect of thermal stress and is more robust to thermal variations. The actuation area is defined

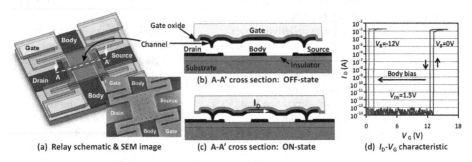

(a) Relay schematic & SEM image   (b) A-A' cross section: OFF-state   (c) A-A' cross section: ON-state   (d) I_D-V_G characteristic

Fig. 11.5 (a) Schematic illustrations of the 4-terminal-relay structure and plan-view SEM image. (b) Cross-sectional view along the channel (A–A′) in the off-state. (c) Cross-sectional view along the channel (A–A′) in the on-state. (d) Measured relay I_D–V_G characteristics with <1 V hysteresis. With body biasing, V_{PI} can be substantially reduced to <1 V with the relay still able to turn off reliably.

by the plate in the center. Flexure dimensions (i.e., the spring restoring force), is decoupled and can be optimized separately from the actuation area (i.e., electrostatic force). For structural strength, the plate is square while the conducting channel is attached via an insulating dielectric to the underneath the plate. This design is demonstrated to be structurally robust, switching well over 10^9 cycles without structural failure.

11.3   Relay process technology

11.3.1   Materials selection and process integration challenges

Developing an integrated device fabrication process involves co-optimization to meet the requirements of process integration and the application. A 4T relay structure consists of four critical elements, namely sacrificial, contact electrode, insulating dielectric, and structural materials. An optional coating layer could be applied to the contact to improve reliability. Each of these has its individual requirements, and must also be process-compatible with the other materials (Table 11.1).

While relays are capable of fully replacing CMOS transistors, monolithic integration with CMOS on a single chip allows the advantages of each technology to be leveraged to realize hybrid systems with enhanced performance and functionality [13]. Both "MEMS-first" (MEMS fabrication done before CMOS process steps) and "MEMS-last" (MEMS fabrication done after CMOS process steps) approaches have been proposed. The "MEMS-first" approach requires a customized CMOS process. On the other hand, a "MEMS-last" process allows a standard foundry CMOS process to be used, and also allows for vertical stacking of MEMS on top of electronics, which saves

Table 11.1 Material selection consideration for a 4T relay for digital IC applications.

Lawyer	Process integration requirements	Application specific requirements
Sacrificial	Dry release process Low-temperature deposition ($<425°$C)	Highly uniform, continuous film Well-controlled deposition (nm range precision)
Contact electrode	Resistant to release etch Good adhesion to underlying layer Low-temperature deposition ($<425°$C)	Highly conductive Sufficiently low contact resistance (<10 kΩ) Resistant to wear and plastic deformation Highly uniform, continuous film
Insulating dielectric	Resistant to release etch Well-controlled deposition Low-temperature deposition ($<425°$C)	High electrical breakdown voltage Low leakage current Low residual stress and strain gradient
Structural	Resistant to release etch Low-temperature deposition ($<425°$C)	Electrically conductive Large Young's modulus Minimal residual stress (<100 MPa tensile) Minimal strain gradient ($<1 \times 10^{-4}$ μm^{-1}) Robust against fracture and fatigue
Contract surface coating	Resistant to release etch Ultra-thin (<1 nm) and conformal deposition Low-temperature deposition ($<425°$C)	Low potential barrier to conduction Low surface adhesion forces

area and reduces parasitic interconnect resistance and capacitance for better performance. However, it imposes a thermal budget constraint for MEMS fabrication (425°C for 6 hours for a foundry 0.25 μm CMOS technology) [14]. In this section, materials selection consideration and process integration challenges are discussed, assuming a "MEMS-last" CMOS compatibility requirement. Subsequently, a robust 4T relay process flow is presented.

11.3.1.1 Sacrificial

In selecting a set of relay materials, the sacrificial material is the first to be considered because it determines the etch chemistry that will be used to release the relay at the very end of the process, thus setting a material constraint to every layer in the relay structure exposed to it. From an operational standpoint, the sacrificial layers determine the actuation and contact gap thicknesses, which in turn determine the switching voltages of the relay. It is critical that the gap thickness can be well-controlled and repeatable during processing, and is uniform across the wafer to ensure a uniform operating voltage. The sacrificial layers must also be continuous and without pinholes, which could cause shorts between electrodes. As relay dimensions are scaled down for improved device density and performance, the ability to controllably deposit thin sacrificial films in the nanometer range will also become critical.

Releasing nanoscale gaps has proved to be quite a challenge. MEMS structures released in a typical wet chemical etch process are prone to stiction, caused primarily by surface adhesive forces. These include capillary force, van der Waals force, Casimir force, electrostatic force, and hydrogen-bond force [15–17]. Of these forces, capillary force is typically dominant during release. Surface tension from the liquid etchant can cause adhesion forces stronger than the spring restoring force of the beam, causing the structure to get stuck down. Susceptibility to this type of stiction is determined by surface property and the gap at the contact. The contact angle (θ_C) is a material property that determines the wettability of a surface, and thus the susceptibility of that surface to stiction due to the capillary effect. When $\theta_C < 90°$, the surface is hydrophilic, while $\theta_C < 90°$ indicates a hydrophobic surface. The surface interaction energy due to capillary forces (E_{cap}) is given by [16]

$$E_{cap} = 2\gamma_1 \cos\theta_C, \qquad \text{for} \quad d \leq d_{cap}, \qquad (11.13)$$

$$E_{cap} = 0, \qquad \text{for} \quad d > d_{cap}, \qquad (11.14)$$

where γ_1 is the surface tension of the liquid (usually water) and θ_C is the contact angle of water on the surface.

Capillary condensation of water occurs when the gap between the solid plates (d) is smaller than the characteristic distance for capillary condensation (d_{cap}) given by

$$d_{cap} = \frac{2\gamma_1 v \cos\theta_C}{RT \log(RH)}, \qquad (11.15)$$

where v is the liquid molar volume, R is the universal gas constant, T is the absolute temperature, and RH is relative humidity.

Solutions such as critical-point drying following a wet etch to eliminate surface tension have been proposed [18] but add to process complexity and do not solve the problem completely. A dry release process is therefore desirable because it eliminates the presence of fluid during the etch process and thereby mitigates stiction. Examples of dry release process are vapor phase HF etch for SiO_2 and XeF_2 etch for Si. The SiO_2/HF vapor combination is particularly attractive as a sacrificial/release etch because SiO_2 can be deposited via low-pressure chemical vapor deposition (LPCVD) at 400°C with good uniformity and conformality, and it allows conventional polycrystalline Si or SiGe to be used as structural material.

11.3.1.2 Contact electrode

The electrode material makes up the source, drain, channel, and body electrodes. Materials that are highly conductive provide good performance, but are prone to reliability issues. This trade-off should be carefully optimized based on the application.

From a performance perspective, the contact determines relay on-resistance (R_{on}). For an optimal digital IC design using relays, R_{on} can be 10–100 kΩ to guarantee that the electrical charging delay will be much smaller than the mechanical switching delay (100 ns), for typical load capacitances of 10–100 fF [19]. In a 4T relay, R_{on} consists of several components (Fig. 11.7):

$$R_{on} = R_{source} + R_{drain} + R_{channel} + 2R_{contact}, \qquad (11.16)$$

where R_{source} is the resistance of the source electrode, R_{drain} is the resistance of the drain electrode, $R_{channel}$ is the resistance of the channel electrode, and $R_{contact}$ is the contact

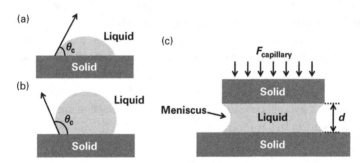

Fig. 11.6 (a) The solid–liquid interface for a hydrophilic and (b) a hydrophobic surface. The contact angle (θ_C) determines wettability of the surface. (c) Illustration of capillary force when liquid is present between two solid plates.

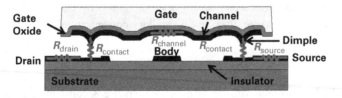

Fig. 11.7 On-resistance components of a 4-terminal relay design.

resistance between the channel and source or drain. Out of these components, the contact resistance is dominant, so that $R_{on} \cong 2R_{contact}$. Therefore, carefully examining contact properties is critical.

A resistance model for an electromechanical contact has been developed by R. Holm [20]:

$$R_{contact} = \frac{4\rho\lambda}{3A_r}, \qquad (11.17)$$

where ρ is the resistivity of the material, λ is the electron mean free path in the material, and A_r is the effective contact area.

The effective contact area is given by

$$A_r \approx \frac{F_{elec}}{\xi H}, \qquad (11.18)$$

where F_{elec} is the electrostatic force that makes the contact (which gives a measure of the loading force), ξ is the deformation coefficient, and H is the hardness of the material.

The contacting surface is always rough [21], so that physical contact is only made at local asperities [22] as illustrated in Fig. 11.8. The real contact area is only a fraction of the apparent contact area and is a function of the applied load and the material hardness. The applied load should be kept low enough to keep the material in its elastic domain and avoid plastic deformation.

From a contact failure perspective, reliability issues include: (1) stiction due to microwelding and surface adhesive forces; and (2) wear and plastic deformation. While stiction during release is dominated by surface adhesion forces, stiction during operation is primarily caused by microwelding due to Joule heating at the contacts when a high level of current flows. When the field of MEMS emerged more than 20 years ago, relays were among the first devices to be investigated. These were built mainly for RF switching applications, which require very low on-resistance to minimize insertion loss. As a result, RF switches are typically built with soft metals such as gold ($H = 0.2$–0.7 GPa), which can achieve very low contact resistance (<1 Ω) [23]. Due to high current

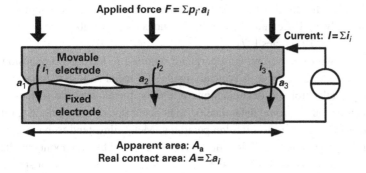

Fig. 11.8 Cross-sectional schematic of an EM contact, showing three asperities where current can flow. Total load P and current I correspond to the sum of the contributions from every asperity.

levels, welding-induced failure is a major concern for this application. In contrast, low on-resistance is not a requirement for relay-based digital logic, where R_{on} can be as high as 10–100 kΩ. Logic relays should instead be designed for high reliability to ensure they meet the industry standard operating lifetime of 10 years. As a reasonable benchmark, in a relay-based microcontroller for embedded sensor applications operating at 100 MHz with 0.01 average transition probability, relays experience ~3 × 10^{14} on/off cycles over 10 years. A contact reliability model based on atomic diffusion was developed and experimentally verified to predict failure due to welding at the contact [24]. This model shows that hard refractory metals are more suitable for the contact electrode material, predicting endurance >10^{15} on/off cycles for relays operating at 1 V.

Metal films are highly conductive and can be deposited using physical vapor deposition (PVD), such as sputtering, which easily meets the thermal budget constraint. Tungsten (W), for example, is an excellent material to prevent contact failure from wear, plastic deformation, and welding-induced stiction. Relays with tungsten contacts have been shown experimentally to endure >10^9 cycles without failure [5, 12]. Tungsten has one of the highest hardnesses of all metals (Mohs hardness of ~7.5 and Vickers hardness ~3.43 GPa). Of all pure metals in the periodic table, tungsten has the highest melting point, 3422°C, for good resistance to welding-induced failure due to Joule heating at the contact. Intrinsically, tungsten has sufficiently low contact resistance (<1 kΩ) to comfortably meet the R_{on} requirement. However, hard materials are more prone to chemical reaction than noble metals (e.g., gold). Tungsten readily oxidizes in air to form an insulating native oxide (W_xO_y), which not only makes R_{on} unstable cycle-to-cycle but increases the contact resistance beyond an acceptable level over time. To improve the contact resistance stability, hermetic sealing technology [25] can be employed. An extensive study of R_{on} evolution with cycling on W contacts shows that this relay technology with W contacts can switch up to ~10^8 on/off cycles under vacuum (5 µTorr) before the contact resistance increases above an acceptable level for IC applications (taken as 10 kΩ for the study) [26]. Auger electron spectroscopy (AES) spectra measurements confirm that the increasing contact resistance is indeed caused by oxidation of the contacting surfaces. Ruthenium is another attractive contact material candidate because it forms a conductive oxide (RuO_2) in air, while still maintaining a relatively high hardness (Mohs hardness of ~6.5), albeit not as high as tungsten. A recent study of Ru contact relays has demonstrated a stable contact resistance of ~9 kΩ for >3 × 10^7 cycles [27].

Structures with released moving parts are required to have good adhesion between layers. The contact electrode is typically made of a metal sitting on top of a dielectric, and is especially prone to delamination. While the metal/dielectric surface interaction energy can be rather strong, delamination may occur due to stress [28, 29]; this issue is exacerbated by exposure to high temperatures. Metal films tend to have high residual stress and strain gradient [24]. Residual stress consists of two components: the intrinsic stress of the film that depends on the microstructure of the film (grain size, geometry, and crystal orientation), and thermal stress caused by a difference in the thermal expansion coefficients of the film and underlying substrate when deposited or annealed at high temperatures. The choice of contact electrode material will impose a strict process temperature limit. For example, the use of tungsten as an electrode material imposes a processing temperature limit of 550°C. Adhesion layers can also be employed to prevent delamination. Ruthenium, for example, has very poor adhesion to SiO_2 even at room temperature. A thin layer of TiO_2 can serve as an adhesion layer to solve delamination issues [27].

11.3.1.3 Insulating dielectric

In a 4T relay, a dielectric is needed in two places: (1) the insulating layer between the electrodes and the silicon substrate (substrate dielectric), and (2) insulating layer between the channel and the structural electrode (gate dielectric). In relays, the sole purpose of the dielectric layers is electrical insulation. Unlike a CMOS transistor, a relay does not depend on good capacitive coupling between the gate and the channel to operate with steep turn-on/off characteristics, so there is no need for very low equivalent gate oxide thickness. In fact, a low-permittivity dielectric material is more desirable to reduce capacitive loading. The most important criteria for dielectric would be a high electrical breakdown voltage to withstand the higher operating voltage often associated with MEMS devices, and minimal gate leakage current to maintain low static power dissipation and good reliability. Mechanical stress is a consideration because the dielectric is part of the structural stack and can affect the curvature (and hence the actuation and contact gaps) of the released structure. Finally, process integration and thermal budget requirements are considerations. It is crucial that the release etch does not degrade the dielectric integrity.

Al_2O_3 is an excellent dielectric material candidate with favorable electrical, mechanical, and chemical properties [30–32]. First, atomic level control and precision can be achieved via atomic layer deposition (ALD). ALD Al_2O_3 can be deposited at 300°C with growth rate of ~0.1 nm/cycle, which easily satisfies the thermal budget requirement. A 300 nm film deposited at 300°C has intrinsic tensile stress of ~200 MPa [30], which is within the acceptable range. A film deposited at 250°C shows leakage current density of $1\,\mu A/cm^2$ at 2 MV/cm, with a breakdown at 7.5 MV/cm [30], which gives a breakdown voltage of over 30 V for typical thicknesses (~40–50 nm). In a device, breakdown is expected to occur at a lower voltage than for a blanket film, due to a concentrated electric field at the corners formed as a result of topography. Finally, despite being attacked in an HF liquid [31], Al_2O_3 is known to be highly resistant to HF vapor [32], to ensure no degradation in quality during release.

11.3.1.4 Structural

The choice of structural material is an important one since it determines relay operating voltage and mechanical reliability. First, an electrically conductive material is required so that it can be electrically biased and actuated. A sufficiently large Young's modulus ensures that it will be stiff enough to turn off (overcome surface adhesive forces) and return to its original position when the bias is removed. For reliable operation over many cycles, a material robust against fracture and fatigue is desired. Stress-induced structural warping (out-of-plane deflection) due to stress causes variation in actuation voltage, reduced reliability, and failure. Minimizing residual stress (<100 MPa tensile) and strain gradient ($<1 \times 10^{-4}$ μm^{-1}) is the key challenge in structural material development. Residual stress represents the average stress in the film due to intrinsic stress related to the film microstructure and thermally induced stress from high-temperature treatments. Tensile films want to contract while compressive films want to expand to relieve the stress and reach equilibrium. The strain gradient represents changes in stress through the thickness of the beam (film growth direction). Following usual convention, positive (+) stress represents tensile stress while negative (–) represents compressive stress. A positive gradient indicates that the film is more tensile towards the top. In a cantilever structure after release, positive strain gradient results in upward bending because the bottom of the film expands more than the top of the film. A negative strain gradient causes downward bending since the top of the film expands more than the bottom of the film.

Polycrystalline silicon (poly-Si) is the conventional structural material used in surface-micromachined MEMS devices [33]. Its mechanical and stress properties have been well characterized for many years [34, 35]. It is a strong structural material with Young's modulus in the 120–180 GPa range and fracture strength between 1 and 3 GPa [36–38], and most importantly, a low residual stress (<25 MPa) with negligible strain gradient can be achieved [35]. However, poly-Si is typically deposited via LPCVD at temperatures $>600°C$. Furthermore, high-temperature ($>900°C$) annealing is typically required to achieve the desired low stress films [33, 35]. This imposes a severe constraint on the choice of contact electrode materials (typically metals) and does not meet the CMOS compatibility requirement. Metal films are potentially attractive due to their high electrical conductivity and low thermal process budget. However, a study investigating Al, Ni, Ti, and TiN show that sputtered metals have very high residual stress, ranging from hundreds of MPa to 1 GPa [39]. Al shows low tensile residual stress within the acceptable range (~10 MPa) but is hampered by a very large strain gradient (~3×10^{-3} μm^{-1}). Hence, metals will only be an attractive structural material if a low stress film can be achieved. Polycrystalline silicon germanium (poly-Si$_{1-x}$Ge$_x$, $x \leq 0.6$) has been recently studied as a low thermal budget structural material for MEMS [40]. Poly-SiGe retains the favorable mechanical properties of poly-Si, and can be deposited by LPCVD at temperatures much lower than poly-Si ($<450°C$). While it tends to have a larger strain gradient than poly-Si, it is still within an acceptable range when carefully optimized (~1×10^{-4} μm^{-1}) for films ~1 μm thick.

11.3.1.5   Contact surface coating

A surface coating layer can greatly improve reliability and stability of the contact. It is important that this layer does not degrade contact resistance significantly and has favorable low surface adhesion properties. The ability to deposit an ultra-thin layer (<1 nm) conformally is especially critical. An attractive candidate to coat tungsten contacts is TiO$_2$ [41], which can be deposited with atomic level control and precision through ALD at 275°C. Since ALD is extremely conformal, deposition can be performed at the very last step (post-release), making process integration simple. TiO$_2$ may help to alleviate both stiction during release and stiction during operation. First, TiO$_2$ helps make the contacting surface less hydrophilic ($\theta_{CTiO2} > 80°$ [42]) to reduce capillary forces. A pure tungsten contact readily forms native oxide (WO_3) at its surface in air ($\theta_{CWO3} < 10°$ [43, 44]). Additionally, TiO_2 improves contact stability by serving as an oxidation barrier to slow down the formation of tungsten native oxide. Unlike WO_x, TiO_2 forms a relatively low potential barrier (~0.8 eV) for electrons to flow from W to TiO_2 to keep contact resistance more stable within the acceptable range (<100 kΩ). In fact, the slight increase in contact resistance due to TiO_2 helps limit the amount of current flow compared to a pure tungsten contact, which is beneficial to mitigate microwelding.

11.3.2   Relay process flow

Figure 11.9 presents a robust 4-terminal relay process flow developed at UC Berkeley in 2009 [12], using the 4T relay design discussed in Fig. 11.5. Excellent repeatability and yield (>95%) are obtained. On a starting silicon wafer substrate, a layer of Al_2O_3 is deposited at 300°C by atomic layer deposition (ALD) to form an insulating substrate surface. Tungsten deposited by DC magnetron sputtering is used as the material for the source, drain, body, and channel electrodes. Sputtering easily meets the thermal budget constraint (<425°C), since the wafer is kept at room temperature. Silicon dioxide (SiO_2) deposited at low temperature (400°C) by low-pressure chemical vapor deposition

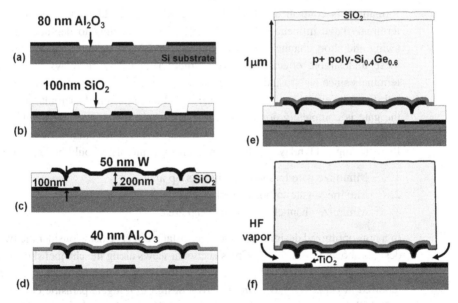

Fig. 11.9 Illustration of the four-mask process used to fabricate the first prototype four-terminal relays, shown along cross section A–A' of Fig. 11.5. (a) 80 nm Al_2O_3 layer deposited and 50 nm W layer deposited and patterned to form source, drain, and body electrodes. (b) First sacrificial SiO_2 layer (100 nm) deposited and source/drain contact regions defined. (c) Second sacrificial SiO_2 layer (100 nm) deposited and 50 nm W layer deposited and patterned to form channel. (d) 40 nm Al_2O_3 gate dielectric layer deposited. (e) 1 μm p$^+$ poly-$Si_{0.4}Ge_{0.6}$ deposited and patterned using LTO hard mask to form gate electrode. (f) Gate-stack released in HF vapor and coated with 3 Å TiO_2.

(LPCVD) was used as the sacrificial material. Note that two layers of sacrificial are used. The first sacrificial determines the thickness of the contact dimple. The second sacrificial determines the contact gap at the source/drain. Both sacrificial layers combined give the actuation gap. This scheme enables the g_d / g_0 ratio to be controlled to the determined mode of operation (pull-in vs. non pull-in). In this particular case, the contact gap is one-half the actuation gap for the most energy-efficient operation [4]. In-situ boron-doped polycrystalline silicon-germanium (poly-Si$_{0.4}$Ge$_{0.6}$) deposited at 410°C by LPCVD is used as a high-quality, low-thermal-budget structural material [8] for post-CMOS integration capability. Lastly, an ultra-thin (~3 Å thick) coating of titanium dioxide (TiO$_2$) is deposited at 275°C by ALD after the gate-stack structure has been released, to improve contact reliability. This process has been further optimized in subsequent publications to lower the operating voltages, scale down the relay dimensions, and improve contact reliability [45–46].

11.4   Relay design optimization for digital logic

11.4.1   Design for improved electrostatics

In an ideal transistor, the modulation of potential barrier at the channel (i.e., channel formation) is controlled entirely by the gate terminal. In reality, the drain and body terminals have influence on the channel as well, leading to degraded sub-threshold swing and short channel effects among others. CMOS design is optimized to mitigate these undesirable effects. Similarly, structural actuation in relays is affected by all terminals since electrostatic force exists between any two electrodes of different potential. For best electrostatic control, relays need to be designed such that the influence of the gate is completely dominant. The electrostatic force (F_{elec}) between two plates is proportional to the capacitance and therefore the overlap area of the plates ($F_{elec} \propto C \propto A$). In designing a 4T relay, the following general principles should be kept in mind:

1.   Minimize gate-to-source/drain overlap area.
2.   Maximize gate-to-body overlap area.
3.   Minimize channel-to-body overlap area.

In a non-optimized design, parasitic electrostatic effects exist. Consider the two 4T relay designs shown in Fig. 11.10. Cross-sectional views along the channel show capacitance components associated with these relays. Relay A is the first prototype relay design reported in [12]. Relay B is an improved design to mitigate parasitic effects [47].

Recall that an important motivation for the 4T relay design is to fix the gate switching voltage with respect to the body terminal (i.e., only gate-to-body voltage ($|V_{GB}|$) influences V_{PI}). In relay A, the gate-to-source/drain overlap area is comparable to the gate-to-body overlap area (~67%). In relay B, the gate-to-source/drain overlap area is now insignificant compared to the gate-to-body overlap area (~2.4%). Figure 11.11 shows how the drain and source bias voltages (V_D and V_S) affect V_{PI} and V_{RL} for relays A and B. Ideally, the source and drain should have minimal effect to the switching voltages

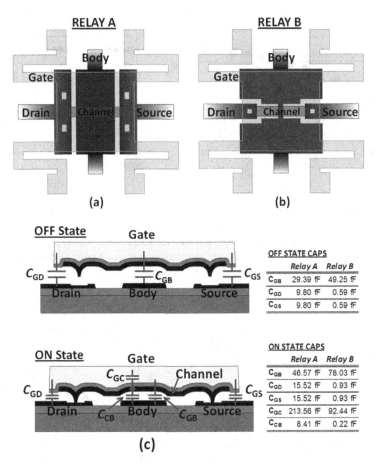

Fig. 11.10 4T relay designs and their capacitance components. (a) Layout view of the first relay prototype [12] (relay A) and (b) an optimized relay design [47] (relay B), showing electrode areas. (c) Off-state and on-state cross-sectional views along the channel showing the associated capacitances. Their values are given in the tables.

(i.e., slope = 0). Significant influence from the source/drain electrodes is observed in relay A, but effectively suppressed in relay B.

The ability to utilize body biasing to tune V_{PI} post process is another key advantage of the 4T relay. It is important that the effect of body biasing can be predictable and well controlled. Figure 11.12 shows how the switching voltages shift with applied body bias. In the ideal case, V_{PI} and V_{RL} should shift the same amount as the applied V_B (i.e., slope = 1). In relay A, only approximately half the actuation (gate) area overlaps the fixed (body) electrode. As a result, the body is only half as effective in modulating the switching voltages (slope ~ 0.5). Note that since the electrostatic force is ambipolar, ideally both the movable electrode and fixed electrode should be equally effective in actuating the relay. Thus, either electrode could be designated as the gate and the other one as the body without changing V_{PI} and V_{RL}. In a non-optimized design (relay A), V_{PI} could be significantly higher when the fixed electrode is used as gate.

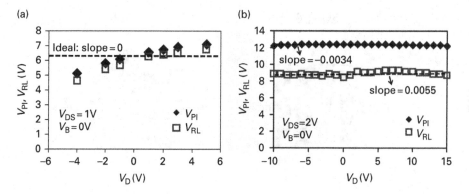

Fig. 11.11 Dependence of the pull-in voltage (V_{PI}) and the release voltage (V_{RL}) on drain bias (V_D) comparing (a) the first prototype (relay A) and (b) an improved design (relay B). In relay A, parasitic electrostatic force between the gate and the source/drain results in a shift in V_{PI} and V_{RL}. Relay B has eliminated source/drain parasitic actuation effect, having slopes close to ideal (i.e., slope = 0). Gate is the movable (SiGe) electrode.

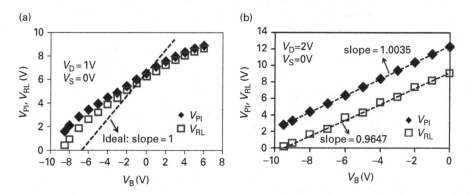

Fig. 11.12 Dependence of the pull-in voltage (V_{PI}) and the release voltage (V_{RL}) on body bias (V_B) comparing (a) the first prototype (relay A) and (b) an improved design (relay B). For a given V_G, more negative V_B results in larger V_{GB} (larger electrostatic force) and so reduces V_{PI} and V_{RL}, and vice versa. For relay B, slopes are close to the ideal case (i.e., slope = 1). Gate is the movable (SiGe) electrode.

Lastly, when the channel region that overlaps the body is large (as in relay A), significant electrostatic force may exist between the channel and the body terminal during the on-state. This additional electrostatic force makes it harder to turn off the relay (decreases V_{RL}) and undesirably increase the hysteresis voltage. When the channel is at high enough potential, this channel electrostatic force may be large enough to overpower the spring restoring force even after gate bias is removed (V_{RL} drops to below 0 V). In a well-designed relay, the channel-to-body overlap should be minimized, if not completely eliminated, which is accomplished in relay B by removing the body electrode under the channel (a minimally sized strip remains to electrically connect the top and bottom halves of the fixed electrode).

11.4.2 Design for highly compact circuits

For a relay-based digital IC, it is desirable to implement functions in a single stage (i.e., one mechanical delay) and minimize the device count to reduce area. The unique properties of mechanical structures can be leveraged to achieve equivalent functions with significantly fewer devices compared to CMOS implementation. Multi-source/drain and multi-gate relays are proposed as the way forward to achieve highly compact circuits.

Figure 11.13 shows a dual-source/drain or 6-terminal (6T) relay. The two pairs of source/drain electrodes can both be utilized to increase circuit functionality without incurring additional area. Alternatively only one pair can be used with the other pair left floating for a standard 4T relay operation. Notice that having the source and drain electrodes on the same side of the actuation plate as opposed to opposite sides (as in relays A and B) is optimal for electrostatic integrity as well. This way, the channel-to-body overlap is completely eliminated without the need to remove fixed electrode area (as in relay B in Fig. 11.10), which in turn maximizes the gate-to-body overlap area. Another pair of source/drain at the opposite side of the actuation plate is necessary to keep symmetry of the electrostatic force so that the relay does not tilt to one side during operation. Relays can also be designed with more than two pairs of source/drain electrodes, albeit at the expense of electrostatic control as the source/drain regions will be a larger percentage of the total actuation area.

Since V_{PI} is dependent on the electrostatic force strength, logic gates can be implemented by modulating the total relay actuation electrode area. The actuation area can be altered by both altering the electrode dimensions and the number of driving input electrodes. This concept allows logic gate designs unique to relays. For example, by carefully designing the beam dimensions, the number of driving input electrodes required to actuate a relay can be adjusted to implement two-input AND, OR [48],

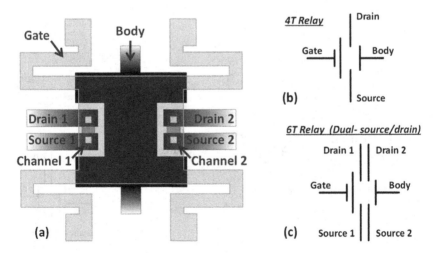

Fig. 11.13 (a) Layout view illustration of a dual-source/drain relay structure, indicating the terminals. (b) Circuit symbol when only one source/drain pair is utilized (4T relay operation) and when (c) both source/drain pairs are utilized (6T relay operation).

and NAND [49] gates. Multi-gate relay design architecture utilizes this concept to implement logic gates in a single device for significant reduction of device count. The fixed electrode (designated as gate) can be subdivided into multiple separate electrodes that can be independently biased. This way, the strength of electrostatic force (therefore actuation) can be controlled by the number of input electrodes (gate) that are driven. For a given process technology, there are several ways to design the relay to implement different logic functions:

1.   *Subdivision of the fixed electrode.* Since electrostatic force depends on the total actuating electrode area, the fixed electrode can be replaced with smaller electrodes of equal area to accommodate multiple inputs – or of different areas to accommodate multiple inputs of varying weight (i.e., influence) – to implement different logic functions.

2.   *Relay dimensions.* The flexure and plate dimensions determine the total electrostatic force needed to turn on the relay. For example, by adjusting flexure length (i.e., stiffness), a relay can be made to turn on with either one or two driven inputs.

3.   *Body biasing.* Given a certain input voltage range, relays can be set to turn on when a specific number of input electrodes is driven by appropriately setting the level of body bias. In this manner, the same relay structure can be used to achieve different logic functions.

A dual-gate (two-input), dual-source/drain (two-output) relay structure is shown in Fig. 11.14. Note that the gate electrodes have equal area and are inter-digitated to ensure that they have equal influence. Figure 11.15 shows V_{PI} and V_{RL} of the dual-gate relay when $V_B = 0\,\mathrm{V}$. The first case, [0, 1], is when only gate 2 is driven, and the second case, [1, 0], is when only gate 1 is driven. The non-driven gate is kept at 0 V. Switching

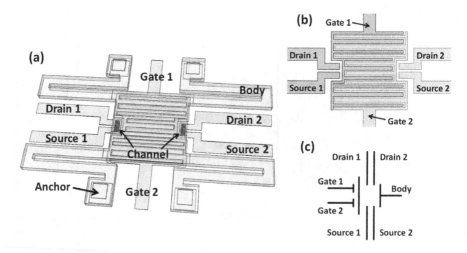

Fig. 11.14 Schematic views of a dual-gate, dual-source/drain relay. (a) Isometric view. (b) Bottom electrode layout, showing inter-digitated gate electrodes to ensure that they have equal influence on the body. (c) Circuit symbol.

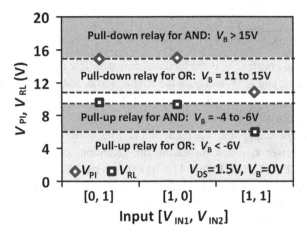

Fig. 11.15 Measured pull-in (V_{PI}) and release (V_{RL}) voltages of a dual-gate relay, for various input combinations. "1" ≡ V_G. Body bias requirements to achieve AND and OR functions are indicated assuming $V_G = 8$ V.

voltages are equal for both cases, confirming that gates 1 and 2 have equal influence. The third case, [1, 1], is when both gates are driven simultaneously to actuate the relay. As expected, the switching voltages are lower in this case due to stronger electrostatic force for the same voltage since the effective actuation area is doubled in this case. With this property, different levels of body bias can be applied to make the relay switch using either one gate or two gate electrodes for the same voltage. For example, an operating voltage of 12 V requires both gates to be driven to turn the relay on with $V_B = 0$ V (AND function). When $V_B = -5$ V is applied, the switching voltages are lowered, so that driving just one of the gates is sufficient to turn the relay on (OR function). In Fig. 11.15, body biasing requirements to achieve AND and OR function are indicated for $V_G = 8$ V.

11.4.3 Design for low-voltage operation

The operating voltages of relays tend to be relatively high (~10 V) for the current technology. In designing relays for digital logic, V_{PI} needs to be reduced for reliable circuit operation and the ultimate goal of ultra-low power operation. This can be done through scaling and design optimization.

Recall from Section 11.1 that V_{PI} in pull-in mode depends on

$$V_{PI} \propto \sqrt{\frac{k_{eff} g_0^3}{A_{eff}}}, \tag{11.19}$$

where g_0 and g_d are the actuation and dimple gaps respectively. A_{eff} can be approximated as the overlap area between the movable and fixed electrode. For this particular relay design, the effective spring constant (k_{eff}) consists of flexural and torsional components, and can be approximated as follows [5]:

$$\frac{1}{k_{\text{eff}}} \cong \left(\gamma_{\text{f}} \frac{EWh^3}{L^3} \right)^{-1} + \left(\gamma_{\text{t}} \frac{GWh^3}{L} \right)^{-1}, \tag{11.20}$$

where γ_{f} is the flexural constant; γ_{t} is the torsional constant; E is the Young's modulus; G is the shear modulus; and L, W, and h are the length, width, and height of the flexures respectively. While W and L are defined by layout, h is the structural SiGe thickness (T_{SiGe}) defined by process.

From a relay design standpoint, V_{PI} reduction can be achieved by optimizing the flexures (W, L) to lower k_{eff}, and increasing the actuation area (A_{eff}) for a given process technology (Fig. 11.16). Figure 11.17(a) shows the effect of flexure lengths (L) on V_{PI} and the hysteresis voltage for different flexure width (W). First, V_{PI} is found to be relatively insensitive to flexure length, while V_{RL} drops with increasing L, widening hysteresis. Since turn-on is brought about by electrostatic force while turn-off is brought about by spring restoring force, L has more significant impact on V_{RL}. V_{PI} starts to increase when flexure lengths are long ($L > 30\,\mu\text{m}$) due to increased warping from strain gradient. As W gets smaller, reductions in V_{PI} and hysteresis are observed. It should be noted that reduction in V_{PI} through decreasing k_{eff} comes with reliability trade-off. Having weaker spring restoring force, the relay becomes more prone to stiction (both during release and operation), so W could not be reduced indefinitely.

Increasing the actuation plate size is an effective way to maximize A_{eff}, but the increase in device footprint is highly undesirable. In this design, device footprint is

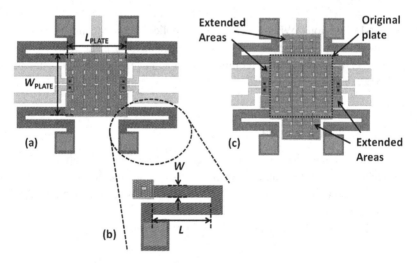

Fig. 11.16 Plate and flexure design parameter definition. (a) Plate size for the original design is defined by plate width (W_{plate}) and length (L_{plate}). (b) Definition of flexure width (W) and length (L). W is the width of the beam which is uniform throughout in all designs. L is defined for one beam section only, not the total beam length. (c) Extended actuation plate designs provide for larger actuation area without any increase in the total device area, because they make use of empty spaces between anchors and flexures. Each design is defined by the size of the original plate ($L_{\text{plate}} \times W_{\text{plate}}$) + total area of the extensions as a percentage of the original plate area.

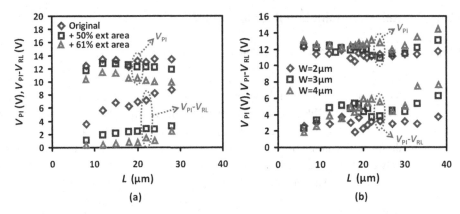

Fig. 11.17 (a) Effect of flexure width (W) on relay operating voltage. Relay V_{PI} and hysteresis voltage vs. flexure lengths (L) compared for various W. (b) Effect of extended actuation plate area on relay operating voltage. Relay V_{PI} and hysteresis voltage vs. flexure lengths (L) compared for relays without and with extended actuation plate. Relays studied are single-gate, dual-source/drain design. Plate size = 30 μm × 30 μm. $V_D = 1.5$ V, $V_S = V_B = 0$ V. Tungsten fixed electrode is biased as gate.

set by the anchor placement (vertically) and the flexure lengths (horizontally). Extensions can be incorporated to the movable plate to increase the effective actuation area without increasing the total device footprint (Fig. 11.16(c)). A study of extended actuation plate effects is presented in Fig. 11.17(b). In general, V_{PI} and hysteresis reduction is observed with larger extensions. However, reliability degrades significantly when the extensions are excessively large, likely caused by strain gradient. Interferometer measurements confirm that negative strain gradient causes the extensions to bend downwards. With long extensions, this bending can be greater than the contact gap so that the channel touches the source/drain electrodes prior to being actuated. Devices that are not stuck down are prone to stiction after a few cycles. Devices with ~60% extra plate area are found to be reliable with >15% reduction in V_{PI} and hysteresis <1 V, and appears to be the optimal point for maximum V_{PI} reduction vs. reliability trade-off. As the plate is scaled down, larger percentage of extension may be allowable as bending due to strain gradient is reduced with scaling.

Note that the hysteresis voltage sets a lower limit for supply voltage scaling, even with body biasing. (To ensure that the relay is able to turn off, V_{RL} can be reduced to near 0 V at best, which means the lowest V_{PI} achievable is equal to the hysteresis voltage.) It is therefore crucial to lower the hysteresis voltage for ultra-low power operation. With relays having <1 V hysteresis, sub-1 V operation can be achieved via body biasing. Ultimately, design optimization is an inexpensive way to tune V_{PI} and the hysteresis voltage, but not as a driver for ultra-low voltage operation. An optimized design can potentially achieve ~20–30% reduction in V_{PI} for a given process. However, sub-1 V V_{PI} can only be achieved through technology scaling. Scaling of the gap sizes (g_0 and g_d) and structural layer thickness (T_{SiGe}) are not trivial, as will be discussed in Section 11.5.

11.5   Relay combinational logic circuits

11.5.1   Complementary relay inverter circuit

Relays can be employed as direct replacements for transistors, as CMOS circuit design techniques are compatible with relays. A 4T relay can be biased to mimic either an n-channel MOSFET (turning on at sufficient positive V_{GS}) or a p-channel MOSFET (turning on at sufficient negative V_{GS}). Thus, an inverter circuit with zero static power dissipation can be formed by connecting two complementary 4T relays in series between the power supply and ground, similarly to the transistors in a CMOS inverter (Fig. 11.18). The inverter is considered to be the basis for all logic circuits, as its operation principle is applicable to logic circuits in general.

Note that in a CMOS inverter the transistors have gradual switching behavior (due to the fact that the sub-threshold swing is fundamentally limited to be no steeper than 60 mV/dec at room temperature) so that the CMOS inverter voltage transfer characteristic (VTC) shows a gradual transition between states, with non-zero "crow-bar current" (I_{dd}, flowing directly from the power supply to ground) in the transition region. In contrast, a relay inverter VTC can show abrupt transitions between states, with zero crow-bar current, if the relay switching voltages are tuned appropriately (ideally, $V_{PI,N} \geq V_{RL,P}$ and $V_{PI,P} \leq V_{RL,N}$ [50]). A nearly symmetric voltage transfer characteristics (Fig. 11.18(b)) is achieved by carefully tuning the V_{PI} values for the two relays via body biasing to achieve complementary switching characteristics for $V_{dd} = 2$ V. Additionally, to achieve the maximum static noise margin, switching from high-to-low and low-to-high should be symmetric about $V_{dd}/2$, and hysteresis should be minimized [50]. Dynamic inverter operation at 50 Hz is shown in Fig. 11.18(c), for a square-wave input signal.

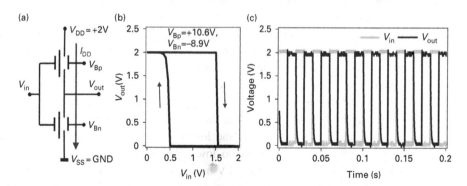

Fig. 11.18 Demonstration of a complementary relay inverter [12]. (a) Circuit schematic. Two relays are individually probed and connected externally. V_{dd} is set to 2 V. Body bias of the P-relay (V_{BP}) and N-relay (V_{BN}) is adjusted to achieve symmetric switching at low voltage. (b) Static inverter measurement with $V_{BP} = +10.6$ V and $V_{BN} = -8.9$ V. (c) Dynamic inverter measurement of the same relay circuit shown in (b), performed at a frequency of 50 Hz. V_{in} is a square function, oscillating between 0 V and 2 V.

(a)

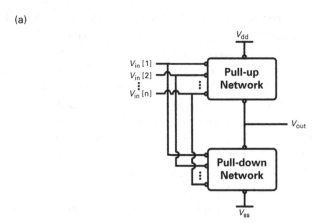

(b)

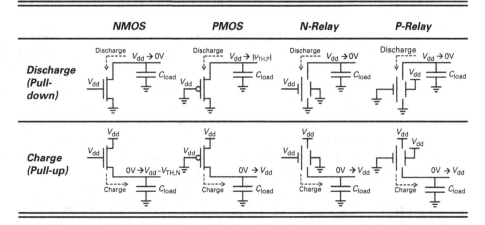

Fig. 11.19 (a) A generic static complementary CMOS logic circuit, consisting of "pull-up" network and a "pull-down" network. Complementary relay logic can be implemented in the same way. (b) Comparison of NMOS, PMOS, N-relay, and P-relay when used as a "pull-down" and "pull-up" device.

11.5.2 CMOS-like circuit design

A static complementary CMOS gate is essentially an extension of the complementary inverter circuit. It consists of a "pull-up network" and a "pull-down network" (Fig. 11.19) [51]. When the output of the logic function is "1," the "pull-up" network provides a connection between the output and V_{dd}. Similarly, when the logic function outputs a "0," the "pull-down" network provides a connection between the output and GND. At steady state, only one of these networks will be on (conducting) at any given time. Thus, the output will always have a connection to either V_{dd} or GND and static current flowing directly from V_{dd} to GND ideally should not exist.

Complementary relay logic gates can be implemented in a similar fashion with one major difference. In CMOS, the "pull-up" network is made up of PMOS transistors, while the "pull-down" network is made up of NMOS transistors. Consider Fig. 11.20. An NMOS turns on when $V_{GS} = V_{th,N}$, while a PMOS turns on when $|V_{GS}| = |V_{th,P}|$. An NMOS is therefore able to discharge ("pull down") a node all the way down to 0 V, while it can only charge ("pull up") a node to $V_{dd} - V_{th,N}$. A PMOS could charge a node all the way up to V_{dd}, but is only capable of discharging a node down to $|V_{th,P}|$. As a result, complementary CMOS circuits are always inverting. Functions such as NAND, NOR, and XNOR can be implemented in one stage, but functions like AND, OR, and XOR are formed by connecting an extra inverter to the inverting functions (thus incurring additional area and delay). A relay, on the other hand, turns on when $|V_{GB}| = V_{PI}$, since electrostatic force is ambipolar. Hence, a relay can charge all the way to V_{dd} and discharge all the way to 0 V regardless of N-relay or P-relay operation mode. A complementary relay circuit can implement both inverting and non-inverting logic in a single stage. In fact, an optimal relay circuit should consist of a complex gate that performs all computation in a single stage (i.e., only one mechanical delay), due to the large ratio between the mechanical (switching) delay and the electrical (capacitive charging) delays [19].

Alternatively, CMOS circuit can be designed using the pass-gate logic style to reduce device count. For example, an AND pass-gate logic could be implemented with only two transistors (plus another two transistors to invert the signal to get its complement, for a total of four). However, an NMOS-only approach is not able to pass V_{dd}, while a PMOS-only approach is not able to pass 0 V, for the same reason that PMOS makes a better "pull-up" device while NMOS makes a better "pull-down" device explained earlier. Without the ability to pass voltages rail-to-rail (to strongly turn on and off driven devices), slower transition (i.e., speed) of the driven gate is expected and static power dissipation potentially rises when devices are not completely off. This problem can be solved by using a transmission gate (having NMOS and PMOS together in parallel), but with increased device count. Relays, on the other hand, do not have such an issue. N-relay or P-relay only implementations are both effective to pass voltages rail-to-rail. Thus, a large reduction in device count can be achieved using relays with the pass-gate architecture.

Since an N-relay or P-relay is implemented with an identical relay structure, a relay logic gate can be configured to implement different functions by simply changing the bias conditions. An example in Fig. 11.20(a) shows a dynamically configurable AND/OR/NAND/NOR gate implemented using 4T relays by applying different biasing conditions. Thus, both inverting and non-inverting logic is achieved in a single stage. Dual-source/drain relay design can be utilized to add functionality to the logic gate by using the second pair of source/drain as a second output. The second output can implement either the complementary signal (to get differential output), the same signal with a different voltage level, or a different function altogether. An example in Fig. 11.20(b) shows a relay logic gate that implements two different functions (AND/OR, OR/AND, NAND/NOR, and NOR/NAND) at its outputs depending on the bias configuration.

A second way to use the dual-source/drain relay design is to reduce the number of devices necessary to implement a certain function. For example, take a CMOS-like

(a)

(b)

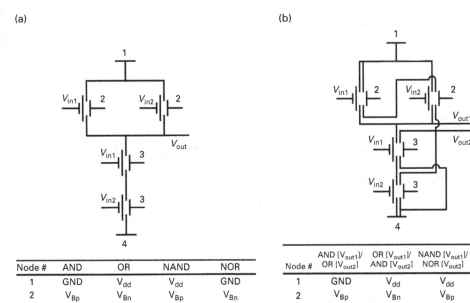

Node #	AND	OR	NAND	NOR
1	GND	V_{dd}	V_{dd}	GND
2	V_{Bp}	V_{Bn}	V_{Bp}	V_{Bn}
3	V_{Bn}	V_{Bp}	V_{Bn}	V_{Bp}
4	V_{dd}	GND	GND	V_{dd}

Node #	AND $[V_{out1}]$/ OR $[V_{out2}]$	OR $[V_{out1}]$/ AND $[V_{out2}]$	NAND $[V_{out1}]$/ NOR $[V_{out2}]$	NOR $[V_{out1}]$/ NAND $[V_{out2}]$
1	GND	V_{dd}	V_{dd}	GND
2	V_{Bp}	V_{Bn}	V_{Bp}	V_{Bn}
3	V_{Bn}	V_{Bp}	V_{Bn}	V_{Bp}
4	V_{dd}	GND	GND	V_{dd}

Fig. 11.20 (a) A dynamically configurable complementary relay AND/OR/NAND/NOR gate implemented with 4T relays with its bias configurations. (b) A dynamically configurable complementary relay AND/OR/NAND/NOR gate implemented with dual-source/drain relays to achieve multiple functionalities with the same number of devices as the 4T implementation.

implementation of an XOR/XNOR gate using 4T relays (Fig. 11.21(a)). Although functional, an eight-relay implementation consumes large chip area. By employing dual-source/drain relays, the same circuit can be implemented with four relays (Fig. 11.21(b)). In this case, two 4T relays with the same gate and body biases are replaced with a single dual-source/drain relay, to reduce the device count by half.

11.5.3 Multi-input/multi-output design

The multi-gate relay design allows a relay device to accommodate multiple inputs. In principle, almost any logic function can be implemented by adjusting the body bias levels and carefully designing the gate electrode areas to adjust the amount of electrostatic force each gate contributes to actuation. The ultimate, most compact digital logic gate comprises only two switches: one "pull-up" switch that connects the output to the power supply when it is turned on, and one "pull-down" switch that connects the output to ground when it is turned on (Fig. 11.22). Each input signal is connected to one input electrode of the pull-up switch and also to one input electrode of the pull-down switch, and only one of these switches is on at any given time, i.e., complementary operation. Note that each relay can comprise multiple pairs of source/drain electrodes as well, to provide for greater functionality, e.g., output signals at various voltage levels

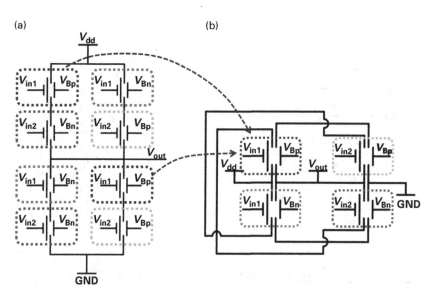

Fig. 11.21 (a) A complementary relay XOR/XNOR gate implemented with 4T relays. XOR biasing condition is shown. XNOR function can be achieved by simply switching V_{dd} and GND. (b) A complementary relay XOR/XNOR gate implemented with dual-source/drain relays to reduce device count for more compact logic. The two relays that have identical gate and body biases in (a) are replaced with one dual-source/drain relay, reducing the device count by half.

(as indicated in the figure) or differential output signals. Two-input logic gates are presented next to demonstrate the feasibility of this approach.

Figure 11.23(a) shows two single-gate, dual-source/drain relays connected together to form a dynamically configurable one-input, two-output complementary relay circuit, which can function either as an inverter/buffer or XOR/XNOR gate, depending on the electrode biasing configuration. First, consider the inverter/buffer bias configuration. When the input voltage (V_{in}) is high the top relay is off and the bottom relay is on, and vice versa. By connecting the sources to V_{dd} or GND, the left/right-hand side source biases are V_{dd}/GND and GND/V_{dd} for the top and bottom relay respectively, such that complementary signals are achieved at the two sides. The relays used in this circuit have <1 V hysteresis, so that low voltage (1 V) inverter/buffer operation is achieved with body biasing. Next, consider the XOR/XNOR bias configuration, which makes use of the ambipolar nature of electrostatic force (actuation depends only on the magnitude of V_{GB}, not its polarity). When the gate and body voltages are complementary, the top relay will turn on. When they are the same, the bottom relay will turn on. Since the body is used as an input electrode in this case, the input voltage range cannot be reduced by body biasing.

The second example (Fig. 11.23(b)) shows two dual-gate, dual-source/drain relays connected together to implement a dynamically configurable two-input, two-output complementary relay circuit, which can function either as an AND/NAND or OR/NOR

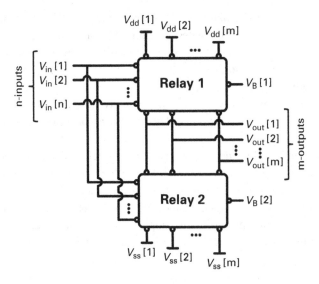

Fig. 11.22 A generic multi-input, multi-output relay combinational logic circuit.

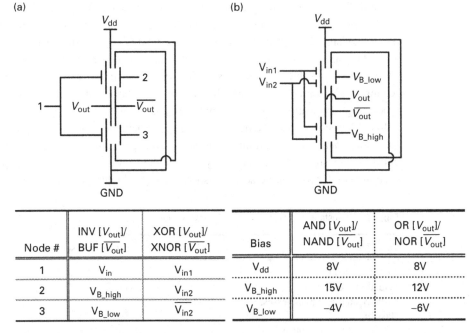

Node #	INV [V_{out}]/ BUF [$\overline{V_{out}}$]	XOR [V_{out}]/ XNOR [$\overline{V_{out}}$]	Bias	AND [V_{out}]/ NAND [$\overline{V_{out}}$]	OR [V_{out}]/ NOR [$\overline{V_{out}}$]
1	V_{in}	V_{in1}	V_{dd}	8V	8V
2	V_{B_high}	V_{in2}	V_{B_high}	15V	12V
3	V_{B_low}	$\overline{V_{in2}}$	V_{B_low}	−4V	−6V

Fig. 11.23 (a) Dynamically configurable complementary relay logic circuit utilizing two single-gate, dual-source/drain relays. The same circuit can achieve INV/BUF or XOR/XNOR functionality depending on the bias configuration. (b) Dynamically configurable complementary relay logic circuit utilizing two dual-gate, dual-source/drain relays. The same circuit can achieve AND/NAND or OR/NOR functionality depending on the body bias voltages. Device operating voltages are given in Fig. 11.15.

gate, depending on the electrode bias configurations. In the AND/NAND configuration, the top relay turns on only when both gates are high. The bottom relay turns on when at least one gate is low. Again, switching is complementary and the source biases (V_{dd} or GND) determine whether that source will act as a "pull-up" or a "pull-down" connection. By simply adjusting the body bias voltages, the same circuit can implement OR/NOR functions. In this configuration, the top relay now turns on when at least one gate is high. Similarly, the bottom relay now turns on only when both gates are high.

The multi-input/multi-output design approach has therefore demonstrated all two-input logic gates using only two relay devices. Further extensions to relay designs comprising greater than two input electrodes [52] and/or greater than two sets of source/drain electrodes are possible, to implement more complex logic functions. Hence, multi-input/multi-input relays are promising to implement multi-functional, dynamically configurable, and highly compact relay-based integrated circuits.

11.6   Relay scaling perspective

Miniaturization of relays results in reduced operating voltage, increased performance, and a higher degree of integration for improved cost and functionality. Similar to the classic MOSFET scaling theory [1], constant field scaling theory for relays have been reported. Using this methodology, the electric field is kept constant across the actuation gap while all device dimensions are reduced by a factor $1/\kappa$. The implications are summarized in Table 11.2 [53], showing the performance and power benefits with scaling. Next, the challenges and limitations for relay scaling are discussed from the processing, application, and functionality points of view.

Sub-1 V relay operation requires scaling of the actuation and contact gap sizes to the sub-10 nm range. First, forming nanometer-scale gaps requires highly controlled and uniform deposition of sacrificial layers and the ability to release the gaps without stiction. The former can be accomplished through a highly optimized CVD process or

Table 11.2 Constant field scaling theory for relays [53]

Variables	Constant-field scaling
Spring constant	$1/\kappa$
Actuation area	$1/\kappa^2$
As-fabricated gap thicknesses, g, g_d	$1/\kappa$
Mass, m	$1/\kappa^3$
Supply and pull-in voltages	$1/\kappa$
Delay	$1/\kappa$
Switching energy	$1/\kappa^3$
Density	κ^2
Power	$1/\kappa^2$
Power density	1

ALD to achieve nanometer-scale precision. The latter is a bigger challenge to achieve. Gaps as small as 15 nm are demonstrated using critical point drying technique [3]. Even a dry release process, such as HF vapor for an SiO_2 sacrificial structure, will be prone to stiction at highly scaled dimensions due to water being an etch by-product. A 13 nm gap has been successfully released by an anhydrous HF vapor system at reduced pressure [54]. Gaps <10 nm require more novel gap formation techniques, such as spacers [55] and silicidation [56].

It is also necessary to reduce the effective spring constant flexures to operate at such low voltages. Thinning down the sacrificial structure could accomplish this but is mainly hampered by a worsened strain gradient. The strain gradient causes an out-of-plane deflection, which is detrimental to relay performance, functionality, and reliability. In a highly scaled dimension, a deflection of just a few nanometers results in a large percentage change in the gap thicknesses. Depending on the bending direction, relays could be stuck down from the start or have significantly higher V_{PI}. At best, V_{PI} becomes highly unpredictable. Hence, structures with strain gradient $<1 \times 10^{-4} \mu m^{-1}$ are required to address this challenge. Relays with poly-SiGe structural layers that have been reported [12, 45] use poly-SiGe processes developed for thicknesses in the micrometer range [40] and suffer from a dramatically increased strain gradient at thicknesses $<1 \mu m$. Alternatively, multi-layered structural materials can be utilized to lower the overall strain gradient of the structure by stress compensation [46]. Ultimately, a material with low spring constant and low strain gradient that can be deposited at sufficiently low temperatures is desired, but this is challenging to achieve.

Another processing challenge in relay scaling is to make very small contact dimples. The contact regions are often the minimum feature size of the structure and forming nanometer-scale holes is a lithography challenge. While e-beam lithography can do the job, throughput is low and not suitable for large-scale integrated relay circuits. Scaling down the dimple size is crucial because it dictates the minimum size of the source/drain regions. When the source/drain areas cannot be scaled proportionately to the rest of the structure, gate electrostatic control degrades (recall Section 11.3.1).

Scalability is also more limited for the multi-gate relay approach described in Section 11.3.2, due to electrostatic control. When the fixed electrode is subdivided, there is a minimum separation distance between the electrodes, which is limited by surface leakage. This minimum separation distance is in the hundreds of nm range and also gets reduced as the operating voltages are scaled. As the size of the relay is scaled down, the amount of fixed electrode area lost due to electrode-to-electrode separation becomes a larger portion of the total area. As a result, actuation becomes less effective. Despite limited scalability, multi-electrode design remains attractive as it potentially enables any logic function to be implemented with only two relays, such that an overall area savings can be achieved from a system standpoint.

From the application standpoint, scaling would be limited by contact resistance. Note that in the multiple asperity model at the relay contact discussed previously (Fig. 11.8), the effective contact area for the contact resistance is not a function of the apparent contact area, but rather the applied load and material properties (Eq. (11.18)). This model assumes that the contact surface is large such that the real contact area is made of a finite number of asperities. As the contact is scaled to dimensions comparable to the asperities ($<100\,nm$), the contact region will eventually consist of just a single asperity. In this regime, the apparent contact area will start to have a significant influence on the contact resistance. Thus, the $10\,k\Omega$ contact resistance limit requirement for relay digital circuit [19] imposes a minimum limit to contact dimple scaling, which in turn dictates the lower limit to relay lateral dimensions that still have good electrostatic gate control as discussed previously.

Finally, since relays rely on the spring restoring force to break contact once electrostatic force is removed, the ultimate limit for relay functionality will be dictated by the ability to turn the relay off [53]. While lowered spring constant and gap size scaling are necessary for ultra-low power operation, the relay must still be able to overcome the surface adhesive forces at the contact to avoid being permanently stuck down:

$$F_{spring} = k_{eff}\, g_d \geq F_{adhesion}. \tag{11.21}$$

With scaling of the dimple size, $F_{adhesion}$ is also expected to scale [5], as the surface area of the contact is reduced. At the same time, processing improvements to reduce surface adhesion forces, such as surface treatment or contact coating, could help reduce this limit. It should also be noted that, in general, lowering the spring restoring force sacrifices reliability. Stiffer flexures tend to achieve higher yield and endurance, albeit with a higher operating voltage. Given the 10^{14} on/off cycles requirement for relay digital logic [57], reliability and yield may be the practical limit to scaling.

References

[1] R. H. Dennard, F. H. Gaensslen, V. L. Rideout, E. Bassous, & A. R. LeBlanc, "Design of ion-implanted MOSFETs with very small physical dimensions." *Journal of Solid-State Circuits*, **9**, 256–268 (1974).

[2] S. D. Senturia, *Microsystem Design* (Boston, MA: Kluwer Academic, 2001).

[3] W. W. Jang, J. O. Lee, J.-B. Yoon et al., "Fabrication and characterization of a nanoelectromechanical switch with 15-nm-thick suspension air gap." *Applied Physics Letters*, **92**, 103110 (2008).

[4] J.-O. Lee, M.-W. Kim, S.-D. Ko et al., "3-terminal nanoelectromechanical switching device in insulating liquid media for low voltage operation and reliability improvement." In *Electron Devices Meeting (IEDM), 2009 IEEE International*, pp. 227–230 (2009).

[5] H. Kam, V. Pott, R. Nathanael, J. Jeon, E. Alon, & T.-J. King Liu, "Design and reliability of a micro-relay technology for zero-standby-power digital logic applications." In *Electron Devices Meeting (IEDM), 2009 IEEE International, Technical Digest*, pp. 809–812 (2009).

[6] S. Chong, K. Akarvardar, R. Parsa et al., "Nanoelectromechanical (NEM) relays integrated with CMOS SRAM for improved stability and low leakage." In *Computer-Aided Design, International Conference on*, pp. 478–484 (2009).

[7] D. A. Czaplewski, G. A. Patrizi, G. M. Kraus et al., "A nanomechanical switch for integration with CMOS logic." *Journal of Micromechanics and Microengineering*, **19**, 085003 (2009).

[8] S. Chong, B. Lee, K. B. Parizi et al., "Integration of nanoelectromechanical (NEM) relays with silicon CMOS with functional CMOS-NEM circuit." In *Electron Devices Meeting (IEDM), 2011 IEEE International, Technical Digest*, pp. 701–704 (2011).

[9] R. Parsa, M. Shavezipur, W. S. Lee *et al.*, "Nanoelectromechanical relays with decoupled electrode and suspension." In *Proceedings of the Microelectromechical Systems Conference*, pp. 1361–1364 (2011).

[10] W. S. Lee, S. Chong, R. Parsa *et al.*, "Dual sidewall lateral nanoelectromechanical relays with beam isolation." In *Solid-State Sensors, Actuators and Microsystems Conference (TRANSDUCERS), 2011 16th International*, pp. 2606–2609 (2011).

[11] S. Chong, B. Lee, S. Mitra, R. T. Howe, & H.-S. P. Wong, "Integration of nanoelectromechanical relays with silicon nMOS." *IEEE Transactions on Electron Devices*, **59**(1), 255–258 (2012).

[12] R. Nathanael, V. Pott, H. Kam, J. Jeon & T.-J. K. Liu, "4-terminal relay technology for complementary logic." *IEEE International Electron Devices Meeting Technical Digest*, pp. 223–226 (2009).

[13] G. K. Fedder, R. T. Howe, T.-J. King Liu, & E. Quevy, "Technologies for cofabricating MEMS and electronics." *Proceedings of the IEEE*, **96**, 306–322 (2008).

[14] H. Takeuchi, A. Wun, X. Sun, R. T. Howe, & T.-J. King Liu, "Thermal budget limits of quarter-micrometer foundry CMOS for post-processing MEMS devices." *IEEE Transactions on Electron Devices*, **52**, 2081–2086 (2005).

[15] V. K. Khanna, "Adhesion–delamination phenomena at the surfaces and interfaces in microelectronics and MEMS structures and packaged devices." *Journal of Physics D: Applied Physics*, **44**(3) (2011).

[16] W. M. van Spengen, R. Puers, & I. De Wolf. "A physical model to predict stiction in MEMS." *Journal of Micromechanics and Microengineering*, **12**(5), 702–713 (2002).

[17] F. M. Serry, D. Walliser, & G. J. Maclay, "The role of the Casimir effect in the static deflection and stiction of membrane strips in MEMS." *Journal of Applied Physics*, **84**(50), 2501–2506 (1998).

[18] P. J. Resnick & P. J. Clews, "Whole wafer critical point drying of MEMS devices." In *Reliability, Testing, and Characterization of MEMS/MOEMS, SPIE Proceedings*, vol. **4558**, pp. 189–196 (2001).

[19] F. Chen, H. Kam, D. Markovic, T. King-Liu, V. Stojanovic, & E. Alon, "Integrated circuit design with NEM relays." *IEEE/ACM International Conference on Computer-Aided Design*, pp 750–757 (2008).

[20] R. Holm, *Electric Contacts: Theory and Applications* (Berlin: Springer-Verlag, 1967).

[21] M. P. de Boer, J. A. Knapp, T. M. Mayer, & T. A. Michalske, "The role of interfacial properties on MEMS performance and reliability." In *Microsystems Metrology and Inspection, SPIE Proceedings*, vol. **3825**, pp. 2–15 (1999).

[22] L. Kogut & K. Komvopoulos, "Electrical contact resistance theory for conductive rough surfaces." *Journal of Applied Physics*, **94**, 3153–3162, 2003.

[23] G. M. Rebeiz, *RF MEMS: Theory, Design, and Technology* (New York: Wiley, 2003).

[24] H. Kam, E. Alon, & T.-J. K. Liu, "A predictive contact reliability model for MEM logic switches." In *Electron Devices Meeting, 2010 IEEE International, Technical Digest*, pp. 399–402 (2010).

[25] R. Candler, W. Park, H. Li, G. Yama, A. Partridge, M. Lutz, & T. Kenny, "Single wafer encapsulation of MEMS devices." *IEEE Transactions on Advanced Packaging*, **26**(3), 227–232 (2003).

[26] Y. Chen, R. Nathanael, J. Jeon, J. Yaung, L. Hutin, & T.-J. K. Liu, "Characterization of contact resistance stability in MEM relays with tungsten electrodes." *IEEE/ASME Journal of Microelectromechanical Systems*, **21**(3), 511–513 (2012).

[27] I.-R. Chen, Y. P. Chen, L. Hutin, V. Pott, R. Nathanael, & T.-J. King Liu, "Stable ruthenium-contact relay technology for low-power logic." Accepted to *The 17th International Conference on Solid-State Sensors, Actuators and Microsystems (TRANS-DUCERS)*, Barcelona, Spain (2013).

[28] V. K. Khanna, "Adhesion–delamination phenomena at the surfaces and interfaces in micro-electronics and MEMS structures and packaged devices." *Journal of Physics D: Applied Physics*, **44**(3) (2011).

[29] R. H. Dauskardt, M. Lane, Q. Ma, & N. Krishna. "Adhesion and debonding of multi-layer thin film structures." *Engineering Fracture Mechanics*, **61**(1), 141–162 (1998).

[30] R. L. Puurunen, J. Saarilahti, & H. Kattelus, "Implementing ALD Layers in MEMS processing." *Electrochemical Society Transactions*, **11**(7), 3–14 (2007).

[31] K. Williams, K. Gupta, & M. Wasilik, "Etch rates for micromachining processing – part II." *Journal of Microelectromechanical Systems*, **12**(6), 761–778 (2003).

[32] T. Bakke, J. Schmidt, M. Friedrichs, & B. Völker, "Etch stop materials for release by vapor HF etching." In *Proceedings of the 16th Workshop on Workshop on Micromachining, Micromechanics, and Microsystems*, pp. 103–106 (2005).

[33] R. T. Howe, B. E. Boser, & A. P. Pisano, "Polysilicon integrated microsystems: technologies and applications." *Sensors and Actuators A: Physical*, **56**(1), 167–177 (1996).

[34] R. T. Howe & R. S. Muller, "Polycrystalline and amorphous silicon micromechanical beams: annealing and mechanical properties." *Sensors and Actuators*, **4**, 447–454 (1983).

[35] M. Biebl, G. T. Mulhem & R. T. Howe, "Low in situ phosphorus doped polysilicon for integrated MEMS." In *Solid State Sensors and Actuators (Transducers 95), Technical Digest, 8th International Conference*, vol. **I**, pp. 198–201 (1995).

[36] J. Bagdahn, W. N. Sharpe Jr, & O. Jadaan, "Fracture strength of polysilicon at stress concentrations." *Journal of Microelectromechanical Systems*, **12**(3), 302–312 (2003).

[37] H. Kapels, R. Aigner, & J. Binder, "Fracture strength and fatigue of polysilicon determined by a novel thermal actuator [MEMS]." *IEEE Transactions on Electron Devices*, **47**(7), 1522–1528 (2000).

[38] R. Modlinski, A. Witvrouw, A. Verbist, R. Puers, & I. De Wolf, "Mechanical characterization of poly-SiGe layers for CMOS–MEMS integrated application." *Journal of Micromechanics and Microengineering*, **20**(1) (2009).

[39] J. Lai, "Novel processes and structures for low temperature fabrication of integrated circuit devices." Ph.D. Dissertation, University of California, Berkeley, CA (2008).

[40] C. W. Low, T.-J. King Liu, & R. T. Howe, "Characterization of polycrystalline silicon-germanium film deposition for modularly integrated MEMS applications." *Journal of Microelectromechanical Systems*, **16**(1), 68–77 (2007).

[41] V. Pott, H. Kam, J. Jeon, & T.-J. King Liu, "Improvement in mechanical contact reliability with ALD TiO_2 coating." In *AVS Conference, Proceedings*, pp. 208–209 (2009).

[42] G. Triani, J. A. Campbell, P. J. Evans, J. Davis, B. A. Latella, & R. P. Burford, "Low temperature atomic layer deposition of titania thin films." *Thin Solid Films*, **518**(12), 3182–3189 (2010).

[43] R. Azimirad, N. Naseri, O. Akhavan, & A. Z. Moshfegh, "Hydrophilicity variation of WO_3 thin films with annealing temperature." *Journal of Physics D: Applied Physics*, **40**(4), 1134–1137 (2007).

[44] M. Miyauchi, A. Nakajima, T. Watanabe, & K. Hashimoto, "Photocatalysis and photo-induced hydrophilicity of various metal oxide thin films." *Chemistry of Materials*, **14**(6), 2812–2816 (2002).

[45] R. Nathanael, J. Jeon, I.-R. Chen *et al.*, "Multi-input/multi-output relay design for more compact and versatile implementation of digital logic with zero leakage." Presented at the 19th International Symposium on VLSI Technology, Systems and Applications (2012).

[46] I.-R. Chen, L. Hutin, C. Park *et al.*, "Scaled micro-relay structure with low strain gradient for reduced operating voltage." Presented at the 221st ECS Meeting (2012).

[47] M. Spencer, F. Chen, C. Wang *et al.*, "Demonstration of integrated micro-electro-mechanical relay circuits for VLSI applications." *IEEE Journal of Solid-State Circuits*, **46**(1), 308–320 (2011).

[48] A. Hirata, K. Machida, H. Kyuragi, & M. Maeda, "A electrostatic micromechanical switch for logic operation in multichip modules on Si." *Sensors and Actuators A*, **80**, 119–125 (2000).

[49] K. Akarvardar, D. Elata, R. Parsa *et al.*, "Design considerations for complementary nano-electromechanical logic gates." In *Electron Devices Meeting, 2007 (IEDM 2007), IEEE International*, pp. 299–302 (2007).

[50] R. Nathanael, V. Pott, H. Kam, J. Jeon, E. Alon, & T.-J. K. Liu, "Four-terminal-relay body-biasing schemes for complementary logic circuits." *IEEE Electron Device Letters*, **31**(8), 890–892 (2010).

[51] J. M. Rabaey, A. P. Chandrakasan, & B. Nikolic, *Digital Integrated Circuits* (Englewood Cliffs, NJ: Prentice-Hall, 2003).

[52] J. Jeon, L. Hutin, R. Jevtic *et al.*, "Multi-input relay design for more compact implementation of digital logic circuits." *IEEE Electron Device Letters*, **33**(2), 281–283 (2012).

[53] V. Pott, H. Kam, R. Nathanael, J. Jeon, E. Alon, & T.-J. K. Liu, "Mechanical computing redux: relays for integrated circuit applications." *Proceedings of the IEEE*, **98**(12), 2076–2094 (2010).

[54] W. Kwon, J. Jeon, L. Hutin, & T.-J. K. Liu, "Electromechanical diode cell for cross-point nonvolatile memory arrays." *IEEE Electron Device Letters*, **33**(2), 131–133 (2012).

[55] J. O. Lee, Y.-H. Song, M.-W. Kim *et al.*, "A sub-1-volt nanoelectromechanical switching device." *Nature Nanotechnology*, **8**, 36–40 (2013).

[56] L.-W. Hung & C. T.-C. Nguyen, "Silicide-based release of high aspect-ratio microstructures." In *Micro Electro Mechanical Systems (MEMS), 2010 IEEE 23rd International Conference on*, pp. 120–123 (2010).

[57] T.-J. K. Liu, J. Jeon, R. Nathanael, H. Kam, V. Pott, & E. Alon, "Prospects for MEM logic switch technology." In *Electron Devices Meeting (IEDM), 2011 IEEE International, Technical Digest*, pp. 424–427 (2010).

Section IV

Spin-based devices

12 Nanomagnetic logic: from magnetic ordering to magnetic computing

György Csaba, Gary H. Bernstein, Alexei Orlov, Michael T. Niemier, X. Sharon Hu, and Wolfgang Porod

12.1 Introduction and motivation

12.1.1 Magnetic computing defined

Magnetic computing – in the broadest sense – is about using magnetic signals (nanomagnets, domain walls) to represent and process information. Nowadays, when "information processing" and "electronics" is synonymous, this concept sounds rather exotic. However, before the triumphant era of CMOS logic devices, non-charge based computers were serious candidates for information processing – for example, ingenious magnetic computing circuits were invented by R. J. Spain [1–3]. It was Cowburn [4] who first realized that the properties of nanoscale, single-domain magnets – which are very different from large, multi-domain magnets – are well suited for digital computing.

This chapter deals with one approach to magnetic computing, nanomagnet logic (or NML) [5, 6]. In NML devices, binary information is represented by the state (magnetization direction) of single domain nanomagnets and the magnetically represented information is propagated and processed by magnetic dipole–dipole interactions. From the circuit architecture point of view, NML builds on the concept of "quantum-dot cellular automata" [7] – they both share the idea of representing binary signals by bistable nanosystems and processing them through field-interactions. For this reason, nanomagnet logic was formerly called "magnetic quantum-dot cellular automata" (QCA), or field-coupled computing.

Magnetoelectronics (or spintronics) has generated a lot of attention in recent years, and using the spin degree of freedom (besides voltages, currents, and charges) as an information carrier is an intriguing possibility to enhance the functionality of electronic devices. Most proposed spintronic devices, however, use spin only as an additional variable to complement electronic functionality. NML and the closely related domain-wall logic [8] are unique in the sense that they use magnetization (spin configuration) as the primary carrier of information.

12.1.2 Qualitative description

Here we give a simple qualitative description of the idea underlying NML. For this discussion we assume that (1) nanoscale magnets behave as ideal compass-like

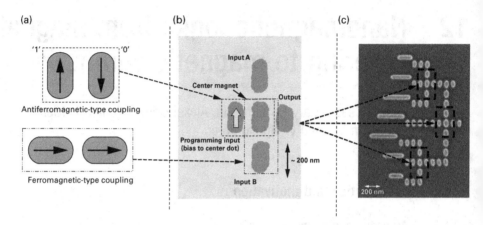

Fig. 12.1 (a) In NML, information is propagated by ferromagnetic and antiferromagnetic interactions. The outcome of competing interactions can result in majority gate functionality – the majority gate of (b) contains both ferromagnetically and antiferromagnetically coupled pairs [9]. Majority gates can build up complex logic circuits, like the (c) full adder [10]. Dotted lines show the three majority gates that build up the adder.

devices,[1] and their magnetization direction (the vector pointing from their north to south pole) always points up or down along their longer geometrical axis, and (2) the magnetization directions settle in such a way that the interaction energies between the magnets are minimized. This means that in the vicinity of north poles there are predominantly south poles and vice versa.

Magnetization direction of a stand-alone nanomagnet straightforwardly represents a single bit of binary information, as shown in Fig. 12.1(a). A pair of coupled nano-magnets can be in antiferromagnetically coupled state or ferromagnetically coupled state (see Fig. 12.1(a)), depending on their relative position. One can look at the antiferromagnetically coupled pair as an elementary inverter, with the left magnet being the input and the right magnet being the output of the gate.

Logic gates are composed from elementary ferromagnetically/antiferromagnetically coupled units. A nanomagnet interacting with three neighbors (such as the one denoted as the center magnet in Fig. 12.1(b)) cannot always be in its lowest energy state with respect to *all* of its neighbors. But it can minimize its energy by going to the ground state with respect to the *majority* of its inputs. This arrangement of magnets realizes the majority gate functionality [7], which is a universal logic gate and, in principle, can be used to realize any complex logic mapping. An example of such more complex magnetic circuits is given in Fig. 12.1(c), where a one-bit full adder, relying entirely on magnetic ordering, is shown. These figures show how elementary interactions can be engineered to perform complex functions. We describe these structures in detail in Section 12.4.

---

[1] We use the word "compass" as a simple illustration for a rotating magnetic moment. When magnets switch, their spins turn to different directions without any actual mechanical rotation.

12.1.3   Why NML? Benefits and challenges

Excessive power dissipation is the single most critical roadblock facing microelectronic scaling [11]. Most of this dissipation comes from leakage currents flowing through deeply scaled MOS transistors. NML does the computation by magnetic ordering, and once the ordering process takes place, no currents need to flow. Such lack of any static power dissipation is a major benefit. Dynamic power dissipation is also potentially very low; it has been shown that the switching of nanomagnets dissipates power close to the theoretical lowest limit of computation [12].

Magnets are intrinsically good for information storage. NML circuits, if properly designed, hold their computational state indefinitely. Such non-volatile logic has special advantages, allowing close integration of memory and logic functions, and instant start-up from an unpowered state. Nanomagnets store information in a robust way: they are entirely insensitive to radiation damage [13] and can be switched virtually unlimited times without degradation [14].

On the downside, the speed of magnetic switching is intrinsically limited by the ferromagnetic resonance frequency, and magnetic switching occurs typically on the nanosecond time scale. This limits the operating frequency of an NML circuit to below a gigahertz.

As we will point out later in detail, NML devices require on-chip magnetic fields to operate. The practicality of NML devices eventually hinges on how efficiently these fields can be generated.

12.1.4   Outline of the chapter

The goal of this review is to show that starting from elementary magnetic behaviors such as the switching of single-domain particles and their interaction via dipole fields, one can arrive at complex, engineered structures, where potentially millions of interacting nano-magnets order in an entirely controlled way while performing computing functions. We therefore start by understanding the switching and the domain properties of single-domain nanomagnets (Section 12.2) and their interactions (Section 12.3). Armed with this knowledge, we design simple logic gates (Section 12.4). We point out in Section 12.5 that spontaneous (uncontrolled) ordering in a larger nanomagnet-array will always lead to erroneous ordering, often referred to as "frustration" in the physics literature.

Larger-scale circuits must have a carefully designed clocking apparatus in order to control the ordering and avoid such errors – methods of clocking are described in Section 12.6. Inputs, outputs, and components to build a complete system are discussed in Section 12.7. We present case studies of more complex devices in Section 12.9 while Section 12.10 demonstrates how circuit design paradigms can be applied to NML.

Throughout this chapter, focus will be placed on permalloy-based, in-plane magnet-ized NML devices. The only exception will be Section 12.8, where NML devices with spatially varying out-of-plane anisotropy will be discussed.

This review is written to be accessible to readers not familiar with the terminology of magnetism. We try to describe even the more advanced topics in layman's terms. For more formal treatment, the reader is referred to the extensive list of references. The fabrication of NML magnetic systems is not the subject of this review, but the interested reader will find a wealth of information on this topic in the cited experimental papers. There are a number of review papers on NML (see, e.g., [15, 16]). In this work we present some of the latest results and focus more on challenges related to devices and less on system integration, where other reviews (such as [17]) are available.

12.2   Single-domain nanomagnets as binary switching elements

In order to use nanoscale magnets as binary switching elements, we need to assign a logic "1" or "0" to two distinct magnetization directions of the magnet. As we show below, this can be done for an elongated and sufficiently small nanomagnet.

The domain size of magnetic materials typically lies in the 100 nm $< l_{\text{domain}} <$1 μm range. So if one reduces the physical size of magnets to this (or even smaller) size range, then only a single domain will fit inside the magnet, and the magnetization of a particle can be characterized in terms of a single vector, $\mathbf{m}$ ($|\mathbf{m}|=1$).[2] This vector points in a direction that minimizes the free energy of the nanomagnet. For a stand-alone, isotropic magnetic particle (such as one made of permalloy) the free energy depends on the particle shape only.[3] Figure 12.2 shows how the magnetostatic energy (demagnetization energy) of rectangular nanomagnets depends on its magnetization direction.[4] There are two equivalent ground states along the longer axis, i.e., the magnet prefers to point "up" or "down" in that direction. This direction is also called the "easy axis" of the nanomagnet. Demagnetization energy is highest if $\mathbf{m}$ points along the shortest axes, called the "hard axes." One can see that there is an energy barrier, $E_{\text{barrier}} \approx 7.8$ eV $\approx 300kT_{@T=300\text{K}}$, between the two ground states (units of $kT_{@T=300\text{K}} \approx 0.0259$ eV are assumed). The fundamentally lowest energy required to process information is $E_{\text{min}} = \ln 2kT$ and processes involving energy barriers of several tens/hundreds kT have a long-term stability against thermal noise. For the rest of the chapter, energies will be expressed in kT units, where $T = 300$ K is assumed. For comparison, the energy scales involved in the switching of sub-micron CMOS gates are $E_{\text{diss}} \approx 10^4$–$10^6 kT$.

The calculation leading to Fig. 12.2 assumes that the particle stays in the uniformly magnetized, single-domain state, and the magnet energies are governed by elementary magnetostatics. This is valid only for sufficiently small or high aspect ratio particles, where the energy cost of splitting into multiple domains would be prohibitively high. Figure 12.3 quantifies what "sufficiently small" means. The simplest domain configuration of nanomagnets is a vortex state, in which the non-uniform magnetization (flux

---

[2] For an excellent, thorough treatment of micromagnetics and domain theory the reader is referred to [18] and [19].

[3] Permalloy is a nickel-iron alloy, with negligible magnetic anisotropy. The magnetic energy of a permalloy nanomagnet does not depend on the angle between its magnetization and crystallographic axes.

[4] For isotropic, single-domain particles only the magnetostatic energy will depend on the magnetization direction.

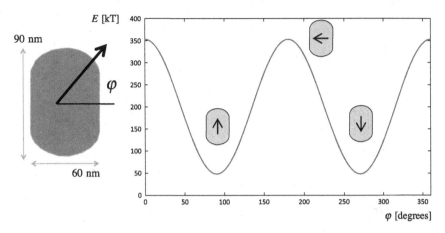

Fig. 12.2 Energy landscape of a single-domain nanomagnet. The magnet prefers to point up ($\phi = 90$ degree) or down ($\phi = 270$ degree)–this results in a bistable system that can represent one bit of information. The two (degenerate) ground states are separated by a $\Delta E = 300kT$ energy barrier, providing long-term stability. For details of the calculation, see [20]. The energy units in the y-axis are $kT_{@T=300K} \approx 0.0259$ eV.

closure state) reduces magnetostatic energy at the expense of creating domain walls (i.e., increasing the exchange energy).

Figure 12.3 shows a phase diagram of a 20 nm thick permalloy particle, i.e., the energy difference between the single-domain and vortex states. Like most simulation results in this review it was calculated by the widely used OOMMF code [21].[5]

According to the phase diagram of Fig. 12.3, magnets that are shorter and narrower than 50 nm remain in the single-domain state regardless of their shape. Larger magnets stay single-domain only if their aspect ratio is sufficiently high. Using magnets with "extreme" aspect ratios is impractical in NML – these magnets display complex switching behaviors despite their simple ground state. Typical magnet dimensions we used in experiments are 60×90 nm and 100×200 nm.

The theoretically calculated phase diagram of Fig. 12.3 fits well to experimental data, such as the magnetic force microscopy (MFM) data of Fig. 12.4. MFM is a common technique used for evaluation of magnetic systems and most experimental results on NML circuits were obtained using MFM. MFM maps the magnetic field above the sample: bright or dark spots in an MFM image indicate magnetic poles. In Fig. 12.4, a pair of bright/dark spots indicate a uniformly magnetized single-domain state with a single north and south pole. For larger or lower aspect ratio magnets, an internal domain structure appears and this is indicated by contrast inside the magnets, along the domain wall boundaries.

---

[5] The Object Oriented MicroMagnetic Framework (OOMMF) is a widely used, public-domain micromagnetic software micromagnetic program, see http://math.nist.gov/oommf/. See also [21]. We also used a non-zero temperature extension of the simulator, developed by the group of R. Wiesendanger: www.nanoscience.de/group_r/stm-spstm/projects/temperature/download.shtml.

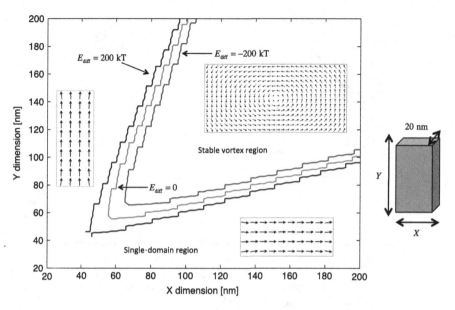

Fig. 12.3 Phase diagram of the single-domain/multi-domain transition of a single-domain nanomagnet [20], showing the size domain where nanomagnets can be viewed as ideal switches. We calculate the energy corresponding to the single domain and vortex states and draw the contour lines separating the two phases. Single-domain configuration is the lowest-energy state only for very small ($<$ 50 nm) or high aspect ratio nanomagnets.

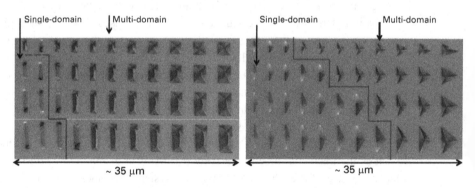

Fig. 12.4 MFM image of nanomagnets with different shapes. A magnet displaying a single bright and dark spot is in the single-domain state and domain walls (lines inside the magnet) indicate a more complicate multi-domain structure. The transition from single-domain to multi-domain state is clearly visible [22] as the magnet size grows. Short and high aspect ratio magnets on the left-hand side of the figure remain in single-domain state regardless of shape.

12.3   Behavior of coupled nanomagnets

In order to quantify the coupling strength of two single-domain nanomagnets, we plot (see Fig. 12.5) the energy landscape of a nanomagnet in the presence of an antiferro-magnetically coupled neighbor. It is seen from Fig. 12.5 that the nanomagnet now has a

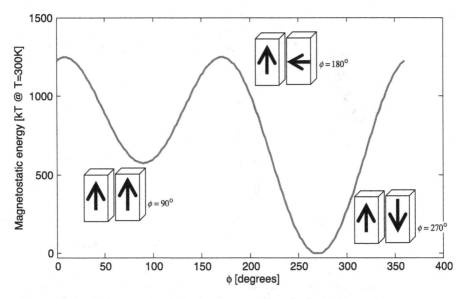

Fig. 12.5 Energy landscape of a single-domain nanomagnet in the presence of an antiferromagnetically coupled neighbor. The x-axis is the angle of magnetization, the y-axis is the energy – the left-hand magnet is assumed to be pinned in an upward-pointing state. The coupling prefers an anti-parallel alignment – there is a $600kT$ energy difference between the two (formerly degenerate) easy-axis states. The size of the magnets is 60 nm × 90 nm × 20 nm and the gap between them is 20 nm wide. (Based on [15]). The energy units in the y-axis are $kT_{@T=300\text{K}} \approx 0.0259$ eV.

unique ground state; i.e., the coupled system prefers to be aligned in an anti-parallel way. Notably, the coupling energy ΔE is comparable to the energy barrier height E_{barrier} coming from shape anisotropy, and is in the few hundreds of kT regime. This indicates that one should expect to see robust ordering phenomena.

In order to experimentally quantify magnetic coupling, we describe below an experiment that measures the hysteresis curve in an unusual configuration. The M_y (B_x) hysteresis of coupled nanomagnets provides deep insight into the switching and coupling properties of single-domain of nanomagnets.

Figure 12.6(a) gives a sketch of the experimental setup, along with the definition of the x and y directions. We refer to the horizontal magnet as the "driver" magnet, and the vertically oriented magnet as the "driven" magnet. External magnetic field is applied along the x-axis, which is the easy axis of the horizontal magnet, but the hard axis of the vertical (driven) one. There is dipole coupling between the driver and driven magnets, which we denote by B_{cpl}. This coupling field is acting upwards in the scenario of Fig. 12.6(a).

If the external field is applied along (or on a small angle to) the $+x$-axis, then, as the field is reduced towards zero, the B_{cpl} field will turn the driven magnet upwards. The two-magnet system ends up in its ground state at $B_x = 0$. The black curve of Fig. 12.6(b), shows this case. However, if the external field is applied along an angle, then the driven magnet experiences the superposition of two fields along its easy axis.

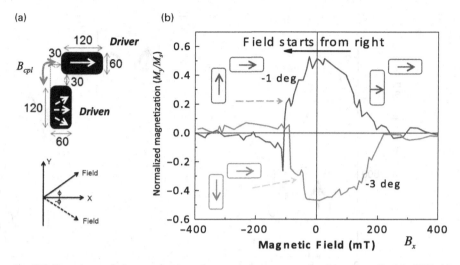

Fig. 12.6 Experimental characterization of a coupled nanomagnet pair, as described in [23]. (a) Geometry of the magnets. Magnetic fields are applied along the x-axis or at a small angle to it. The vertical magnet responds to the superposition of the coupling field and the small B_y^{ext} vertical component of the external field. These fields compete with each other (b); for $\phi=-1°$ the coupling wins and for $\phi=-3°$ the external field wins. These data were obtained using a vibrating sample magnetometer (VSM) and a sample comprising several million nominally identical nanomagnet pairs.

For $\phi < 0$ (bottom of Fig. 12.6(a)) the coupling field B_{cpl} will act against the y component of the external field, which is $B_y^{ext} = |B| \sin \phi$. Whether B_{cpl} or $B_y^{ext} = |B| \sin \phi$ wins determines whether the driven dot will eventually point up or down.

Figure 12.6(b) shows that at $\phi \approx -3°$ the magnet turns downwards – it turns out that this angle is actually the tipping point. So at $\phi \approx -3°$ and $| B_{ext} | \approx 150$ mT the vertical components of the external and coupling fields are roughly equal: $B_{cpl} \approx B_y^{ext}$. This results in volume-averaged coupling field of $B_{cpl} \approx 7$ mT. The coupling energy can be straightforwardly calculated from the coupling field, yielding $\Delta E = 600kT$. This barrier is higher than the one shown in Fig. 12.5, because the magnet sizes in this experiment were larger.

12.4   Engineering the coupling: gates and concatenated gates

The two-magnet coupling scheme above described can be straightforwardly extended to construct nanomagnet wires, i.e., chains of coupled single-domain nanomagnets that transmit information via magnetostatic coupling, as shown in Fig. 12.7. The magnetization state of the driver is propagated along multiple magnets, either via ferromagnetic coupling (Fig. 12.7(a)) or antiferromagnetic coupling (Fig. 12.7(b)).

Perhaps the most important component of NML circuits is the magnetic majority gate, shown in Fig. 12.8(a) [9]. Its functionality could be best understood by considering the "programming input" in Fig. 12.8(a) to be fixed pointing upwards. In that case,

(a)　　　　　　　　　　　　　　　　　　　(b)

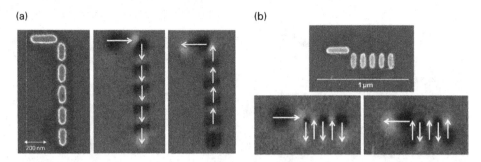

Fig. 12.7 Experimental characterization of nanomagnet wires. (a) Ferromagnetic coupling and (b) antiferromagnetic coupling is used to propagate information [24]. Both SEM images and MFM images are shown. Images are courtesy of E. Varga, see also [24].

(a)　　　　　　　　　　　　　　　　　　　(b)

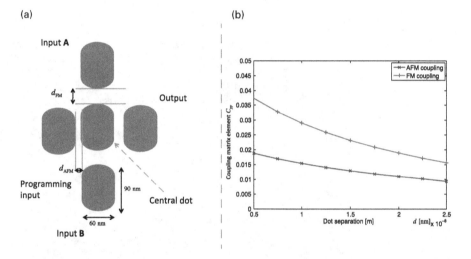

Fig. 12.8 (a) Majority gate with input and output dots. (b) Strength of ferromagnetic and and antiferromagnetic couplings in the gate [20]. The coupling magnetic fields of (b) are normalized to the saturation magnetization of permalloy ($M_S = 8.6 \times 10^5$ A/m). The graph of (b) is used to design a balanced majority gate, where $d_{AFM} > d_{FM}$ and all inputs have an equal weight in determining the state of the central dot.

the center dot is biased downwards due to antiferromagnetic coupling, so the center dot points upwards only if both input A *and* input B are pointing up. The output inverts the central dot's magnetization state, so the gate performs the NAND(A,B) operation.

There are many subtle details to consider for the design of the majority gate. For example, the center dot has to weight its inputs equally, i.e., the coupling fields from inputs A, B and the programming input should be identical. Since antiferromagnetic coupling is weaker than ferromagnetic coupling, inputs A and B must be placed somewhat farther than the programming input. Figure 12.8(b) shows how the coupling fields change with distance, which can be used for finding the optimal distance. A more sophisticated design also has to take into account parasitic (non-nearest neighbor) interactions between the magnets [25].

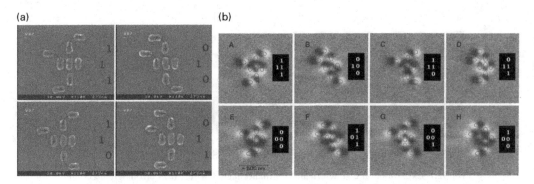

Fig. 12.9 Experimental demonstration of majority gate operation [9]. (a) SEM and (b) MFM images of four majority gate designs, each magnetized in opposite directions.

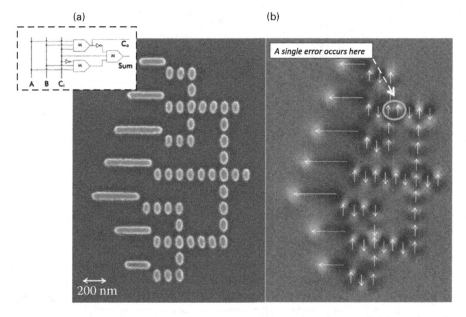

Fig. 12.10 Experimental realization of a full adder – the most complex coupled dot array so far built [10]. (a) SEM image of the adder and (b) MFM image after magnetic fields were applied to drive it into ground state. The logic diagram is shown in the inset, based on [27]. The level of ordering is promising, but not perfect: one error occurs at the output of a majority gate (see encircled region).

The majority gate was first demonstrated experimentally by our group in [9] – SEM and MFM images verifying its operation are shown in Fig. 12.9. In this demonstration the inputs were "hard wired": driver magnets placed at different positions provided the different input combinations to the gate.

Cascading NAND gates/majority gates enable any Boolean mapping. One needs auxiliary configurations (such as fanouts [26]), which are demonstrated as well. Properly designed structures show remarkably robust ordering. As an example, experimental realization of a 1-bit full adder is shown in Fig. 12.10.

The driver magnets of Fig. 12.10 are more elongated than the magnets building up the rest of the gate. Higher aspect ratio magnets have higher switching fields so the magnetization of drivers can be fixed before the rest of the circuit relaxes to ground state – this way the operation of the NML circuit can be tested for arbitrary input combinations without the need to fabricate electrical inputs.

The degree of ordering is remarkably high in the gate of Fig. 12.10, but a watchful eye can see that this adder is not working properly. An ordering error occurred at the output of one majority gate, as indicated by a circle in the MFM image.

Experimental evidence (and basic physics insights) show that a larger array of nanomagnets is extremely unlikely to order perfectly in its ground state. Ordering errors (commonly referred to as frustration) will inevitably occur. One needs to control the ordering dynamics in order to eliminate these errors.

12.5   Errors in magnetic ordering

Analysis of MFM images indicates that dots close to a strong input (such as the ones on the left-hand side of Fig. 12.10) almost always switch into the ground state dictated by their neighbors. But nanomagnets farther away from the input often switch into seemingly random states. These magnets do not feel a strong input and their magnetization is set by "random" magnetic fields. The sources of these random fields are:[6]

(1)   *Fabrication variations*: due to variations in shape and misalignment in easy-axis directions, magnets may experience a built-in upward/downward bias field, which may overwhelm the coupling.

(2)   *Non-nearest neighbor couplings*: non-nearest neighbors may drive the circuit into a state different from the desired computational ground state [25].

(3)   *End domain states*: small deviations from the single-domain state may occur that make magnetic coupling fields dependent on the magnetization history of the magnet.

(4)   *Temperature fluctuations*: the effect of non-zero temperature appears as a random (white-noise like) magnetic field that causes switching field variations and switching into random directions [29].

Note that the effect of item 1 can be reduced by optimizing the technology, while the impact of item 2 is reduced by proper design. Item 3 becomes less important with shrinking magnet sizes. Item 4 is the most fundamental, as it prescribes that the coupling field should always be larger than the field caused by thermal fluctuations in order to avoid unacceptably high error rates.

Figure 12.11(b) is an experimentally measured (VSM) hysteresis curve of a large (several-million magnet) array of $100\,\text{nm} \times 200\,\text{nm} \times 20\,\text{nm}$ size nanomagnets at

---

[6] From the physics point of view, the nanomagnet wire of Fig. 12.7 is similar to an Ising chain – a well-established model system in physics known to order with errors at non-zero temperature. It was recognized early that such ordering errors are a main obstacle in the operation of NML devices [28].

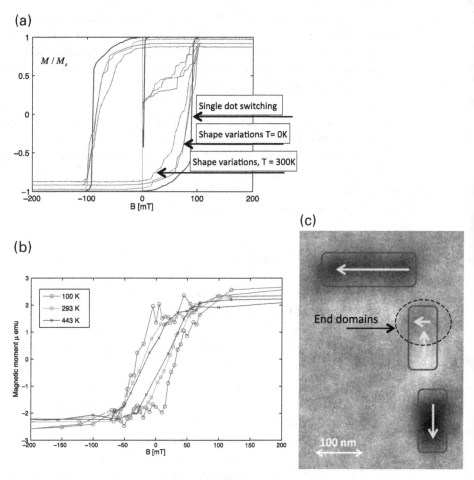

Fig. 12.11 Experimental characterization of nanomagnet variability. (a) Simulation showing how temperature fluctuations and fabrication variations will yield to a switching field distribution [30]. The hysteresis loop of an "ideal" nanomagnet is square-shaped. (b) The experimental data of [30] confirm this. (c) SEMPA image of nanomagnets after switching [30]. The lack of a clear magnetic pole indicates an end-domain structure in the middle magnet.

various temperatures [30]. Such-shaped nanomagnets should have a well-defined switching field and a perfectly square hysteresis loop. The slope of the hysteresis curve indicates a switching field distribution. Comparing the measurements and simulation results, Fig. 12.11(a) clearly shows that both shape variations and thermal effects contribute about equally to the broadening of the switching field.

Figure 12.11(c) is a scanning electron microscopy with polarization analysis (SEMPA) image of coupled nanomagnets – the high-resolution and non-invasive nature of this technique reveals the end-domain states that may make the coupling and switching fields dependent on the history of the magnetic switching – this exemplifies the mechanism mentioned as item above.

We discussed earlier that large coupling energies (typically exceeding $100kT$) should guarantee robust ordering. However, criteria for temperature stability of coupled n-magnet systems is a more complex problem. An n-magnet system is characterized by an at least n-dimensional energy landscape; this landscape may have many possible transition paths. Temperature fluctuations may switch the magnet system to an error state via a path through these transition states.

To fully characterize the energetics of a pair of coupled single-domain nanomagnets, a two-dimensional energy surface (like the one shown in Fig. 12.12) can be constructed. The independent variables are the ϕ_1, ϕ_2 magnetization angles of the magnets. If the nanomagnets are high aspect ratio and not very close (like in Fig. 12.12(a)), then there is a sufficiently high barrier between the various stable states; however, for other geometries (like in Fig. 12.12(b)), a path opens between the ground state and the metastable state if both magnets switch simultaneously via a $\phi_1 = \phi_2$ ferromagnetically coupled state. This is indicated by a $\leftarrow\leftarrow$ symbol on the figure. If there is a path in the energy landscape below $E_{\text{barrier}} = 40kT$ height, then the long-term stability becomes questionable; if E_{barrier} falls in the $\approx 10kT$ range, then magnetic ordering may not be observable at all.

12.6   Controlling magnetic ordering: clocking of nanomagnets

12.6.1   Control of speed and error rates using hard-axis clocking

Errors can be minimized or entirely eliminated by controlling the switching dynamics of nanomagnets. This can be done by applying hard-axis fields that drive out metastable states by first switching the nanomagnets to a "neutral" hard-axis logic state. Subsequently the hard-axis field is gradually removed. As the hard-axis field decreases, the magnets choose either the up- or down-pointing easy-axis state. Small coupling fields along the easy axis can tip the magnetization upwards or downwards. Upon the removal of the external field, dots close to the drivers initiate the ordering, and subsequently all the farther-lying magnets relax to their computational ground state. This is the extensively studied "hard-axis clocking" scheme [31] for NML (illustrated in Fig. 12.13).

Hard-axis clocking alone cannot eliminate all ordering errors. Magnets lying far from the driver may still choose an arbitrary state (due to random fields) before the ordering "wave" can reach them. To illustrate this we simulated a hard-axis clocking scenario assuming that thermal noise is the only source of random magnetic fields [32]. The results are shown in Fig. 12.14(a): short wire segments (fewer than five nanomagnets) almost always work correctly, but longer wire segments have diminishing probability for working properly. Faster switching times also lead to more errors.

However, if only short NML segments are clocked together, errors can be avoided. This is shown in Fig. 12.14(b), where the clocking fields are applied to five magnets at once – once this wire segment is ordered, clocking of the next block starts. The graph of Fig. 12.14(b) shows the time-dependent switching process of the magnets and the

(a)

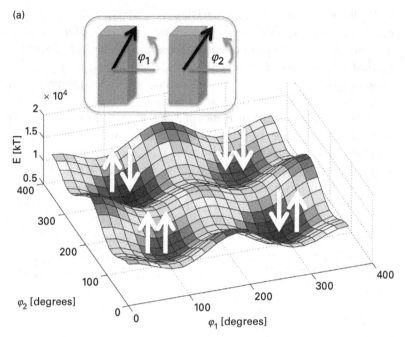

100 nm × 200 nm × 20 nm magnets, 50 nm apart

(b)

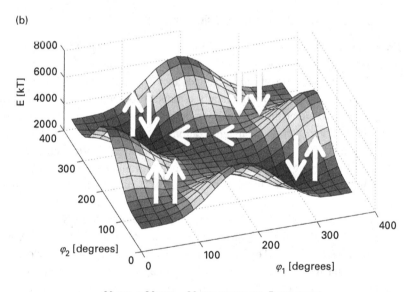

60 nm × 90 nm × 20 nm magnets, 5 nm apart

Fig. 12.12 Energy landscape of two-magnet system for two different geometries. Depending on the aspect ratios and the coupling strength, different transition pathways may exist between the ground state and the metastable states. In (a), there is a barrier in the center of the energy landscape, while in (b) a transition state (denoted as ←←) becomes available. The energy units in the y-axis are $kT_{@T=300K} \approx 0.0259$ eV.

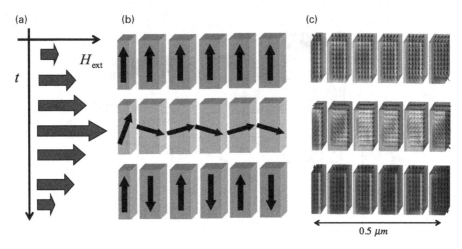

Fig. 12.13 Schematic of the hard-axis switching process. (a) Field sequence. (b) Cartoon showing that the increasing then decreasing hard-axis field switches the nanomagnets from an "all up" metastable state to the computational (up/down) state via an intermediate (hard-axis) state. The left-most magnet is narrower than the others and it keeps its upward-pointing magnetization state. (c) OOMMF simulation result [31].

"noise" on the m_y magnetization component is the actual thermal noise, simulated according to [20, 32]. Switching of a five-magnet zone was performed in $T = 10$ ns.

12.6.2   Energetics of NML clocking

In order to understand power flow in an NML system, it is instructive to study the effect of the hard-axis clocking process on the energy landscape, as shown in Fig. 12.15. The hard-axis magnetic field flattens the potential barrier between the two easy-axis states – in this state a very small field from the neighboring magnet (i.e., a small "kink" energy propagating in the magnetic signal) is sufficient to switch the magnet toward the up or down state. When the field is released (the potential barrier between the two states rises back) the logic state of the magnet stabilizes.

In this process, there is a significant energy exchange with the external magnetic field – this is required to manipulate the energy barrier. However, this field energy is not dissipated and could be recycled by appropriate driving circuitry. If the magnetic field is generated by current-carrying wires or coils (i.e., inductances), then coupling to the nanomagnets appears as an additional inductive load on the wires. Energy flows from the clocking apparatus toward the magnets when the magnets are biased in the high-energy, hard-axis biased state. Most of this energy will flow back from the inductor when the magnets order to their ground state. This is somewhat similar to the energy recycling in (transistor-based) adiabatic circuits. It is also clear that the clock signal has twofold functionality: it does not only control the magnetization dynamics, but also supplies power to the NML computation.

Dissipation in the computing magnets circuit will have two sources. If the clocking process is done fast (i.e., the magnet is allowed to quickly slide between states 1–7) then

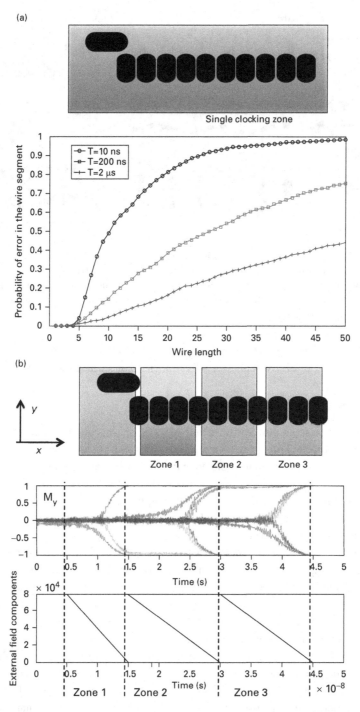

Fig. 12.14 (a) Probability of error as a function of wire length. If a long chain of magnets is placed in a single clocking zone (as represented in (a)), then thermally induced error rates become unacceptably high. Error rates are somewhat lower if clocking is done slower (for larger *T*). (b) By splitting the nanomagnet wire into shorter clocking zones, errors are eliminated [20].

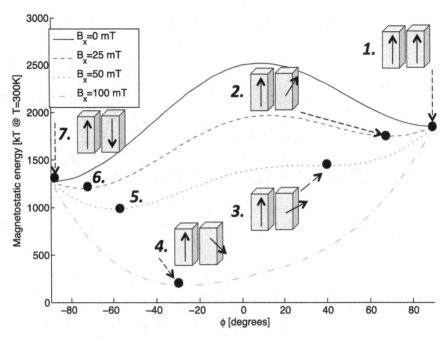

Fig. 12.15 Hard-axis clocking of nanomagnets on the energy landscape. Initially the magnet is in a metastable state (stage 1, $\phi = 90°$). A magnetic field along the x-axis direction (B_x) flattens the potential barrier separating the two easy-axis states (stages 2 and 3). Once the barrier completely disappears, the magnetization flips downwards (stage 4). When the field is removed and the barrier re-emerges, the magnetization settles to its ground state (stages 5–7: $\phi = -90°$). The $B_x = 0$ curve is identical to the corresponding section of the curve of Fig. 12.12, apart from an arbitrary constant energy term.

a large fraction of the magnetostatic energy (in the hard-axis state) will be converted into heat. In practice, this energy will fall in the few kT range [12, 33] if $t \approx 100$ ns. Also, some energy in the magnetic signal will have to be lost, since the energy propagating in the signal must be sufficiently above $\approx kT$ to stand out from thermal noise. Still, these considerations suggest that nanomagnets will dissipate close to the lowest limit allowed by thermodynamics [34–36]. Even when they are clocked very quickly ($T \approx 1$ ns), the dissipation per switch is in the 100 kT (2.5 eV) range. We will refer to the $E_{\text{base}} = 100kT$ energy as the baseline when estimating net dissipation in an NML system.

In practice, clock field generation will account for the vast majority of power dissipation in NML. If clocking fields are generated by bare wires, then the net power dissipation due to Joule heating will be several orders of magnitude larger than E_{base}.

12.7   Computing systems from NML

12.7.1   Components of the NML system

A high-level sketch of an NML computing system is given in Fig. 12.16. Hard-axis clocking fields are generated by wires running under the magnets. They provide

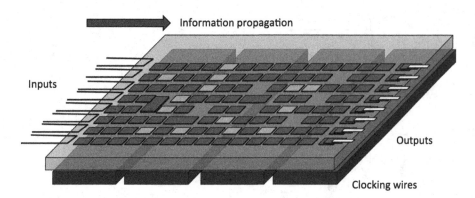

Fig. 12.16 A high-level sketch of the NML system indicating its main components. Inputs and outputs are applied at the edges, and information is processed inside the array in an entirely magnetic way [37].

localized fields for the NML blocks. Per the discussion above, a clocking line may drive only a few nanomagnets in a particular signal path, but a large number of non-consecutive blocks can be clocked by the same line. Not shown in the sketch is the circuitry generating the clock signals – this is assumed to be a "standard" CMOS pulse generator. The clocking lines can be long, and drive several parallel magnetic pathways.

Inputs and outputs are placed at the edges of the structure and provide the "boundary conditions" for the ordering. We assume that there are several NML gates between the inputs and outputs, and the system performance is dominated by the interior of the NML system, not the I/O structures. The practical size of the NML block of Fig. 12.16 depends on the computing task to be performed. As crossing of NML signals is problematical, the layout should consider several constraints. Furthermore, propagation of magnetic signals over a nanomagnet wire is several orders of magnitude slower than propagation of electronic signals would be. An irregular circuit layout with many long-range interconnections should be split into a number of electrically interconnected blocks.

12.7.2   Clocking structures and dissipation by the clocking apparatus

In practice, magnetic fields have to be generated by on-chip wires, which wastes a lot of power. Joule heating in the wires dissipates orders of magnitude more heat than the magnets [38]. Wires that are surrounded by a ferromagnetic cladding layer make field generation more efficient [39].

One obvious strategy to lessen resistive losses is to reduce the magnitude of the required B_{clock} clocking field – reducing B_{clock} by a factor of h reduces the required currents by h and the dissipation by h^2. B_{clock} can be reduced either by decreasing the aspect ratio of the nanomagnets or increasing their size. The room for optimization is small: per Fig. 12.3, increased nanomagnet size will result in undesirable multidomain states, and small aspect ratios and/or strong couplings result in thermal stability problems (per Fig. 12.12). Considering these constraints on the design space, the lower limit of the hard-axis clocking field lies in the $5 \text{ mT} < B_{\text{clock}} < 10 \text{ mT}$ range.

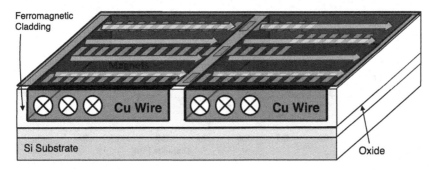

Fig. 12.17 Schematic of an NML system that is clocked by yoked wires placed under the nanomagnets [39].

Using bare wires for magnetic field generation, resistive losses in the clocking apparatus are about five orders of magnitude larger than the E_{base} baseline losses. Having a cladding layer (a magnetic yoke) around the wire reduces the overhead of field generation to about $E_{baseline}$ per switching. A schematic of such an NML system is given in Fig. 12.17.

Using enhanced permeability dielectrics (EPDs) promises further power savings. EPDs are aggregates of paramagnetic particles embedded in a non-magnetic matrix. They concentrate magnetic field lines around on-chip nanomagnets, and were shown to boost the power efficiency of MRAM devices [40]. In NML, they were recently demonstrated to allow a factor of three reduction [41] in clocking field and increase the net power efficiency in the vicinity of $10^3 \cdot E_{baseline}$ per switching. This is already better than advanced, low-power CMOS at iso-performance.

12.7.3   Novel methods of NML clocking

Clocking is the Achilles heel of NML circuitry, since the clocking apparatus accounts for most of the complexity of an NML layout and for the vast majority of dissipated power. This section discusses several proposals in the community that aim to reduce the overhead of clocking.

The idea behind domain wall clocking is to use the stray field of propagating domain walls to provide the clocking field. Domain walls can be propagated by weak external fields, but they generate strong, localized fields – exactly the type of fields required for NML clocking [42, 43]. The simulation of Fig. 12.18(a) shows the magnetic field distribution emanating from a domain wall – this field is very strong ($B \approx 300\,\text{mT}$ next to the wall) and very localized. Experimental results (Fig. 12.18(b)) show that the propagating wall can indeed flip neighboring nanomagnets.

Electric-field control of magnetism is a new, intensively researched area, promising to switch nanomagnets without the need of current flow. Possibilities include multi-ferroic materials [44], piezoelectric stacks [45, 46] or electric field control of magnetic anisotropy [47].

It was also demonstrated that without any external field, thermally activated switching can drive small NML circuits into the ground state [48], leading to speculation about a Brownian computer in NML.

(a)

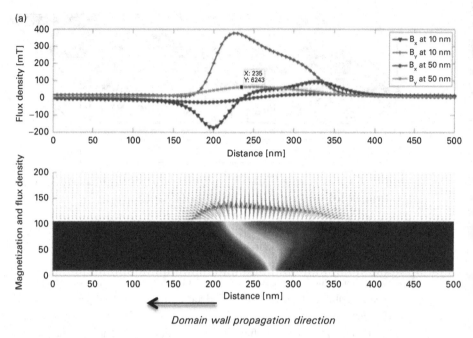

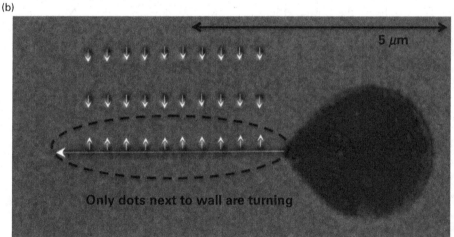

Domain wall propagation direction

(b)

Fig. 12.18 Simulated field distribution and experimental demonstration for domain-wall assisted switching of nanomagnets [43]. (a) Magnetic field distribution of a propagating transverse domain wall, that (in a small volume) significantly exceeds the switching field of nanomagnets. (b) MFM image showing nanomagnets adjacent to a domain-wall conductor. Magnets lying close to the domain wall (and only those) can be switched [43].

12.7.4   Inputs and outputs

In practice, NML circuits will be embedded in a CMOS-based system and require a magnetoelectrical interface (MEI) to communicate with the surrounding circuits. So far, most NML demonstrations have focused only on the study of the magnetic parts alone.

Inputs were imitated by magnets having different switching field from the rest of the NML circuit. In some work current loops or wires were used to set the input dots of NML circuits [49, 50]. Outputs were mostly tested either by MFM or optical methods. Electrical outputs were based on the extraordinary Hall effect (EHE) [51, 52] or Hall effect [53] – these output structures are well suited for laboratory characterization but impractical in a circuit environment.

Magnetic random access memory (MRAM) and spin transfer torque RAM (STT-RAM) [54] both use magnetic multilayer structures to sense the magnetization of a nanomagnet. They are becoming a mainstream technology, and it is straightforward to adopt these technologies for the MEI of NML. An appealing feature of STT devices is that essentially the same physical structure can be used for input and output: a STT-based multilayer can both switch the input magnets and sense them via spin-dependent tunneling in a magnetic tunnel junction (MTJ).

A number of designs have been proposed [55, 56], and there are experimental demonstrations of coupled nanomagnets integrated with read-out structures [57, 58]. However, we are not aware of a functional experimental demonstration of an NML gate with STT-based inputs and outputs. There are a number of design trade-offs that must be considered [59]. For example, STT inputs work well for thin, $d \approx 3\,\mathrm{nm}$ thick free layers, while typical nanomagnets in an NML circuit are in the $d \approx$ 10–20 nm thick range, which maximizes coupling. The multilayer structure should be very precisely compensated to avoid any residual stray fields. Stray fields originating from the fixed layer or the bias caused by effects such as orange peel coupling and exchange bias can easily overwrite the coupling field of the neighboring NML output.

No matter which technology is chosen for the MEI, it will add to the energy footprint, and will significantly complicate fabrication. Magnetic multilayers, tunnel oxides, and exchange biasing layers typically require a much more complex technology than fabrication of a single (or few) magnetic layers, and electric signals must be routed on top/below the magnets.

12.8   Nanomagnet logic in perpendicular media

12.8.1   Properties of magnetic multilayers with perpendicular anisotropy

In the NML devices shown so far, the bistable behavior of nanomagnets originated from shape anisotropy. This results in an in-plane easy axis, which places constraints on the circuit layout. Using the out-of-plane magnetization direction for representing the information would relax these constraints and allow the arrangement NML gates much more freely on the chip surface. The variant of NML that exploits out-of-plane (perpendicular) anisotropy, referred to as pNML, was pioneered by a group at the Technical University of Munich, Germany [59].

An out-of-plane magnetic easy axis can be realized by magnetic multilayers (such as in stacks of alternating cobalt/platinum or cobalt/nickel layers). A typical layer composition that exhibits perpendicular anisotropy is $Pt_{5nm}5 \times [Co_{0.3nm} + Pt_{0.8nm}]Pt_{4.5nm}$. The cobalt layers must be a few atomic layers thick, or the shape anisotropy overwhelms the interfacial out-of-plane anisotropy. The film can be thick overall, consisting of up to 40 bilayers [59]. Depending on the layer composition/thickness, nanomagnets may stay single-domain up to microns in size [60], and the domain structure is fairly insensitive to the shape of the single-domain dot.

The switching properties of perpendicularly magnetized multilayers are different from permalloy nanomagnets, and enable the construction of logic gates with higher functionality. Magnetic multilayers typically switch by domain wall nucleation and propagation: magnetization starts turning at a certain spot (the nucleation center) of the film, and the reversal propagates across the magnets. The switching properties of the multilayer film (or dots made out of it) are dominated by the spatial distribution of the magnetic anisotropy: low-anisotropy regions will be nucleation sites (where switching can start easily) and high-anisotropy regions act as pinning sites, which may stop wall propagation.

The magnetic anisotropy can be altered intentionally: focused ion beam (FIB) irradiation can locally alter the multilayer structure and create artificial nucleation centers (ANCs). In other words, it is possible to define at which point switching will start. Engineering the position of the ANC provides a tool to control the switching and coupling behavior of multilayer-based nanomagnets [61, 62].

In order to study how ANCs can be used to engineer switching, a three-magnet structure was fabricated, as is shown in the AFM image of Fig. 12.19(a). An ANC was created at the left-hand side of the middle magnet (T). The left and right magnets (denoted as D1 and D2, respectively) have no ANC. The magnets were probed optically with magneto-optical Kerr-effect microscopy (MOKE) in order to obtain individual hysteresis loops of the individual dots. An example of these hysteresis loops is given in Fig. 12.19(b). The switching fields of D1 and D2 are determined by their layer composition – they are significantly different due to the non-uniformity of the as-grown film and also due to the rough edges introduced by the etching process. The switching field of the magnet "T" can be precisely tuned by the ANC FIB dose, and is the lowest of three magnets [63].

Most importantly, not only the switching but also the coupling properties of the nanomagnets can be controlled by the FIB irradiation. The measurements of Fig. 12.19(c) show the switching field of the T magnet for all possible magnetization directions of the D1 and D2 magnets. A distribution of switching fields arises from temperature fluctuations [64] and inaccuracies of the measurement setup. However, it is clearly observable that the center magnet is sensitive only to the coupling field of its left neighbor (D1) and entirely insensitive to the field of its right neighbor (D2). This is because the ANC is intentionally placed on the left-hand side of T, right in the stray field of D1, and where the stray field of D2 is negligible. Properly placing the nucleation site can be used to (1) enhance effective coupling between nanomagnets and (2) selectively couple the magnets to certain neighbors and decouple them from others. In NML devices, this can be used to define input and output sides of a dot and direction of signal propagation.

Figure 12.20(b) shows experiments on a pNML inverting majority gate. The center magnet has an ANC. This eliminates the input/output symmetry of the permalloy NML gate. Only the inputs can switch the output, thus preventing the back-propagation of signals from the output toward the inputs.

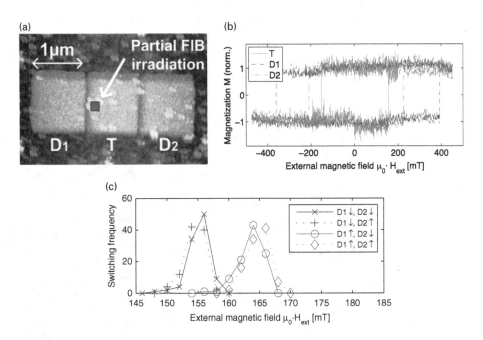

Fig. 12.19 Experimental demonstration of uni-directional coupling between nanomagnets. (a) AFM image of three coupled CoPt nanomagnets, the middle one (T) being irradiated on its left-hand side. (b) MOKE measurement of the hysteresis loop on each of the magnets – only T will switch at small fields. (c) Switching field distribution of magnet T for various states of the neighboring magnets. The switching field of T depends only on the state of its left neighbor (D_1) and not on the right (D_2). For details see [63].

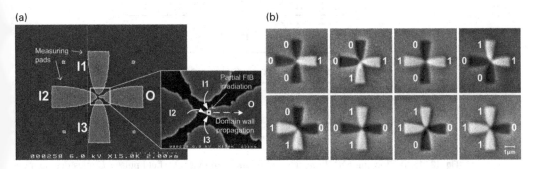

Fig. 12.20 Inverting majority gate in pNML. The ANC in the center dot defines signal propagation direction. (a) SEM micrograph of the gate. (b) MFM image showing correct operation for all input combinations. For details see [65].

12.8.2   Clocking

The fact that each nanomagnet is sensitive to its (say) left neighbor allows an entirely deterministic demagnetization (clocking) scheme using global external fields. Figure 12.21(a) sketches how a pNML wire, starting from an arbitrary initial state, ends up in its computational ground state after the metastable state is deterministically driven out by oscillating field pulses. We refer to this scheme as "easy-axis clocking." The fact that the metastable state always steps toward the right eliminates the "wandering" of error states (present in permalloy-based NML [66]) and results in a perfectly deterministic magnetization dynamics, even using global clocking fields.

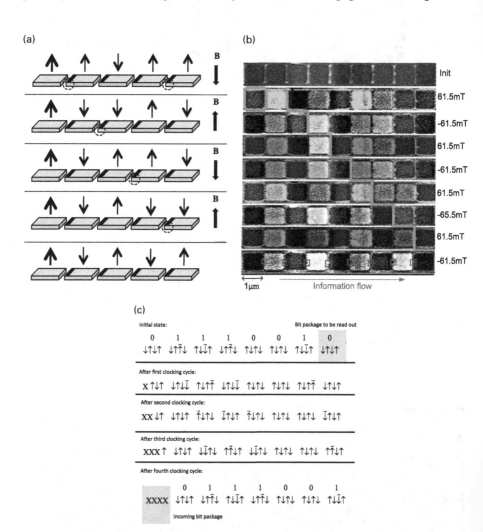

Fig. 12.21 (a) Clocking pNML using spatially uniform magnetic field pulses. Circles denote the metastable states driven out in the next clocking phase [62]. (b) Experimental data: white/black colors correspond to up/down magnetization [67]. (c) Four-bit sequences can use pNML wires as a ratchet, storing, and shifting information [62].

Figure 12.21(b) shows magneto-optical images of data moving through a clocked pNML wire. Experimental results verify the above simple model.

Figure 12.21(a) and (b) essentially depicts a shift register (ratchet). By forming a loop from this shift register, a ring-oscillator-like device can be built [68]. These shift registers can also be used for non-volatile storage. For example, bit values can be represented by four-magnet sequences where the presence or absence of a metastable state indicates the stored bit value. The stepping of such four-bit sequences is exemplified in Fig. 12.21(c).

Perhaps the most important benefit of easy-axis clocking is that it facilitates an efficient geometry for generating magnetic fields. For example, in the geometry described in [69], the nanomagnets are placed in the gaps of an on-chip electromagnet. Two high-permeability magnetic layers below and above the nanomagnet layer act as the poles of the electromagnet. The gap between the poles is only a few to ten nanometers wide, and electrical wires can be several micrometers apart. Preliminary finite element calculations show that this electromagnet can generate several 10 mT fields at a few times 10 MHz frequencies [69]. The energy overhead of clocking is about $10 \times E_{\text{baseline}}$, meaning that pNML systems offer order of magnitude power savings compared to predicted, end-of-roadmap, low-power CMOS at iso-performance.

12.8.3   A full-adder benchmark structure

Just as in the case of in-plane NML, a full adder was used to benchmark a more complex circuit [70]. The structure and MFM images are shown in Fig. 12.22(a) and (b), respectively. The circuit schematics are identical to the one shown in Fig. 12.10, and there are three majority gates that produce the sum, S, and carry-out, C_{out}, signals.

Nanomagnet sizes in Fig. 12.22 are on the micron scale – this is for experimental convenience, as these large-sized magnets enable direct optical probing. There is not yet experimental data on pNML devices with sub-100 nm size dots, but simulations indicate no roadblocks against scaling pNML to this size range [62].

The full adder structure also illustrates that most circuits are easier to lay out in pNML than in permalloy-based NML devices. Thanks to the relatively large single-domain size in thin CoPt or CoNi films, magnets can be arbitrarily shaped up to, typically, a few micrometers in size, simplifying longer-range interconnections. An elongated CoPt stripe (called the domain wall conductor) can be used as the interconnect. Such domain-wall conductors can also be used to cross signals [71], which is a challenge in permalloy-based NML. There are experiment-based compact models [72] that help pNML design and estimate the robustness of the resulting circuits.

12.9   Case study on two circuits

In order to illustrate some challenges related to more complex NML circuits, we show two examples below.

The first design concerns the permalloy-based full adder circuit. The design of Fig. 12.10 consists of long antiferromagnetically coupled nanomagnet-wires that are

a)

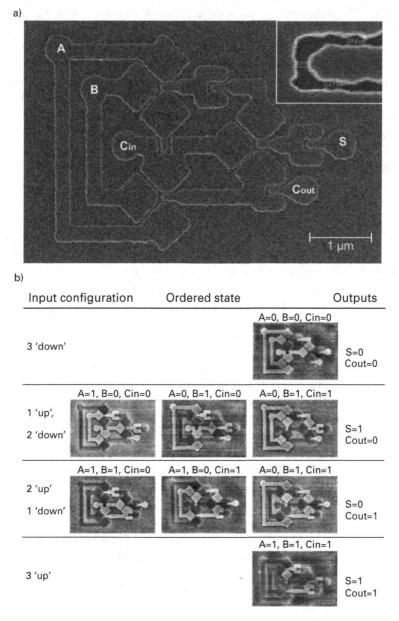

b)

Fig. 12.22 A one-bit full adder realized from Co/Pt multilayers [70]. (a) SEM image. (b) MFM images for various input combinations.

prone to errors and would require to be split into multiple clocking zones in order to operate properly. We simplified the design of circuit of Fig. 12.10 so that it would operate in a single clocking zone.

Slant-shaped magnets [73] act like tilted nanomagnets [74], switching into a definite upwards or downwards state even when their clocking field is applied along their

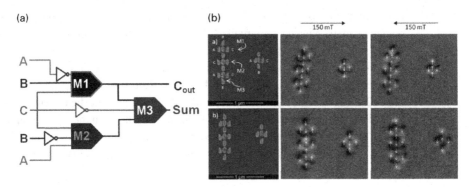

Fig. 12.23 One-bit full adder designed from slant nanomagnets. By using slant magnets and assuming that inputs can be applied at a number of positions the layout greatly simplifies. (a) Circuit schematic. (b) Test of the gate for two different hard-wired input combinations (input *B* has changed). For more details see [75].

geometrical short axis. Biasing dots of majority gates can be substituted by slant-shaped magnets [75], greatly reducing the number of magnets required for certain functions. The generation of the carry and sum bits can be split into two separate circuit parts, eliminating long interconnections. The design shown in Fig. 12.23 was verified to operate correctly for all input combinations. More complex permalloy-based NML circuits will almost certainly require an integrated clocking apparatus per Section 12.6.

Our second example is based on pNML, and exists yet only in simulations. We pointed out earlier that NML devices are the most amenable to circuits where deep pipelining is desirable and no long-range interconnections are required. Systolic circuits are perfect examples for such circuits, and we choose a systolic pattern-matching circuit as an example of larger-scale circuits built with pNML technology.

The purpose of an *n*-bit systolic pattern matcher is to identify *n*-bit-long data sequences in a continuous stream of binary information. It can be built from *n* identical, nearest-neighbor interconnected circuit blocks, as illustrated in Fig. 12.24. The circuit block is challenging to design; it requires inputs and outputs to be placed in proper parts of the circuits (so the circuit is tileable) and timing constraints to be satisfied (for details see [76]). However, once the block is designed and verified, then arbitrarily large pattern matching circuits (i.e., for any *n*) can be put together by simply repeating this block. Input/output and long-range signal routing constraints do not appear in this design, while the benefits of NML circuits such as non-volatility, overhead-free pipelining, no clock-signal routing problems, low power, and massive parallelism are exploited to their fullest extent.

12.10   The circuit paradigm in NML

In electronics, there is a relatively well-established "checklist" of requirements that a new electronic device should satisfy in order to be a useful building block for electronic circuits. For example, a device should exhibit power gain, sufficient non-linearities, and noise immunity to be practicable. In this concluding section, we point out how NML devices can be viewed as circuits, thus assessing their value as building blocks of complex circuits and systems.

Circuit theory deals with "black box" models that interact with their environment via well-defined ports. Such a black box model for nanomagnets is sketched in Fig. 12.25(a), where a nanomagnet controlled by magnetic field generated by wires (current loops) is shown. Physically, current flowing through the wires induces magnetization dynamics (i.e., precession of $\mathbf{M}(\mathbf{r},t)$) resulting in a time-dependent flux, which induces counter electromotive force on the wires. Energy between the wires and the

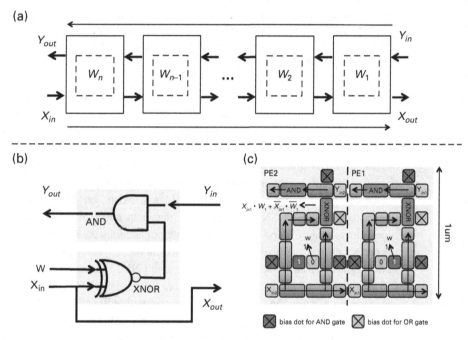

Fig. 12.24 Design of a systolic pattern-matching circuit from pNML [76]. (a) Block diagram – each block is responsible for matching one bit of the pattern in the input stream. (b) Circuit-level schematic of the block and (c) its pNML implementation.

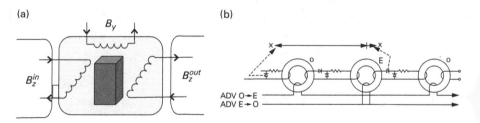

Fig. 12.25 (a) Circuit model of nanomagnets interacting with their environment via magnetic fields [74]. (b) An ancient, ferrite-ring-based clocked shift register from [77]. From a circuit theory point of view this ferrite-ring device is based on principles very similar to that of NML.

magnet is exchanged in the same way as an inductor would exchange energy with its circuit environment.

The equivalent circuit of the nanomagnet can be constructed using three coupled, non-linear inductors as detailed in [74]. The non-linearities depend on the shape of the nanomagnet, and the three inductors are required to represent the components of the $\mathbf{M}$ magnetization vector. The three ports corresponding to the B_x, B_y, B_z magnetic fields act on the magnet.

The equivalent circuit is based on a simplified micromagnetic model (single-domain approximation [74]). But, unlike the micromagnetic model, it facilitates a straightforward analysis of the energy flow in the signal path. It allows direct analysis of the energy exchange between power clock, magnets, and dissipative environment. One can also see that, in the hard-axis clocking scheme, the nanomagnets represent an inductive load to the clocking wires, and most of the energy pumped into the magnets can be (at least in principle) recycled from the clock signal, as happens in adiabatic circuits.

We found that many problems related to the system properties of coupled nano-magnets (such as power gain, dissipation, noise, non-linearities, etc.) boil down to the question of whether it is possible to construct useful circuits from non-linear inductors. Interestingly, there is a definite answer to this in the literature. Computers utilizing non-linear inductors exist [78–80], as do standalone magnetic amplifiers [81, 82], which were the active circuit elements of inductor-based technologies. Figure 12.25(b) shows one of those inductor-based circuits [77]. NML devices use magnetic-field-based interconnections in place of wired interconnections, but use the proven principles of circuit design from non-linear, parametrically controlled inductors. This observation demonstrates that future nanoelectronic devices, such as non-linear inductances realized by nanomagnets, may not replace the transistor as a switch, yet be equally useful as circuit elements.

12.11   Outlook: the future of NML circuits

This chapter has provided an overview of nanomagnetic logic, a computing architecture that exploits magnetic ordering for computation. We described how elementary magnetic phenomena (magnetic switching and magnetic dipole interactions) can be used to engineer a complex computing system that is functionally equivalent to microelectronic circuits.

Since its inception in 2000, NML has reached an advanced state as one of the very few unconventional circuit technologies where functional circuits (well beyond elementary switches) have been demonstrated. Potentially ultra-low-power dissipation and nonvolatility are the strongest reasons that NML is a serious contender for future electronic circuits. NML stands out in applications where massive storage needs to be combined with some logic functionality (such as integrated data storage/processing devices), and where parallelism/pipelining can make up for limited single-device speed. Research on NML devices was also an opportunity for us to appreciate the rich possibilities of non-charge-based computing devices.

Most current NML research focuses on efficient clock field generation in order to demonstrate net ultra-low-power operation of an entire NML computing system. In pNML, where no multi-phase clocking is needed, on-chip electromagnets [69] and possibly three-dimensional arrangements of nanomagnets can minimize the overhead of magnetic field generation. In permalloy-based (in-plane) NML, various approaches to voltage clocking are a promising route. New material systems, such as multilayers that use exchange coupling in addition to dipolar interactions [83], may potentially open new avenues in NML research.

In the foreseeable future NML circuits and their descendants will compete for the title of CMOS replacement technology, and we believe that for certain application spaces they may eventually win that title.

Acknowledgments

Most of the actual work presented in this chapter was done by students of the ND Nanomagnetics Group: Alexandra Imre, Lili Ji, Edit Varga, Faisal Shah, M. Jafar Siddiq, M. Tanvir Alam, Steve Kurz, Shilialng (Shawn) Liu, Aaron Dingler, Peng Li, Himadri Dey, and Katherine Butler. We are grateful to Arpad Csurgay and Joe Nahas for helping us with circuit-related issues.

We worked closely with groups from the Technical University of Munich, Germany, who developed pNML to an experimentally successful technology. Section 12.8 is based on their work. Markus Becherer led the experimental group, and we thank Stephan Breitkreutz, Xueming Ju, Josef Kiermaier, and Irina Eichwald for their successful work, and Doris Schmitt-Landsiedel and Paolo Lugli for continuously supporting this work.

We are grateful to collaboration with Jeff Bokor and Sayeef Salahuddin (UCB), Stuart Parkin (IBM Almaden), and Eugene Chen (Samsung).

Support from our sponsors enabled this work to grow from a small pilot project to larger-breadth experimental work. We greatly acknowledge the support by the Defense Advanced Research Projects Agency (DARPA) grant "Nonvolatile Logic," the SRC-NRI center Midwest Institute for Nanoelectronics Discovery (MIND), National Science Foundation (NSF), the Office of Naval Research (ONR), and the W. M. Keck foundation.

References

[1] R. J. Spain, "Controlled domain tip propagation – part I." *Journal of Applied Physics*, **37**, 2572 (1966)

[2] R. J. Spain, "Controlled domain tip propagation – part II." *Journal of Applied Physics*, **37**, 2584 (1966)

[3] R. J. Spain, H. I. Jauvtis, & F. T. Duben, "DOT memory systems." In *Proceedings of the National Computer Conference and Exposition, May 6–10, 1974*, pp. 841–846 (1974).

[4] R. P. Cowburn & M. E. Welland, "Room temperature magnetic quantum cellular automata." *Science*, **287** (2000).

[5] G. Csaba, A. Imre, G. H. Bernstein, W. Porod, & V. Metlushko, "Nanocomputing by field-coupled nanomagnets." *IEEE Transactions on Nanotechnology*, **1**(4), 209–213 (2002).

[6] G. Csaba, W. Porod, & A. I. Csurgay, "A computing architecture composed of field-coupled single domain nanomagnets clocked by magnetic field." *International Journal of Circuit Theory and Applications*, **31**, 67–82 (2003).

[7] C. Lent, P. D. Tougaw, W. Porod, & G. H. Bernstein, "Quantum cellular automata." *Nanotechnology*, **4**(1), 49 (1993).

[8] D. A. Allwood, G. Xiong, C. C. Faulkner, D. Atkinson, D. Petit, & R. P. Cowburn, "Magnetic domain-wall logic." *Science*, **309**(5741), 1688–1692 (2005).

[9] A. Imre, G. Csaba, L. Ji, A. Orlov, G. H. Bernstein, & W. Porod, "Majority logic gate for magnetic quantum-dot cellular automata." *Science*, **311**(5758), 205–208 (2006).

[10] E. Varga, G. Csaba, G. H. Bernstein, & W. Porod. "Implementation of a nanomagnetic full adder circuit." In *Nanotechnology (IEEE-NANO), 2011 IEEE Conference on*, pp. 1244–1247 (2011).

[11] See the "Emerging Research Devices" section of the International Technology Roadmap for Semiconductors, www.itrs.net/.

[12] G. Csaba, P. Lugli, A. Csurgay, & W. Porod, "Simulation of power gain and dissipation in field-coupled nanomagnets." *Journal of Computational Electronics*, **4**(1–2), 105 (2005).

[13] M. Key, "Material and quantum cellular magnetic automata radiation effects characterization report/test plan." Internal report, 2010, Naval Surface Warfare Center, Crane, IN.

[14] J. Akerman, P. Brown, M. DeHerrera *et al.*, "Demonstrated reliability of 4-mb MRAM." *IEEE Transactions on Device and Materials Reliability*, **4**(3), 428–435 (2004).

[15] A. Orlov, A. Imre, G. Csaba, L. Ji, W. Porod, & G. H. Bernstein, "Magnetic quantum-dot cellular automata: recent developments and prospects." *Journal of Nanoelectronics and Optoelectronics*, **3**(1), 55–68 (2008).

[16] G. H. Bernstein, A. Imre, V. Metlushko *et al.*, "Magnetic QCA systems." *Microelectronics Journal*, **36**(7), 619–624 (2005).

[17] M. T. Niemier, G. H. Bernstein, G. Csaba *et al.*, "Nanomagnet logic: progress toward system-level integration." *Journal of Physics: Condensed Matter*, **23**(49), 493202 (2011).

[18] A. Hubert & R. Schaefer, *Magnetic Domains: The Analysis of Magnetic Microstructures*, corrected edition (New York: Springer, 2008).

[19] A. Aharoni, *Introduction to the Theory of Ferromagnetism*, 2nd edn (New York: Oxford University Press, 2001).

[20] G. Csaba, M. Becherer, & W. Porod. "Development of CAD tools for nanomagnetic logic devices." *International Journal of Circuit Theory and Applications* (2012).

[21] M. J. Donahue and D. G. Porter, OOMMF User's Guide, Version 1.0, Interagency Report NISTIR 6376, National Institute of Standards and Technology, Gaithersburg, MD (1999).

[22] A. Imre, *Experimental study of nanomagnets for magnetic quantum- dot cellular automata (MQCA) logic applications*, PhD thesis, University of Notre Dame (2005).

[23] P. Li, G. Csaba, V. K. Sankar, X. Ju *et al.*, "Direct measurement of magnetic coupling between nanomagnets for nanomagnetic logic applications." *IEEE Transactions on Magnetics*, **48**(11), 4402–4405 (2012).

[24] E. Varga, G. Csaba, G. H. Bernstein, & W. Porod, "Implementation of a nanomagnetic full adder circuit." In *Nanoelectronic Device Applications Handbook*, eds. J. E. Morris & K. Iniewski (Boca Raton, FL: CRC Press), pp. 765–779.

[25] S. Liu, G. Csaba, X. S. Hu *et al.*, "Minimum-energy state guided physical design for nanomagnet logic." In *Proceedings of the 50th Annual Design Automation Conference*, p. 106 (2013).

[26] E. Varga, A. Orlov, M. T. Niemier, X. S. Hu, G. H. Bernstein, & W. Porod, "Experimental demonstration of fanout for nanomagnetic logic.", *IEEE Transactions on Nanotechnology*, **9**(6), 668–670 (2010).

[27] Wei, Wang, K. Walus, & G. A. Jullien, "Quantum-dot cellular automata adders." In *Nanotechnology, 2003. IEEE-NANO 2003. 2003 Third IEEE Conference on*, vol. **1**, pp. 461–464 (2003).

[28] R. P. Cowburn, "Probing antiferromagnetic coupling between nanomagnets." *Physical Review B*, **65**(9), 9 (2002).

[29] D. Carlton, B, Lambson, A. Scholl *et al.*, "Investigation of defects and errors in nanomagnetic logic circuits." *IEEE Transactions on Nanotechnology*, **11**(4), 760–762 (2012).

[30] P. Li, G. Csaba, V. K. Sankar *et al.*, "Switching behavior of lithographically fabricated nanomagnets for logic applications." *Journal of Applied Physics*, **111**(7), 07B911–07B911 (2012).

[31] G. Csaba & W. Porod, "Simulation of field coupled computing architectures based on magnetic dot arrays." *Journal of Computational Electronics*, **1**(1–2), 87–89 (2002).

[32] G. Csaba & W. Porod, "Behavior of nanomagnet logic in the presence of thermal noise. In *Proceedings of the 14th International Workshop on Computational Electronics (IEEE-IWCE)*, pp. 26–29 (2010).

[33] G. Csaba, P. Lugli, & W. Porod, "Power dissipation in nanomagnetic logic devices." In *Nanotechnology, 2004. 4th IEEE Conference on*, pp. 346–348 (2004).

[34] W. Porod, R. O. Grondin, D. K. Ferry, & G. Porod. "Dissipation in computation." *Physical Review Letters*, **52**(3), 232–235 (1984).

[35] T. Toffoli, "Comment on 'Dissipation in computation'," *Physical Review Letters* **53**(12) 1204–1204 (1984).

[36] B. Lambson, D. Carlton, & J. Bokor, "Exploring the thermodynamic limits of computation in integrated systems: magnetic memory, nanomagnetic logic, and the Landauer limit." *Physical Review Letters*, **107**(1), 010604 (2011).

[37] G. Csaba, A. Imre, G. H. Bernstein, W. Porod, & V. Metlushko, "Signal processing with coupled ferromagnetic dots." In *Nanotechnology, 2002. IEEE-NANO 2002. Proceedings of the 2002 2nd IEEE Conference on*, pp. 59–62 (2002).

[38] A. Dingler, M. T. Niemier, X. S. Hu, & E. Lent, "Performance and energy impact of locally controlled NML circuits." *Journal of Emergent Technology in Computer Systems*, **7**(1), 1–24 (2011).

[39] M. T. Alam, M. J. Siddiq, G. H. Bernstein, M. Niemier, W. Porod, & X. S. Hu, "On-chip clocking for nanomagnet logic devices." *IEEE Transactions on Nanotechnology*, **9**(3), 348–351 (2011).

[40] S. V. Pietambaram, N. D. Rizzo, R. W. Dave *et al.*, "Low-power switching in magneto-resistive random access memory bits using enhanced permeability dielectric films." *Applied Physics Letters*, **90**(14) 143510–143510 (2007).

[41] L. Peng, V. K. Sankar, G. Csaba *et al.*, "Magnetic properties of enhanced permeability dielectrics for nanomagnetic logic circuits." *IEEE Transactions on Magnetics*, **48**(11), 3292–3295 (2012).

[42] G. Csaba, J. Kiermaier, M. Becherer *et al.*, "Clocking magnetic field-coupled devices by domain walls." *Journal of Applied Physics*, **111**(7), 07E337–07E337-3 (2012).

[43] E. Varga, G. Csaba, G. H. Bernstein, & W. Porod. "Domain-wall assisted switching of single-domain nanomagnets." *IEEE Transactions on Magnetics*, **48**(11) 3563–3566 (2012).

[44] S. W. Cheong & M. Mostovoy, "Multiferroics: a magnetic twist for ferroelectricity." *Nature Materials*, **6**, 20 (2007).

[45] F. M. Saleh, K. Roy, J. Atulasimha, & S. Bandyopata, "Magnetization dynamics, Bennett clocking and associated energy dissipation in multiferroic logic." *Nanotechnology*, **22**, 155201 (2011).

[46] A. Khitun, B. Mingqiang, & K. L. Wang, "Spinwave magnetic nanofabric: a new approach to spin-based logic." *IEEE Transactions on Magnetics*, **44**(9), 2141–2152 (2008).

[47] U. Bauer, M. Przybylski, J. Kirschner, & G. S. D. Beach, "Magnetoelectric charge trap memory." *Nano letters*, **12**(3), 1437–1442 (2012).

[48] D. B. Carlton, B. Lambson, A. Scholl *et al.*, "Computing in thermal equilibrium with dipole-coupled nanomagnets." *IEEE Transactions on Nanotechnology*, **10**(6), 1401–1404 (2011).

[49] J. Kiermaier, S. Breitkreutz, G. Csaba, D. Schmitt-Landsiedel, & M. Becherer, "Electrical input structures for nanomagnetic logic devices." *Journal of Applied Physics*, **111**(7), 07E341–07E341 (2012).

[50] M. A. Siddiq, M. T. Niemier, G. Csaba, X. S. Hu, W. Porod, & G. H. Bernstein, "Demonstration of field-coupled input scheme on line of nanomagnets." *IEEE Transactions on Magnetics*, **49**(7), 4460–4463 (2013).

[51] J. Kiermaier, S. Breitkreutz, X. Ju, G. Csaba, D. Schmitt-Landsiedel, & M. Becherer, "Field-coupled computing: investigating the properties of ferromagnetic nanodots." *Solid-State Electronics*, **65**, 240–245 (2011).

[52] M. Becherer, J. Kiermaier, S. Breitkreutz *et al.*, "On-chip extraordinary Hall-effect sensors for characterization of nanomagnetic logic devices." *Solid-State Electronics*, **54**(9), 1027–1032 (2010).

[53] D. Kanungo, A. I. Pratyush, W. Bin *et al.*, "Gated hybrid Hall effect device on silicon." *Microelectronics Journal*, **36**(3), 294–297 (2005).

[54] E. Chen, D. Apalkov, Z. Diao *et al.*, "Advances and future prospects of spin-transfer torque random access memory." *IEEE Transactions on Magnetics*, **46**(6), 1873–1878 (2010).

[55] X. Shiliang Liu, S. Hu, J. J. Nahas, M. Niemier, W. Porod, & G. H. Bernstein, "Magnetic-electrical interface for nanomagnet logic." *IEEE Transactions on Nanotechnology*, **10**(4), 757–763 (2011).

[56] S. Liu, X. Hu, M. T. Niemier *et al.*, "Exploring the design of the magnetic-electrical interface for nanomagnet logic." *IEEE Transactions on Nanotechnology*, **12**(2), 203–214 (2013).

[57] A. Lyle, A. Klemm, J. Harms *et al.*, "Probing dipole coupled nanomagnets using magnetoresistance read." *Applied Physics Letters*, **98**, 092502 (2011).

[58] Stuart Parkin, private communication.

[59] M. Becherer, G. Csaba, W. Porod, R. Emling, P. Lugli, & D. Schmitt-Landsiedel, "Magnetic ordering of focused-ion-beam structured cobalt-platinum dots for field-coupled computing." *IEEE Transactions on Nanotechnology*, **7**(3), 316–320 (2008).

[60] O. Hellwig, A. Berger, J. B. Kortright, & E. E. Fullerton, "Domain structure and magnetization reversal of antiferromagnetically coupled perpendicular anisotropy films." *Journal of Magnetism and Magnetic Materials*, **319**(1), 13–55 (2007).

[61] M. Becherer, G. Csaba, R. Emling, W. Porod, P. Lugli, & D. Schmitt-Landsiedel, "Field-coupled nanomagnets for interconnect-free, nonvolatile computing" *IEEE International Solid-State Circuits Conference (ISSCC), Digest, Technical Papers*, pp. 474–475 (2009).

[62] X. Ju, S. Wartenburg, J. Rezgani *et al.*, "Nanomagnet logic from partially irradiated Co/Pt nanomagnets." *IEEE Transactions on Nanotechnology*, **11**(1), 97–104 (2012).

[63] S. Breitkreutz, J. Kiermaier, X. Ju, G. Csaba, D. Schmitt-Landsiedel, & Markus Becherer. "Nanomagnetic logic: demonstration of directed signal flow for field-coupled computing devices." In *Solid-State Device Research Conference (ESSDERC), 2011 Proceedings of the European*, pp. 323–326 (2011).

[64] M. P. Sharrock, "Time-dependent magnetic phenomena and particle-size effects in recording media." *IEEE Transactions on Magnetics*, **26**, 1 (1990).

[65] S. Breitkreutz, J. Kiermaier, I. Eichwald *et al.*, "Majority gate for nanomagnetic logic with perpendicular magnetic anisotropy." *IEEE Transactions on Magnetics*, **48**(11), 4336–4339 (2012).

[66] B. Lambson, G. Zheng, D. Carlton *et al.*, "Cascade-like signal propagation in chains of concave nanomagnets." *Applied Physics Letters*, **100**(15), 152406–152406 (2012).

[67] I. Eichwald, A. Bartel, J. Kiermaier *et al.*, "Nanomagnetic logic: error-free, directed signal transmission by an inverter chain." *IEEE Transactions on Magnetics*, **48**(11), 4332–4335 (2012).

[68] J. Kiermaier, S. Breitkreutz, I. Eichwald *et al.*, "Information transport in field-coupled nanomagnetic logic devices." *Journal of Applied Physics* **113**(17) 17B902–17B902 (2013).

[69] M. Becherer, J. Kiermaier, S. Breitkreutz, I. Eichwald, G. Csaba, & D. Schmitt-Landsiedel, "Nanomagnetic logic clocked in the MHz regime." In *ESSDERC* (2013).

[70] S. Breitkreutz, J. Kiermaier, I. Eichwald *et al.*, "Experimental demonstration of a 1-bit full adder in perpendicular nanomagnetic logic." *IEEE Transactions on Magnetics*, **49**(7), 4464–4467 (2013).

[71] I. Eichwald, J. Kiermaier, S. Breitkreutz *et al.*, "Towards a signal crossing in double-layer nanomagnetic logic." *IEEE Transactions on Magnetics*, **49**(7), 4468–4471 (2013).

[72] S. Breitkreutz, J. Kiermaier, C. Yilmaz *et al.*, "Nanomagnetic logic: compact modeling of field-coupled computing devices for system investigations." *Journal of Computational Electronics*, **10**(4), 352–359 (2011).

[73] M. Niemier, E. Varga, G. Bernstein *et al.*, "Shape engineering for controlled switching with nanomagnet logic." *IEEE Transactions on Nanotechnology*, **35**(3), 281–293 (2007).

[74] G. Csaba, W. Porod, P. Lugli, & A. Csurgay, "Activity in field-coupled nanomagnet arrays." *International Journal of Circuit Theory and Applications*, **35**, 281–293 (2007).

[75] E. Varga, M. T. Niemier, G. Csaba, G. H. Bernstein, & W. Porod, "Experimental realization of a nanomagnet full adder using slanted-edge magnets." *IEEE Transactions on Magnetics*, **49**(7), 4452–4455 (2013).

[76] X. Ju, M. T. Niemier, M. Becherer, W. Porod, P. Lugli, & G. Csaba, "Systolic pattern matching hardware with out-of-plane nanomagnet logic devices." *IEEE Transactions on Nanotechnology*, **12**(3), 399–407 (2013).

[77] H. D. Crane, "A high-speed logic system using magnetic elements and connecting wire only." *Proceedings of the IRE*, **47**(1), 63–73 (1959).

[78] H. W. Gschwind, *Design of Digital Computers* (New York: Springer, 1965).

[79] E. L. Braun, *Digital Computer Design* (New York/London: Academic Press, 1963).

[80] H. J. Gray, *Digital Computer Engineering* (Englewood Cliffs, NJ: Prentice-Hall, 1963).

[81] W. A. Geyger, *Magnetic-amplifier Circuits*, 2nd edn. (New York: McGraw-Hill, 1957).

[82] G. M. Ettinger, *Magnetic Amplifiers* (New York: Wiley, 1957).

[83] R. Lavrijsen, J.-H. Lee, F.-P. Amalio, D. Petit, R. Mansell, & R. P. Cowburn, "Magnetic ratchet for three-dimensional spintronic memory and logic." *Nature*, **493**(7434), 647–650 (2013).

13 Spin torque majority gate logic

Dmitri E. Nikonov and George I. Bourianoff

13.1 Introduction

Nanomagnetic or spintronic circuits hold the promise of non-volatile and reconfigurable logic with low switching energy. One such circuit is the magnetic majority gate formed by concatenating several magnetic tunnel junctions together in such a manner that they interact with each other through a common ferromagnetic free layer to achieve the desired functionality. A key advantage of this configuration is that multiple majority gates can be concatenated together entirely in the magnetic domain without conversion to electric signals. The magnetic majority gates can in turn be concatenated together to form more complex circuits, such as a full magnetic adder circuit described here and simulated with a micromagnetic solver. The dynamics of magnetic polarization propagate through the adder circuit via the motion of magnetic domain walls and correspond exactly to the propagation of information through a ripple adder circuit. The switching speed and energy of the fundamental magnetic switching operation in the magnetic adder is comparable to the same fundamental switching operation in single magnetic gates or nanomagnetic memories. It provides a basis for estimating the operational speed and energy of the more complex magnetic circuits. A non-linear transfer characteristic ensures noise margin and signal restoration after every operation critical for Boolean logic.

The most common applications of spintronic devices in production today are non-volatile memories, namely magnetic random access memory (MRAM), which employ field induced switching of magnetic polarization. More recently, however, a much more efficient magnetic switching mechanism, based on current-induced switching, has been introduced and used to fabricate spin transfer torque RAM (STTRAM) memories [1]. It is natural to consider extending the physics of STTRAM to other magnetic logic functions [2], including the spin torque majority gate (STMG) described here. One obvious benefit of magnetic logic circuits is they are non-volatile, and hence do not suffer from standby power dissipation. A related benefit is that they can be turned on instantly since the circuit is non-volatile in the absence of input signals. In spite of these obvious advantages and the fact that numerous spintronic logic devices have been proposed, few of them have been fabricated and none have been demonstrated to function in an integrated circuit.

In this chapter, we focus on STMG [3, 4], and describe the magnetization dynamics of this device, the operation of a one-bit full adder, and make performance projections for it. These results support the feasibility of an experimental implementation of STMG and demonstrate the possibility of creating extended spintronic circuits (e.g., adders) without the need of spin-to-electrical conversion. We will also provide an argument that such circuits can have performance comparable with the incumbent CMOS technology.

Spintronic logic devices have been extensively studied in the last few years [5] due to their potential for non-volatility and a lower switching energy (though at a slower switching speed) as explained previously. However, there are many barriers to practical implementation, and all of the proposed spintronic devices [6–11] have faced one or more of these obstacles. They include difficulties of spin injection from magnetic metals into semiconductors, clocking by external magnetic fields, inefficiency of spin wave generation and detection, reliability of tunnel barrier materials, high magnetic switching currents, statistical variation of switching parameters, and a host of other issues. In spite of these issues STMGs appear to offer significant performance benefits in certain application areas.

13.2   In-plane magnetization SMG

The structure of the STMG device with in-plane magnetization is shown in Figs. 13.1 and 13.2. The stack of layers is similar to that of a magnetic tunnel junction (MTJ) used in magnetic memories (e.g., STTRAM). A typical MTJ comprises two ferromagnetic (FM) layers separated by a non-magnetic tunneling barrier. One of the FM layers is free, i.e., can be switched. It has two stable directions of magnetization, to the left and to the right, corresponding, e.g., to logical "0" and "1," respectively. Another FM layer is fixed, i.e., has a permanent direction of magnetization. It can be achieved by making this layer thicker than the free layer, but a more reliable way is to place an antiferromagnetic layer in direct contact with it. (An antiferromagnetic layer has spins in its crystal layer in interchanging directions.) The spins at the surface of the

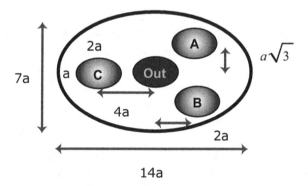

Fig. 13.1 Layout of a spin torque majority gate (STMG). The larger ellipse is the free ferromagnetic (FM) layer. Smaller ellipses are nanopillars containing fixed FM layers. Input nanopillars are labeled "A," "B," and "C," the output nanopillar is "out" in the middle. Minimum lithography width is a. The aspect ratio for all ellipses is 2. The elliptical shapes enforce stable states of magnetization along their long axis.

antiferromagnet hold the spins in the ferromagnet in the same direction via the exchange bias interaction.

Magnetization direction in the free layer is detected via tunneling magnetoresistance (TMR) [12]. The resistance of the stack is higher in the case of anti-parallel magnetizations of the free and fixed layers than in the case of parallel magnetizations, typically by 100% to 200%. This difference of resistance is measured by comparing it to an etalon resistance using a sense amplifier, see Fig. 13.3.

The geometry of STMG is different from STTRAM because it uses separate nanopillars, which are formed for inputs and outputs, with their own fixed FM layers. The

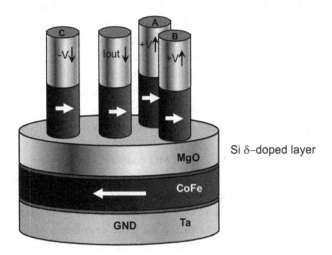

Fig. 13.2 Schematic side view of STMG layers. Magnetization directions in FM layers, CoFe in this case, are designated by arrows. Every nanopillar has its own fixed FM layer and a metal contact on the top. The contacts are labeled with applied voltages ($+V$ or $-V$) or detected current (I_{out}). The common free FM layer, thickness t, below, is separated by a tunneling barrier of MgO. The template layer for growing CoFe, in this case Ta, is on the bottom and connected to "ground."

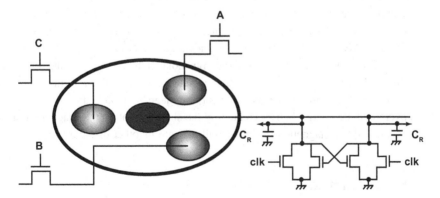

Fig. 13.3 Circuit schematics of a reconfigurable AND/OR gate with STMG. A transistor drives current through each input nanopillar, A, B, and C. Output current flows out of the output nanopillar in the center and is detected by a sense amp.

side view of the structure is shown in Fig. 13.2. The free FM layer is common to all the pillars. The device operates by applying voltage with positive (+) or negative (–) polarity to each input nanopillar, labeled "A," "B," and "C" in Fig. 13.1. The combined action of spin torques [13] resulting from currents in the three input pillars transfers enough torque to switch magnetization first under the nanopillar, and then over the whole common free FM layer (which is a much larger area than the three input nanopillars). Digital inputs with voltage polarities designated as plus (p) or minus (m), determine the direction current flows and the resulting torques. For the three nanopillars that result (e.g., in the following combinations: (ppp) = A+, B+, C+, (ppm) = A+, B+, C–, (pmp) = A+, B–, C+, etc.) the three torques fight to control the magnetization direction in the free layer, which is in the end set by the majority of them. We need to design the gate such that if there is a majority of positive voltages, as in combinations (ppp), (mpp), (pmp), and (ppm), then the magnetization will switch from right to left. If it has been already pointing left, it will remain so. In other words, the feature of the gate is that its final state should be determined just by its inputs and not the prior state, like a STTRAM cell. Conversely, if there is a majority of negative voltages, as in combinations (mmm), (pmm), (mpm), and (mmp), then the magnetization will switch from left to right or remain pointing to the right.

In addition to the three input pillars, the STMG contains a fourth pillar, which is the device output terminal. The output pillar is another MTJ, which displays a high or low resistance depending on whether the fixed magnetic layer is aligned or counter aligned with the free layer. The resistive state of the output MTJ is read with a sensitive sense amp designed so the read current disturbs the magnetic orientation of the free layer (non-destructive read). A single STMG has a useful functionality as a reconfigurable AND/OR gate due to the following feature of the majority gate truth table. If any one of the inputs is set to "1," the logical function of the remaining two is OR. If any one of the inputs is set to "0," the logical function of the remaining two is AND.

Efficient spin injection from normal metals into ferromagnetic materials requires a resistance × area product in the neighborhood of $10\,\Omega\text{-}\mu m^2$. This allows a current greater than the critical current required for switching to flow through the interface at low voltage without excessive loss at the interface. The quantitative values of the critical current and operational voltage will be derived later in this chapter. However, the qualitative nature of the STMG critical current is easily understood in terms of the critical current required for MTJ switching. Simply put, the spin polarized currents flowing through each of the MTJs each exert a torque on the magnetic material in the common free layer. If the vector sum of those torques is sufficient, the magnetic polarization of the free layer will overcome the energy barrier created by the shape anisotropy, a magnetic transition will occur, and the magnetic orientation will flip into its other bistable state.

13.3   Simulation model

The magnetization dynamics were modeled with the OOMMF micromagnetic solver [14]. It produces a time-dependent solution of the magnetization dynamics in the plane

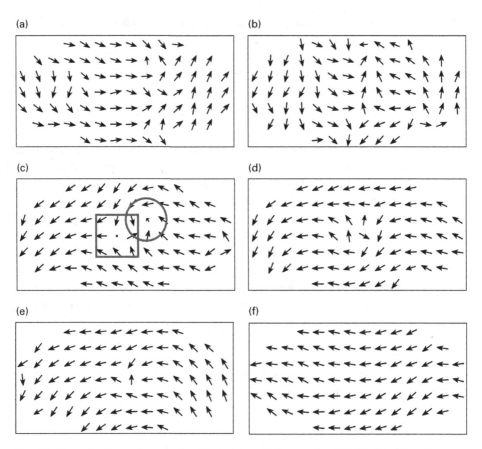

Fig. 13.4 Magnetization patterns in the elliptical free FM layer of an STMG. Polarities of voltages in nanopillars are (ppp), current in each nanopillar is $I = 4\,\text{mA}$, minimum size $a = 24\,\text{nm}$, free FM layer thickness $t = 2\,\text{nm}$. Snapshots (a–f) taken at 0.1 ns and then with time intervals of 0.2 ns in sequence from left to right in rows. A vortex is highlighted by a circle, an antivortex by a square in (c). A transient vortex and an anti-vortex appear and disappear. Magnetization is mostly uniform in the final state.

of the free layer as shown in Fig. 13.4 and on. The mathematical model includes a term describing the spin torque transfer arising from the current passing through fixed layers in the MTJ input pillars. In these simulations, the random thermal fluctuations of magnetization are not modeled, which corresponds to zero effective temperature. However, we set the angle of magnetization in nanopillars to a value of 10° to account for the deflections of this angle due to thermal fluctuation. (This non-zero angle is necessary to initiate switching by spin torque.) We model cases when the initial magnetization is uniformly pointing to the right (average relative value 1, in units of saturation magnetization M_s), along the easy axis of the ellipse. The expected result of the gate operation is for the applied spin torque to switch the direction of the magnetization to more or less pointing to the left (average relative value −1). Switching from right to left looks like an exact mirror image of the above.

The dynamics of magnetization in a nanomagnet is described by the Landau–Lifshitz–Gilbert equation that solves for the amplitude and direction of the magnetization vector **M**. The dimensionless vector of magnetization is obtained by dividing it by the value of saturation magnetization $\mathbf{m} = \mathbf{M}/M_S$. With this substitution, the LLG equation is

$$\frac{d\mathbf{m}}{dt} = -\gamma\mu_0[\mathbf{m} \times \mathbf{H}_{\text{eff}}] + \alpha\left[\mathbf{m} \times \frac{d\mathbf{m}}{dt}\right] + \Gamma, \tag{13.1}$$

where the gyromagnetic factor is $\gamma = g\mu_B / \hbar$ given by the electron g-factor, and the Bohr magneton μ_B. The effective magnetic field is composed of the external magnetic field **H** and the gradient relative to magnetization of all other magnetic energies per unit volume E of the nanoscale circuit

$$\mathbf{H}_{\text{eff}} = \mathbf{H} - \frac{1}{\mu_0}\frac{\delta E}{\delta \mathbf{M}}. \tag{13.2}$$

The second term in Eq. (13.1) corresponds to Gilbert damping with a constant $\alpha \sim 0.01$. The third term describes the spin torque which is proportional to current density J. It has, in its turn, the spin transfer (Slonczewski) term and the field-like term,

$$\Gamma = \frac{\gamma\hbar J}{M_s et}\left(\varepsilon[\mathbf{m} \times [\mathbf{p} \times \mathbf{m}]] + \varepsilon'[\mathbf{p} \times \mathbf{m}]\right). \tag{13.3}$$

The coefficient in Eq. (13.3) is proportional to spin polarization P, and contains the angle-dependent factor

$$\varepsilon = \frac{P}{2g(\theta)}. \tag{13.4}$$

We are also taking into account the magnetic field produced by the current in wires attached to nanomagnets ("Oersted field"). Its dependence in the distance from the center of the wire r and its radius a is

$$H(r) = \begin{cases} \dfrac{Ir}{2\pi a^2}, r < a, \\ \dfrac{I}{2\pi r}, r > a. \end{cases} \tag{13.5}$$

The partial differential equation for the vector field **m** given by Eq. (13.1) is driven by two inhomogeneous terms, Γ and H, given by Eqs. (13.3) and (13.5) respectively. It is helpful to evaluate the scaling of those two terms with domain size and thickness in order to gain a qualitative understanding of the competing structures that are observed. The characteristic size of nanomagnet is taken to be equal to a, and the thickness of nanomagnets is designated by t.

It is readily seen that, for a magnetic pillar of radius a with the current distributed across the entire cross section, H is given by

$$\mu_0 H \propto \frac{\mu_0 I}{2\pi a}. \tag{13.6}$$

For a similar configuration, the spin torque contribution is

$$\Gamma/\gamma \propto \frac{P\hbar I}{2M_s e \pi a^2 t},\tag{13.7}$$

By taking the ratio of the above terms we find that for larger and thicker devices

$$at > \frac{P\hbar}{\mu_0 M_s e} = \frac{0.5 \times 10^{-34}}{1 \times 1.6 \times 10^{-19}} \sim 300\,\text{nm}^2,\tag{13.8}$$

as parameterized by the at product, the Oersted field produced by the current I dominates the spin torque term as shown in Eq. (13.7). In this range, the nanomagnet is more likely to develop magnetic vortices and not switch properly, as will be described in the next section. This mechanism of Oersted field is used for switching of magnetization in MRAM. For the opposite limit of smaller and thinner devices spin torque is dominant. That mechanism of switching is used in STTRAM.

There are three other energy terms where it is useful to evaluate the scaling since their relative magnitudes determine the stability properties of the layer, specifically how likely it is to switch. The first is related to the shape anisotropy and is called the shape anisotropy or, equivalently, the demagnetization energy:

$$E_{\text{dem}} = \frac{\mu_0 M_s^2}{2}(N_{xx}m_x^2 + N_{yy}m_y^2 + N_{zz}m_z^2),\tag{13.9}$$

where $N_{xx} + N_{yy} + N_{zz} = 1$ are components of the demagnetization tensor dependent on the geometry of a nanomagnet, and m_x, m_y, m_z are projections of the dimensionless magnetization vector.

The second term is the magneto-crystalline anisotropy, which is caused by the orientation of the crystal lattice or by strain or deposition conditions of the ferromagnetic film. We will consider only the out-of-plane uniaxial anisotropy:

$$E_{\text{mc}} = K_u(1 - m_z^2),\tag{13.10}$$

where k_u is the anisotropy energy per unit volume.

The third term is the exchange energy, which is simply the energy associated with a particular configuration of individual magnetic dipoles. If neighboring dipoles are counter aligned with each other, the exchange energy will be low. If the shape anisotropy is large compared to the exchange energy, the field will resist switching and stay stable:

$$E_{\text{exc}} = A(|\nabla m_x|^2 + |\nabla m_y|^2 + |\nabla m_z|^2),\tag{13.11}$$

where A is the exchange constant.

As a rule, exchange energy promotes a more uniform magnetization, demagnetization energy promotes magnetization to be parallel to the surface of the nanomagnet, and magneto-crystalline energy aims to align magnetization with the preferred axis (z-axis in this case).

13.4   Patterns of in-plane magnetization switching

The dynamics of magnetization can and does exhibit complex geometrical patterns. The in-plane projections of magnetization are represented in Figs. 13.4–13.8 as a collection of arrows. The out-of-plane projection is negligible in most cases (with the exception of cores of vortices, as will be seen later) and therefore not shown in these graphs. Each represents the direction of average magnetization in a certain area of the elliptical free layer. Note that each of these areas encompasses several points of the simulation grid. In the initial state all arrows are pointing to the right. Our goal is to switch all these arrows to point to the left. However, the switch does not happen as a uniform rotation of magnetization. The ferromagnetic free layer may reach the desired final state without vortex formation (Fig. 13.5) or, under other conditions it will form transient vortices and anti-vortices as shown in Fig. 13.4. In other cases the final state may contain a vortex (Figs. 13.6 and 13.7) or a stable anti-vortex (Fig. 13.8). An anti-vortex may form and then convect off the edge of the free layer (Fig. 13.7). Formation of a stable vortex or anti-vortex in the final magnetization

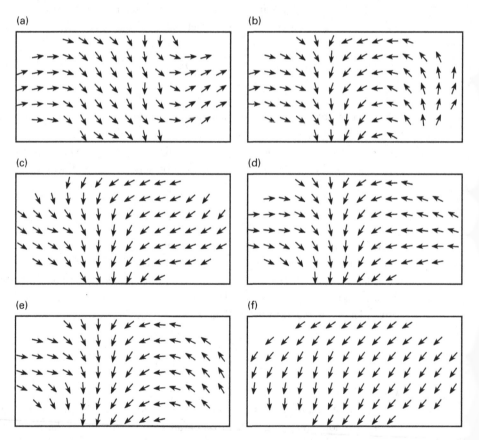

Fig. 13.5 Same as Fig. 13.4, but polarities (ppm). Magnetization is switched via non-uniform bending. No vortices appear.

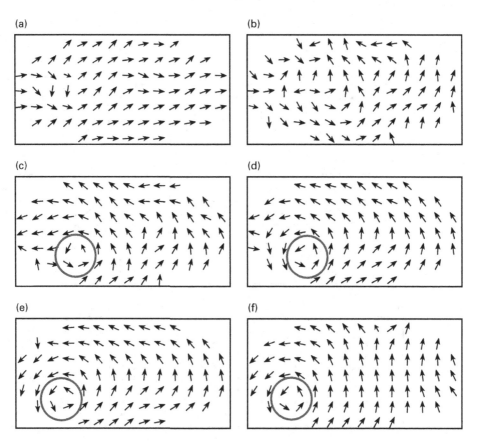

Fig. 13.6 Same as Fig. 13.4, but polarities (pmp). After some time a vortex, highlighted by a circle in (c–f), forms and persists indefinitely. Not a desirable final state.

state will result in a failure of majority logic. In this case the magnetization under the output pillar does not align along the easy axis and will not provide a stable magneto-resistance value for the sense amplifier to sense. Such a situation must and can be avoided by a proper choice of proper geometry, current magnitude and duration, which can be derived from the expressions shown earlier. Finally, in the speed plots shown in Fig. 13.9, failure to switch is indicated as zero switching speed. Fortunately, this happens over a rather limited range of parameters so it is possible to find a broad operation range where normal switching occurs for all polarities of inputs.

In order to analyze the conditions under which a stable vortex will form, it is necessary to estimate energies for two-dimensional cylindrical volumes of radius a and thickness t. The demagnetization energy per unit volume and exchange energy per unit volume is

$$E_{\mathrm{mag}}/V \sim \frac{\mu_0 M_s^2 t}{2a}.$$

$$(13.12)$$

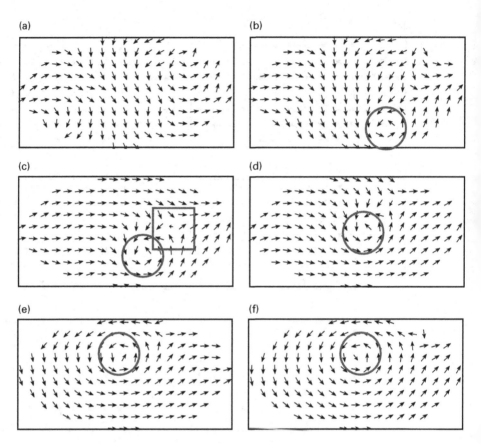

Fig. 13.7 Magnetization patterns with polarities (ppm), $I = 16\,\mathrm{mA}$, $a = 48\,\mathrm{nm}$, $t = 3\,\mathrm{nm}$, at $0.2\,\mathrm{ns}$ and then time intervals of $0.6\,\mathrm{ns}$. A vortex, highlighted by a circle in (b–f), and an anti-vortex, highlighted by a square in (c), form. The anti-vortex exits at an edge; the vortex persists indefinitely. This is not a desirable final state.

The exchange energy per unit volume is

$$E_{\mathrm{ex}}/V = A\left(\frac{\pi}{a}\right)^2. \tag{13.13}$$

The magnetic exchange length is an important parameter because it governs the width of the transition between magnetic domains in ferromagnetic material. It is a function of the relative strength of exchange and demagnetization energies and characterizes the magnetic behavior of a given magnetic material, e.g. permalloy. The exchange length for permalloy is given by [15]

$$\lambda_{\mathrm{e}}^2 = \frac{2A}{\mu_0 M_{\mathrm{s}}^2} \to 5\,\mathrm{nm}. \tag{13.14}$$

The condition that the demagnetization energy exceeds the exchange energy, i.e., when vortex formation becomes energetically favorable

$$at > \pi^2 \lambda_{\mathrm{e}}^2 \approx 250\,\mathrm{nm}^2. \tag{13.15}$$

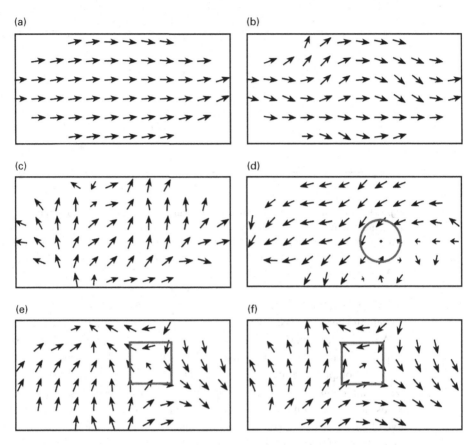

Fig. 13.8 Magnetization patterns with polarities (pmp), $I = 1$ mA, $a = 10$ nm, $t = 2$ nm, at 0.1 ns and then time intervals of 0.2 ns. A vortex, highlighted by a circle in (d), and an anti-vortex, highlighted by a circle in (e) and (f), form. The vortex exits at an edge; the anti-vortex persists indefinitely. This is not a desirable final state.

Note that this condition is very similar to the condition of magnetic field dominance in Eq. (13.7).

Simulations at various current values and size provide values of switching speed as shown in Fig. 13.9. The normalized projection of magnetization on the axis of stable states varies between 1 and −1. We define the switching time as the time between the start of the current pulse (0 ns) and the last time magnetization projection crosses −0.6 (i.e., 80% of switching amplitude). Switching speed is the inverse of switching time.

Similarly to switching of STTRAM, there is an initial time interval when the magnetization almost does not change. In STMG it is set by the time necessary for the domain walls to reach the output arm. This time is roughly inversely proportional to the input current magnitude. At lower values of current, the spin torque is not enough to overcome damping. So STMG has a threshold current similarly to STTRAM. Switching also fails (designated as zero speed) at larger values of $a = 24$ nm and for larger values of current due to a vortex formation. Even in this case, normal switching occurs for a

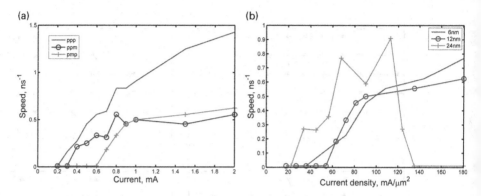

Fig. 13.9 Switching speed of STMG: (left) vs. current for various voltage polarities, $a = 12$ nm, $t = 2$ nm; (right) vs. current density per unit area of free layer for (pmp) polarity at various a. Top left corner is preferred in these and further plots, corresponding to higher switching speed with smaller current. Switching does not occur below certain critical current. Switching speed in general increases with current. Switching fails for larger gates due to vortex formation.

broad range of intermediate values of current. However, in most of the cases shown, the switching speed increases with current density as one expects intuitively.

Another critical parameter to evaluate is noise immunity of this logic. An important requirement for logic is that noise in the input should be suppressed and should not affect the output. Thermal noise in magnetic circuits appears as fluctuations of magnetization direction. We performed simulations of STMG (such as shown in Fig. 13.2) with an assumed polarization configuration of (pmm). The directions of magnetizations in two of the three input layers are kept constant while the direction of magnetization in the input layer of the third input is changed so that the configuration change to the polarization of the common free layer will change from (m) to (p), assuming the other conditions for switching have been met. The transfer characteristic, i.e., the dependence of the magnetization angle with the x-axis at the output on the magnetization in the fixed layer of the third input, is shown in Fig. 13.10. This shows that the transfer function has a sharply non-linear transfer characteristic, similar to that of CMOS inverter. If the direction of the third input is between $0°$ and ~$84°$, the output direction is approximately flat, close to $90°$. (We interpret this as a noise margin of $84°$.) The slope in the mid-point ($90°$) corresponds to a gain of ~15. The combination of solid noise margin and good gain supports the suitability of STMG for logic circuits.

13.5   Perpendicular magnetization SMG

Devices in Figs. 13.1 and 13.2 correspond to cases with in plane magnetic anisotropy which requires elliptical domains to provide anisotropy. However, STMGs can also be fabricated from materials with out-of-plane magnetization (such as FePt, TbCoFe, CoPt multilayers, or CoNi multilayers) to achieve a more compact layout and to simplify the etching process. More importantly, we are not restricted to elliptic shapes and can have,

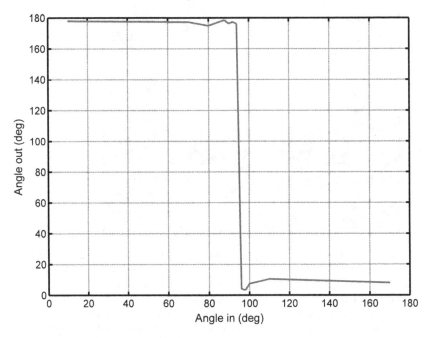

Fig. 13.10 Angle of final magnetization vs. angle of spin polarization in nanopillar A, for polarization (pmp), $I = 1$ mA, $a = 12$ nm, $t = 2$ nm. This dependence is similar to an input–output characteristic of a CMOS inverter. It demonstrates that if the input magnetization deflects from the digital values, 0 and 180 degrees, within a certain range ("noise margin"), the output is still close to the digital values.

e.g., cross shapes with inputs at the edges of the cross. One expects lower switching current with perpendicular magnetization, e.g. see [16], but the structure of the layers (Fig. 13.11) is more complicated. Both the free and the fixed layers are formed of a "synthetic anti-ferromagnet" (SAF) [17] consisting of two or more very thin ferromagnetic layers (such as CoFe) separated by a thin metal layer (such as 0.8 nm of Ru). This ensures that the directions of magnetization in the two co-layers are opposite due to exchange interaction through a thin ruthenium layer.

There is no natural inverter gate in magnetic circuits similar to the classic CMOS inverter built out of an nMOS and a pMOS transistor. One of the ways to create inverter operation is to use the effect of SAF structure in the free layer (Fig. 13.11). An inverter is a structure in an SAF where the ferromagnetic layer is deposited on a tilted plane to connect the top and bottom co-layers. Therefore the magnetization of the top layer is opposite before and after the inverter. This is done to permit a simple passive element to perform an inverter function: if the top of the SAF on one side is connected to the bottom of the SAF on the other side, the magnetizations in the top layer, sensed in TMR, are opposite. Thus the functionality of an inverter can be achieved in a wire without active devices (see [3] for details).

Out-of plane magnetization STMG [3, 4] are laid out as cross shapes of FM wires (Fig. 13.12) with three inputs and one output. STMG has a useful functionality on its

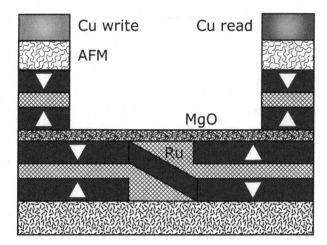

Fig. 13.11 Schematic cross-section of a magnetic inverter device based on perpendicular magnetization. White triangles designate magnetization directions in FM layers. Two lower FM layers with a layer of ruthenium between them form a synthetic antiferromagnet (SAF) free layer. They are separated by a tunneling oxide (MgO) from a similar SAF fixed layer. Its magnetization is pinned by an adjacent antiferromanget (AFM) layer. Two nanopillars (for write and read) are shown and have copper electrodes on top. The inverter structure is shown in the middle.

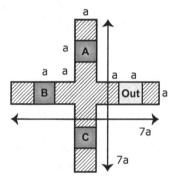

Fig. 13.12 Layout of STMG with out-of–plane magnetization. The squares designated by "A," "B," and "C" are input nanopillars. The square designated by "Out" is the output nanopillar. The minimum width of arms, the size of nanopillars and gaps between them is a. Its operation does not rely on shape anisotropy. Therefore inputs and outputs can be shifted to the periphery.

own: by switching one of the outputs it can be re-configured to either AND or OR logical function of the remaining two outputs. The magnetic computational state does not need to be converted to an electric signal after each gate. Instead one can concatenate them: an output of one gate can drive the input to another. In addition to the STMG, other, more interesting, circuits are obtained by concatenating the majority gates. The first example of an all-magnetic integrated circuit containing only three majority gates is shown in Fig. 13.13. The purpose of this circuit is one bit of a full adder. Three independent electrical signals drive two or three inputs each: A and B are bits of the two numbers to be summed, and C is a carry from a previous bit. Two output signals are

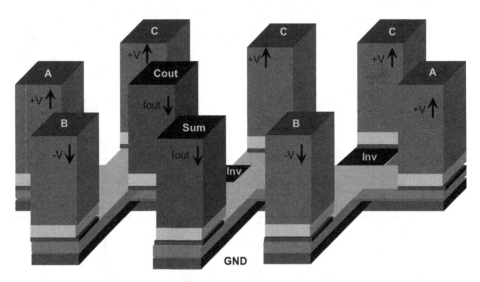

Fig. 13.13 Schematic side view of a STMG adder. Voltages ($+V$ or $-V$) are applied to nanopillars according to the polarity of inputs A and B (the two added bits) and C (carry). Output current (Iout) is detected in the "Sum" and "C_{out}" (for the carry output) nanopillars. Inverter structures "Inv" as in Fig. 13.11 are placed in some of the ferromagnetic wires.

derived from it: C_{out} is the carry signal to be passed to the next bit of the adder, and Sum is the summation of two bits, A and B. In this study we set the width of each FM wire to $a = 20$ nm, and the size of each electrode to 20 nm $\times$ 20 nm, so the total size of a STMG is 140 nm $\times$ 140 nm, and the size of an adder is 340 nm $\times$ 140 nm. The thickness of the free layer is $t = 2$ nm.

13.6   Patterns of perpendicular magnetization switching

For the perpendicularly magnetized materials we take the following set of material parameters (unless stated otherwise): saturation magnetization, $M_s = 400$ kA/m; exchange stiffness, $A = 20$ pJ/m; perpendicular magnetic anisotropy, $K_u = 100$ kJ/m^3; polarization of injected electrons, $P = 0.9$; and Gilbert damping, $\alpha = 0.015$. We simulate the current pulse switched on for not more than 10 ns (unless magnetization reaches the steady state faster). We do not include the random torques associated with thermal fluctuations of magnetization, which corresponds to an effective temperature of zero. However, the random deviations of magnetization from an equilibrium direction (out-of-plane) are modeled by setting its initial angle at 10° away from perpendicular.

First we demonstrate the patterns of magnetization switching of STMG. The in-plane projections of magnetization in small areas of the free layer are still represented by arrows. Lighter gray shadings correspond to positive projections on the z-axis (pointing up). Darker gray shadings correspond to negative projections on the z-axis (pointing down). As before, voltage polarities applied to an STMG are designated as (p) for plus and (m) for minus.

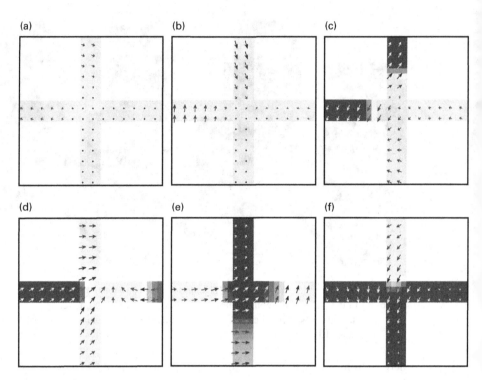

Fig. 13.14 Snapshots of magnetization in STMG of Fig. 13.2 at 2 ns and then every 0.5 ns left to right in each row. Arrows designate in-plane projections. Light grey/white = positive, dark grey/black − negative out-of-plane projection. Voltage polarities are A+, B+, C−. The current in each nanopillar is 0.05 mA, wire width $a = 20$ nm, thickness $t = 2$ nm.

We observe that magnetization switches in a complicated oscillatory manner, as in Fig. 13.14. Switching occurs as motion of domain walls separates areas with magnetization pointing up and down. The dynamics of switching is represented by the motion of domains within the STMG structure. It starts with switching of magnetization in the A+ and B+ arms. As the domain wall reaches the middle of the cross, magnetization overshoots to opposite directions in the B+ and C− arms. The domain wall eventually reaches the Out arm, becoming stationary and carrying negative magnetization there, as needed for the majority gate functionality.

At the same voltage polarity but higher current (Fig. 13.15) the magnetization performs more oscillations before the domain walls reach an equilibrium at the middle of the cross. There the magnetization becomes parallel to the plane of the domain wall, and it stops (similar to the effect of Walker breakdown [18]). Moreover, to move to the output arm, the domain wall needs to significantly increase its length, but the system lacks energy for length extension. As a result, the domain wall gets stuck in the middle of the cross and the gate fails to switch in response to the majority of inputs. The STMG needs to be designed to avoid such regimes.

A counterexample of well-behaved switching for different voltage polarities is shown in Fig. 13.16. As previously, the arms A+ and C+ switch via motion of the

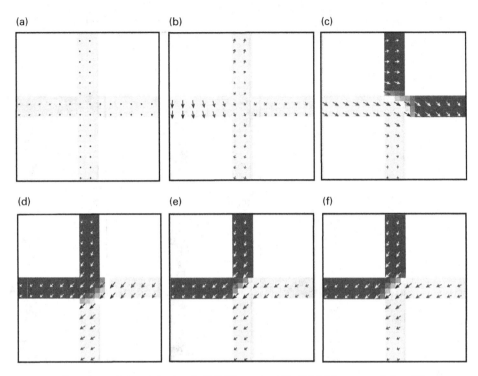

Fig. 13.15 Snapshots of magnetization in STMG, same as Fig. 13.14, but every 0.5 ns. The current in each nanopillar is 0.1 mA.

domain walls to the middle of the cross. Magnetization of the middle switches with overshoot of magnetization in arms A+ and B−. The domain wall moves to switch the Out arm. The B− arm returns to the direction of magnetization dictated by its electrode, and this input remains isolated by a domain wall from the output.

Examples of magnetization evolution in the whole adder device (as shown in Fig. 13.13, but disregarding inverters) are presented in Figs. 13.17 and 13.18 with the polarities as indicated in the figure caption. From them, the reader can see that the general way of switching is similar to a single cross STMG: the domain walls move towards the middles of crosses, where the majority of influences win in the middle, and then the output assumes the direction of the middle. In an adder, the same electrical signal needs to open two or three transistors (not shown) to drive current through corresponding input electrodes. Note that the designation of voltage polarities in an adder is slightly changed: ppp = C+, A+, B+, ppm = C+, A+, B−, pmp = C+, A−, B+. The behavior of domain walls is more complicated than in a single STMG. They reflect from the ends of the magnetic wires and collide at the crosses. Under the influence of demagnetization and the magnetic field of the input current, they sometimes retreat to the unswitched state, but eventually the whole area of the adder (except for the arms with opposing inputs) is switched. On a positive note, there are fewer cases of the domain walls stuck in an adder than in a single STMG. This is probably due to absence of reflection of domain walls from the output arm of each cross.

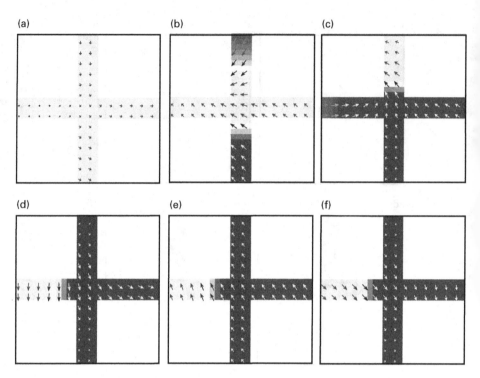

Fig. 13.16 Snapshots of magnetization in STMG, same as Fig. 13.14, but at 1 ns and then every 0.5 ns. Voltage polarities are A+, B−, C+. The current in each nanopillar is 0.1 mA.

The dependence of switching speed on-current, Figs. 13.19–13.21, proves to be non-monotonic and saturating at higher values of current. This first trend is due to timing of reflections that help or delay switching and the second trend is due to the time needed for the domain walls to propagate from inputs to outputs. The results of simulations for various voltage polarities and input currents are summarized in Fig. 13.19. For a wire width of 20 nm, a switching time of ~3 ns can be achieved with a relatively small current of 80 μA.

The gates fulfill a natural requirement for logic circuits in that there is a wide range of current in which they switch at approximately the same speed for all voltage polarities. The threshold current for all polarities is approximately the same, 40 μA, which corresponds to a current density of 10 MA/cm². An exception is a failure of switching at 60 μA for the ppm = C+, A+, B − polarity. Note that switching does not occur over a range of medium and higher currents. This happens when magnetization waves, reflecting from ends of the arms of the cross, converge to near the middle of the cross and form pinned domain walls. In order for the domain wall to move, it needs to become longer, which costs energy. The critical current corresponds to a current density of $J_c = 0.4$ MA/cm² per input, which is related to the total area of the free layer.

From comparing switching of the same layout of devices with different thicknesses, one can see that the switching speeds are comparable (see Fig. 13.20). We also observe that the critical current density increases with the thickness of nanomagnets. The thicker

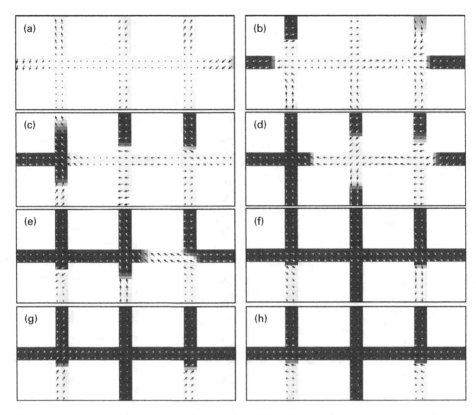

Fig. 13.17 Snapshots of magnetization in the adder of Fig. 13.3 at 1.8 ns and then every 0.6 ns left to right in each row. Inverters are excluded for simplicity. Voltage polarities are C+, A+, B−. The current in each nanopillar is 0.1 mA, thickness $t = 3$ nm.

gate has a higher threshold current due to two factors: (i) a larger volume of magnetic material; (ii) a larger demagnetization contribution to the energy barrier between the stable states.

Since scalability of devices is necessary for logic, we examine the dependence of switching speed on the size of the devices (Fig. 13.21), as set by the width of wires. Current density is used to compare them fairly. It is found that larger devices (with 40 nm wide wires) suffer from domain wall pinning and do not switch at higher values of current, while the problem disappears as they are scaled to smaller sizes. Note that the threshold current density is approximately the same for the width considered. The switching speed of the 10 nm wire adder is the fastest due to the fewest oscillations of magnetization. The speed of the 40 nm wide adder is the slowest and is mostly determined by the duration of the pulse. After it is switched off, the magnetization settles down to the equilibrium state. The case of 20 nm wires is in between: at some values of current the domain wall motion is timed to efficiently pass to the output arm, but at other values of current multiple reflections delay switching. Also the critical current density increases for smaller devices, because they have a different aspect ratio (arm width to thickness) and thus smaller out-of-plane component of shape anisotropy.

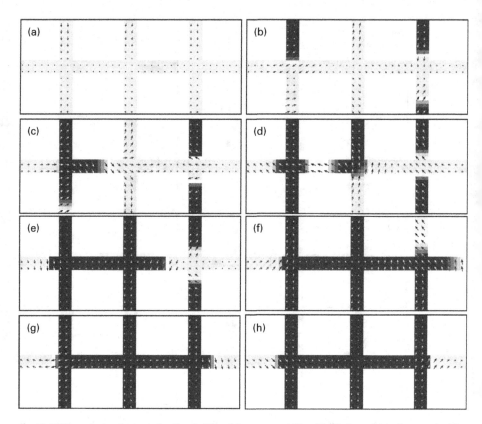

Fig. 13.18 Snapshots of magnetization in the adder, same as Fig. 13.17, but with voltage polarities C+, A−, B+.

13.7 Conclusion

Finally we use the obtained device characteristics to compare the computational performance of STMG adder circuits with CMOS ones. The data for CMOS are derived from ITRS [19]. For the STMG we choose the operating point with current of 75 μA. There are a few factors that are beneficial for STMG circuits. They have a higher density than CMOS, because 28 transistors are replaced by three majority gates. The magnetic circuits are placed between metallization layers, and thus the necessary driver and sense circuits can share the area. STMG circuits can operate at a much smaller drive voltage, because they are not based on raising and lowering the potential barrier. They also switch slower, but correspondingly have lower active power. Moreover, they do not lose their logic states without power, and can be switched off for periods of inactivity, thus resulting in effectively zero standby power.

By choosing an operation point from these plots, we can estimate the circuit's performance. In Table 13.1 we compare an STMG adder, an adder based on standard CMOS [20], and an adder composed of MTJs and CMOS. This last adder is based on a design and parameters in [21] scaled from the generation of 180 nm to 22 nm.

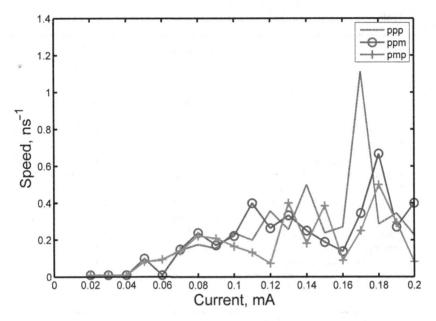

Fig. 13.19 Switching speed in the adder vs. current for various voltage polarities at $a = 20$ nm, $t = 2$ nm.

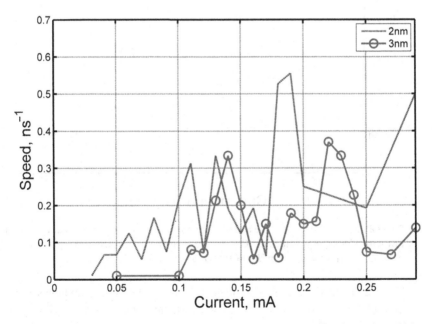

Fig. 13.20 Switching speed in the adder vs. current in each nanopillar for various thicknesses of the free FM layer. Voltage polarities are C+, A−, B+, width $a = 20$ nm, anisotropy $K_u = 110$ kJ/m^3.

Table 13.1 Comparison of adders created using different technologies

	CMOS	STMG	MTJ+CMOS
Process feature, F, nm	22	22	22
Area factor, $F*F$	10413	2727	10124
Area per gate, um^2	5.0	1.3	4.9
Voltage, V	0.81	0.1	0.81
Switching time, ps	16	2826	1.25
Clocking time, ps	250	5651	2000
Switching energy intrinsic, aJ	1382	147703	???
Switching energy with circuits, aJ	17640	257680	326000
Power per gate, active, μW	70.6	45.6	163.0
Power per gate, standby, μW	0.81	0	0
Activity factor	0.01	0.01	0.01
Power per gate, average, μW	1.52	0.46	1.63
Power per unit area, W/cm^2	30.1	34.5	33.3
Throughput, Mops/ns/cm^2	79.4	13.4	10.2

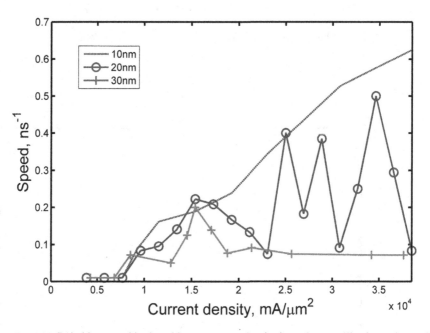

Fig. 13.21 Switching speed in the adder vs. current density in each nanopillar for various widths of FM wires. Voltage polarities are C+, A−, B+, thickness $t = 3$ nm (except for width 10 nm, thickness is 2 nm), anisotropy $K_u = 100$ kJ/m^3.

The scaling of the delay was assumed proportional to the size, and the switching energy proportional to the square of the size. All three circuits were adjusted to work at approximately the same power per unit area.

In conclusion, we propose nanomagnetic logic circuits based on relatively mature STTRAM technology that can be easily implemented. Switching over a wide range of parameters is demonstrated by micromagnetic modeling. Switching speed and current are comparable to existing STTRAM. The magnetization angle transfer characteristic corresponds to circuit gain. Though the switching speed and energy of these circuits are no better than in CMOS for the same process generation and the same activity factor, the dissipated power and computational throughput are within one order of magnitude of CMOS. For applications where the activity factor is much less than the value of 0.01 assumed here, the relative performance of STMG logic will become proportionally greater. The extreme case of that is the ability of magnetic-based logic circuits to maintain their state indefinitely with no power and no resume time.

References

[1] N. M. Hosomi, H. Yamagishi, T. Yamamoto *et al.*, "A novel nonvolatile memory with spin torque transfer magnetization switching: spin-RAM." In *Electron Devices Meeting, 2005. IEDM Technical Digest. IEEE International*, pp. 459–462. (2005).

[2] D. E. Nikonov & G. I. Bourianoff, "Operation and modeling of semiconductor spintronics computing devices." *Journal of Superconductivity and Novel Magnetism*, **21**(8), 479–493 (2008).

[3] D. E. Nikonov & G. I. Bourianoff, "Recent progress, opportunities and challenges for beyond CMOS information processing technologies." *ECS Transactions*, **35**(2), 43–53 (2011).

[4] D. E. Nikonov, G. I. Bourianoff, & T. Ghani, "Proposal of a spin torque majority gate logic." *IEEE Electron Device Letters*, **32**(8), 1128–1130 (2011).

[5] K. Bernstein, R. K. Cavin, III, W. Porod, A. Seabaugh, & J. Welser, "Device and architecture outlook for beyond CMOS switches." *Proceedings of the IEEE*, **92**(12), 2169–2184 (2010).

[6] S. Sugahara and M. Tanaka, "A spin metal–oxide–semiconductor field-effect transistor using half-metallic-ferromagnet contacts for the source and drain." *Applied Physics Letters*, **84**, 2307–2309 (2004).

[7] D. E. Nikonov & G. I. Bourianoff, "Spin gain transistor in ferromagnetic semiconductors: the semiconductor Bloch equations approach." *IEEE Transactions on Nanotechnology*, **4**, 206 (2005).

[8] R. P. Cowburn & M. E. Welland, "Room temperature magnetic quantum cellular automata." *Science*, **287**, 1466 (2000).

[9] D. A. Allwood, G. Xiong, C. C. Faulkner, D. Atkinson, D. Petit, & R. P. Cowburn, "Magnetic domain-wall logic." *Science*, **309**, 1688 (2005).

[10] B. Behin-Aein, D. Datta, S. Salahuddin, & S. Datta, "Proposal for an all-spin logic device with built-in memory." *Nature Nanomaterials*, **5**, 266 (2010).

[11] A. Khitun & K. L. Wang, "Nano scale computational architectures with spin wave bus." *Superlattices and Microstructures*, **38**, 184 (2005).

[12] I. Zutic, J. Fabian, & S. D. Sarma, "Spintronics: fundamentals and applications." *Reviews in Modern Physics*, **76**(2), 323–410 (2004).

[13] J. A. Katine, F. J. Albert, R. A. Buhrman, E. B. Myers, & D. C. Ralph, "Current-driven magnetization reversal and spin-wave excitations in Co/Cu/Co pillars." *Physics Review Letters*, **84**, 3149–3152 (2000).

[14] M. J. Donahue & D. G. Porter, "OOMMF User's Guide, Version 1.0." National Institute of Standards and Technology Report No. NISTIR 6376 (1999).

[15] G. S. Abo, Y.-K. Hong, J. Park, J. Lee, W. Lee, and B.-C. Choi, "Definition of magnetic exchange length." *IEEE Transactions on Magnetics*, **49**, 4937–4939 (2013).

[16] S. Mangin, D. Ravelsona, J. A. Katine, M. J. Carey, B. D. Terris, & E. E. Fullerton, "Current-induced magnetization reversal in nanopillars with perpendicular anisotropy." *Nat. Mater.*, **5**, 210–215 (2006).

[17] J. Hayakawa *et al.*, "Current-induced magnetization switching in MgO barrier magnetic tunnel junctions with CoFeB-based synthetic ferrimagnetic free layers." *IEEE Technology Magazine*, **44**, 1962–1967 (2008).

[18] N. L. Schryer & L. R. Walker, *J. Applied Physics*, **45**, 5406 (1974).

[19] "Emerging research devices." In *International Technology Roadmap for Semiconductors (ITRS)* (2011). Available at: www.itrs.net.

[20] "Process integration and device structure." In *International Technology Roadmap for Semiconductors (ITRS)* (2011). Available at: www.itrs.net.

[21] S. Matsunaga, J. Hayakawa, S. Ikeda *et al.*, "Fabrication of a nonvolatile full adder based on logic-in-memory architecture using magnetic tunnel junctions." *Applied Physics Express*, **1**, 091301 (2008).

14 Spin wave phase logic

Alexander Khitun

14.1 Introduction

A spin wave is a collective oscillation of spins in a spin lattice around the direction of magnetization. Similar to lattice waves (phonons) in solid systems, spin waves appear in magnetically ordered structures, and a quantum of a spin wave is called a "magnon." Magnetic moments in a magnetic lattice are coupled via the exchange and dipole–dipole interaction. Any local change of magnetization (disturbance of magnetic order) results in the collective precession of spins propagating through the lattice as a wave of magnetization – a spin wave. The energy and impulse of the magnons are defined by the frequency and wave vector of the spin wave. Similar to phonons, magnons are bosons obeying Bose–Einstein statistics. Spin waves (magnons) as a physical phenomenon have attracted scientific interest for a long time [1, 2] and a variety of experimental techniques including inelastic neutron scattering, Brillouin scattering, X-ray scattering, and ferromagnetic resonance have been applied to the study of spin waves [3, 4]. Over the past two decades, a great deal of interest has been attracted to spin wave transport in artificial magnetic materials (e.g., composite structures, so-called "magnonic crystals" [5, 6]) and magnetic nanostructures [7–9]. New experimental techniques including time-domain optical and inductive techniques [7] have been developed to study the dynamics of spin wave propagation. In order to comprehend the typical characteristics of the propagating spin wave, we will refer to the results of the time-resolved measurement of propagating spin waves in a 100 nm thick NiFe film presented in [8]. In this experiment, a set of asymmetric coplanar strip (ACPS) transmission lines was fabricated on top of permalloy ($Ni_{81}Fe_{19}$) film. The strips and magnetic layer are separated by an insulating layer. One of the transmission lines was used to excite a spin wave packet in the ferromagnetic film, and the rest of the lines located 10 μm, 20 μm, 30 μm, 40 μm, and 50 μm away from the excitation line were used for detection of the inductive voltage. When excited by the 100 ps pulse, spin waves produce an oscillating inducting voltage, which reveals the local change of magnetization under the line caused by the spin wave propagation. The detected inductive voltage signals reveal the spatially localized spin waves, which are damped and spread in time. For instance, the speed of propagation of the detected magnetostatic spin waves in permalloy is about 10 μm/ns, and the attenuation time at room temperature is about 0.8 ns. The inductive voltage is about 30 mV at the point of excitation and drops to 2 mV as the wave propagates 50 μm. These experimental data obtained for magnetostatic spin waves in permalloy may be used as a benchmark for practically feasible spin wave devices, though the group velocity and the attenuation time may vary significantly for different materials and waveguide dimensions.

The relatively slow group velocity (more than three orders of magnitude slower than the speed of light) and high attenuation (more than six orders of magnitude higher than for photons in a standard optical fiber) explain the lack of interest in spin waves as a potential candidate for information transmission. The situation has changed drastically, however, as the characteristic distance between the devices on the chip has entered the deep-sub-micron range. It has become more important to have fast signal conversion/ modulation, while the short traveling distance compensates for the slow propagation and high attenuation. From this point of view, spin waves possess certain technological advantages: (i) spin waves can be guided by magnetic waveguides similar to optical fibers; (ii) the spin wave signal can be converted into a voltage via inductive coupling; (iii) a magnetic field can be used as an external parameter for spin wave signal modulation. The wavelength of the exchange spin waves can be as short as several nanometers, and the coherence length may exceed tens of microns at room temperature. The latter raises the intriguing possibility of building scalable logic devices utilizing spin waves.

14.2   Computing with spin waves

There are three basic approaches to spin wave logic devices, using the amplitude [10–12], the phase [13], or the frequency [14] of the spin wave signal for information encoding. Each of these approaches possesses certain technological advantages and constraints. The first working spin wave-based logic device was experimentally demonstrated in 2005 by Kostylev et al. [10]. They built a Mach–Zehnder-type current-controlled spin wave interferometer to demonstrate output voltage modulation as a result of spin wave interference. This first working prototype was of great importance for the development of magnonic logic devices. The device operates in the GHz frequency range and at room temperature. Later on, exclusive-NOT-OR and NOT-AND gates were experimentally demonstrated on a similar Mach–Zehnder-type structure [11]. A complete set of logic devices such as NOT, NOR, and AND based on Mach–Zehnder-type spin wave interferometer devices has been proposed [12]. In [10–12], spin wave amplitude defines the logic state of the output. At some point, these amplitude-based devices resemble the classical field effect transistor, where the magnetic field produced by the electric current modulates the propagation of the spin wave – an analog to the electric current.

In the phase-based approach, logic 0 and 1 are assigned to the phase (0 or π) of the propagating spin wave [13]. A phase-based circuit receives spin waves of the same frequency and amplitude, while the phase of the input wave may have only two possible values: 0 or π. The circuit consists of the ferromagnetic junctions and phase shifters so that the phase of the output wave is a function of the input phase combination (e.g., π, π, $\pi \rightarrow \pi$, or π, 0, $\pi \rightarrow 0$) as well as the waveguide configuration. Computing with phases has certain advantages in making special types of logic gates (e.g., MAJ, MOD) [15], some of which have been experimentally demonstrated [16, 17]. A more detailed description of the operation of the phase-based logic circuits is presented in Section 14.4.

Frequency-based magnonic circuits have been recently proposed [14]. The proposed circuits consist of spin torque oscillators communicating via spin waves propagating through the common free layer. This approach is based on the property of nanometer-scale spin torque devices generating spin waves in response to a d.c. electrical current [18, 19]. Electric current passing through a spin torque nano-oscillator (STNO) generates spin transfer torque and induces auto-oscillatory precession of the magnetic moment of the spin valve free layer. The frequency of the precessing magnetization is tunable by the applied d.c. voltage due to the strong non-linearity of the STNO. In the case of two or more STNOs sharing a common free layer, the oscillations can be frequency and phase locked via the spin wave exchange [20, 21]. The unique properties of spin torque oscillators have great promise for magnonic logic circuitry, although the main challenge is associated with the relatively high current required for operation.

14.3   Experimentally demonstrated spin wave components and devices

Over the past decade, there have been a number of works focused on the feasibility study of magnonic circuits exploiting multi-wave interference, as well as the search for more efficient elements for spin wave excitation and detection. Figure 14.1 shows the schematics of a 4-terminal spin wave device used as a prototype MAJ gate [17]. The material structure, starting at the bottom, consists of a silicon substrate, a 300 nm thick silicon oxide layer, a 20 nm thick ferromagnetic layer made of $Ni_{81}Fe_{19}$, a 300 nm

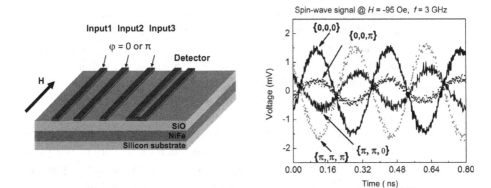

Fig. 14.1 (a) Schematics of the 4-terminal magnonic device. The device structure comprises a silicon substrate, a 20 nm thick layer of permalloy, a layer of silicon dioxide, and a set of five conducting wires on top (three wires to excite three spin waves, the other two wires connected in a loop are to detect the inductive voltage). The initial phase of the excited spin wave (0 or π) is controlled by the direction of the excitation current. (b) Experimental data showing the inductive voltage as a function of time. The different curves correspond to the different combinations of the phases of the interfering spin waves. (Reprinted with permission from A. Khitun *et al.*, "Demonstration of majority logic gate with spin-waves." *Annual report of Western Institute of Nanoelectronics*, Abstract 2.1 (2009).)

thick layer of silicon oxide, and the five conducting wires on top. The distance between the wires is 2 µm. In order to demonstrate a three-input/one-output majority gate, three of the five wires were used as the input ports, and the remaining two wires were connected in a loop to detect the inductive voltage produced by the spin wave interference. An electric current passing through the "input" wire generates a magnetic field, which, in turn, excites spin waves in the ferromagnetic layer. The direction of the current flow (the polarity of the applied voltage) defines the initial spin wave phase. The various curves in Fig. 14.1(b) depict the inductive voltage as a function of time for different combinations of the spin wave phases (e.g., 000, $0\pi0$, $0\pi\pi$, and $\pi\pi\pi$). These results show that the phase of the output inductive voltage corresponds to the majority of the phases of the interfering spin waves. Spin waves produce an inductive voltage output of several mV with a signal to noise ratio of about 10:1. The data are taken for a 3 GHz excitation frequency and at a bias magnetic field of 95 Oe (perpendicular to the spin wave propagation). All measurements were accomplished at room temperature. There are several important features of this device to be emphasized: (i) all input/output ports are located on one ferromagnetic strip; (ii) there are just two working phases 0 and π, which are controlled by the direction of the excitation current; (iii) the whole circuit is built on a silicon substrate.

Later, spin wave interference has been also studied in magnetic cross-junctions [22], where two spin waves were excited in the perpendicular magnetic waveguides. Figure 14.2(a) shows the 4-terminal magnonic device comprising a magnetic cross-junction made of permalloy and four micro-antennas fabricated on top of the waveguides. The micro-antennas are connected to the network analyzer to measure the S parameters corresponding to signal transmission between the micro-antennas. It turned out that the spin wave transport depends significantly on the local junction magnetization. The experimental data in Fig. 14.2(b) show the S_{13} and S_{14} parameters as a function of the angle of the in-plane junction magnetization. The same signal generated by the micro-antenna marked 1 in Fig. 14.2(a) can be redirected between ports 3 and 4 by changing the local magnetic texture of the junction. The data in Fig. 14.2(c) show the result of spin wave interference, where two waves were excited in ports 3 and 4. Thus, one magnetic junction can be utilized as a multi-functional element, which is of great promise for special type logic architecture to be discussed later in the chapter.

The excitation and detection of spin waves in [17, 22] was accomplished by the micro-antennas. Micro-antennas are convenient for laboratory proof-of-concept experiments, though their application in scaled devices has several significant drawbacks, including the strong direct coupling between the input and output antennas via stray fields and inefficient energy conversion. Only a small percentage (>1%) of the input electric power comes to the spin waves via the micro-antenna. A more efficient mechanism of spin wave excitation by multiferroic elements (i.e., magneto-electric, or ME, cells) has been recently demonstrated [23]. Figure 14.3 shows the test structure of two ME cells, where the cell on the left is for spin wave generation, and the cell on the right is a spin wave detector. The structure includes a 40 nm thick Ni/NiFe bilayer (20 nm × 20 nm), patterned in a 4 µm wide and 200 µm long bar (spin wave bus) on the top of the PMN–PT substrate with (001) crystal orientation. The material of the bar was

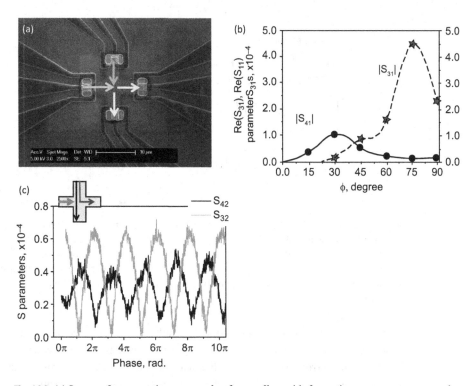

Fig. 14.2 (a) Image of a magnetic cross made of permalloy with four micro-antennas connected to the network analyzer. (b) Experimental data on the signal transmission between ports 1, 3, and 4 as a function of the direction of the external magnetic field. (c) Experimental data on spin wave interference. Two spin waves of 8 GHz were excited at ports 1 and 2. The black and grey curves show the interference detected at ports 3 and 4, respectively. (Reprinted with permission from A. Kozhanov *et al.*, "Spin wave topology and imaging." *Annual report of Western Institute of Nanoelectronics*, Abstract 2.2 (2011).)

chosen to provide both low loss medium for spin wave propagation (NiFe) and magnetostrictive properties for strain-induced anisotropy change (Ni). A radio-frequency voltage was applied to a micron-scale confined region of the PMN–PT substrate underneath the ferromagnetic bar (ME cell) to excite spin waves by rapid anisotropy change due to the local voltage induced strain. The measurements were performed by using a two-port vector network analyzer simultaneously detecting reflected and transmitted signal components. Figure 14.3(b) shows the experimental data on spin wave excitation and detection by ME cells. The various plots show the transmitted amplitude as a function of frequency at the fixed bias magnetic field. The propagating spin waves were detected at distances as far as 40 μm away from the excitation cell. To verify the origin of the detected signal, all experiments were carried out at different bias magnetic field from −500 Oe to 500 Oe applied along the hard axis of the spin wave bus. The obtained data clearly show the Kittel's mode behavior as expected for surface magnetostatic spin waves in permalloy. This experiment confirmed the feasibility of using ME cells for spin wave generation and detection. The utilization

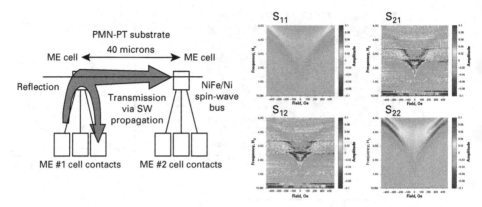

Fig. 14.3 (a) Schematics of the experiment on spin wave excitation and detection by multiferroic elements (ME cells). (b) Collection of experimental data (S_{11}, S_{12}, S_{21}, and S_{22} parameters) obtained at different frequencies and bias magnetic field. (Reprinted with permission from S. Cherepov *et al.*, "Electric-field-induced spin wave generation using multiferroic magnetoelectric cells." *Annual report of Western Institute of Nanoelectronics*, Abstract 2.2 (2011).)

of multiferroic elements has immediately resulted in an enhancement of the excitation efficiency (e.g., of about 1 fJ per wave for a micrometer-scale ME cell). The development of multiferroic elements together with the demonstration of multi-wave interference devices has given impetus to the practical realization of magnonic logic circuits.

14.4   Phase-based logic devices

We consider a phase-based approach to be the most promising [13, 15], taking advantage of a phase for data encoding and processing, and providing an alternative way for logic gate construction. The principle of operation of the phase-based magnonic logic circuit is fundamentally different from the conventional field-modulated amplitude approach. Within this approach, one bit of information is assigned to the phase of the propagating spin wave. A simple computation is associated with the change of the phase of the propagating spin wave, which provides an alternative route to the NOT and majority logic gate construction. The schematics of a phase-based magnonic logic circuit are shown in Fig. 14.4. The circuit comprises the following elements: (i) magneto-electric cells; (ii) magnetic waveguides–spin wave buses; and (iii) a phase shifter. The magneto-electric cell (ME cell) converts the applied voltage into the spin wave as well as to read-out the voltage produced by the spin waves. The operation of the ME cell is based on the effect of the magneto-electric coupling (i.e., multiferroics) enabling magnetization control by applying an electric field and vice versa. The waveguides are simply the strips of ferromagnetic material (e.g., NiFe) used to transmit the spin wave signals. The phase shifter is a passive element (e.g., the same waveguide of different width, a domain wall) providing a π-phase shift to the propagating spin waves.

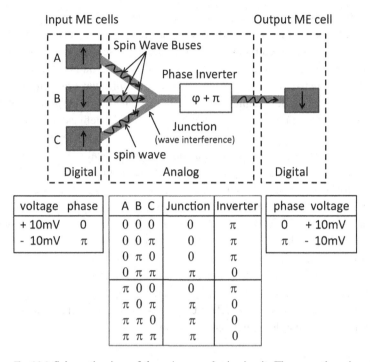

voltage	phase		A B C	Junction	Inverter		phase	voltage
+ 10mV	0		0 0 0	0	π		0	+ 10mV
- 10mV	π		0 0 π	0	π		π	- 10mV
			0 π 0	0	π			
			0 π π	π	0			
			π 0 0	0	π			
			π 0 π	π	0			
			π π 0	π	0			
			π π π	π	0			

Fig. 14.4 Schematic view of the spin wave logic circuit. There are three inputs (A, B, and C) and one output. The inputs and the output are the ME cells connected via the ferromagnetic waveguides–spin wave buses. The input cells generate spin waves of the same amplitude with initial phase 0 or π, corresponding to logic 0 and 1, respectively. The waves propagate through the waveguides and interfere at the point of junction. The phase of the wave after the junction corresponds to the majority of the interfering waves. The phase of the transmitted wave is inverted (e.g., passing the domain wall). The table illustrates the data processing in the phase space. The phase of the transmitted wave defines the final magnetization of the output ME cell. The circuit can operate as NAND or NOR gate for inputs A and B depending on the third input C (NOR if C = 1, NAND if C = 0). Reprinted with permission from A. Khitun, "Multi-frequency magnonic logic circuits for parallel data processing," *Journal of Applied Physics*, 111, 2012. Copyright 2013, American Institute of Physics.

The principle of operation is as follows. Initial information is received in the form of voltage pulses. Inputs 0 and 1 are encoded in the polarity of the voltage applied to the input ME cells (e.g., +10 mV corresponds to logic state 0, and −10 mV corresponds to logic 1). The polarity of the applied voltage defines the initial phase of the spin wave signal (e.g., a positive voltage results in a clockwise magnetization rotation and a negative voltage results in a counter-clockwise magnetization rotation). Thus, the input information is translated into the phase of the excited wave (e.g., initial phase 0 corresponds to logic state 0, and initial phase π corresponds to logic 1). Then, the waves propagate through the magnetic waveguides and interfere at the waveguide junction. For any junction with an odd number of interfering waves, there is a transmitted wave with non-zero amplitude. The phase of the wave passing through the junction always corresponds to the majority of the phases of the interfering waves (e.g., the transmitted wave will have phase 0 if there are two or three waves with initial phase 0; otherwise the wave will have a π-phase). The transmitted wave passes the phase shifter and accumulates an additional π-phase shift (i.e., phase $0 \rightarrow \pi$, and phase $\pi \rightarrow 0$). Finally, the spin wave signal reaches the output ME cell, which has two stable magnetization states. At the moment of spin wave arrival, the output cell is in the metastable state (magnetization is along the hard axis perpendicular to the two stable states). The phase of the incoming spin wave defines the direction of the magnetization relaxation in the output cell [15, 24]. The magnetization change in the output ME cell is associated with a change of electrical polarization in the multiferroic material and can be recognized by the induced voltage across the ME cell (e.g., $+10$ mV corresponds to logic state 0, and -10 mV corresponds to logic 1). The truth table in Fig. 14.4 shows the input/output phase correlation. The waveguide junction works as a majority logic gate. The amplitude of the transmitted wave depends on the number of in-phase waves, while the phase of the transmitted wave always corresponds to the majority of phase inputs. The π-phase shifter works as an inverter in the phase space. As a result of this combination, the three-input/one-output gate in Fig. 14.4 can operate as a NAND or NOR gate for inputs A and B depending on the third input C (NOR if C = 1, NAND if C = 0). Such a gate can be a universal building block for any Boolean logic gate construction.

14.5   Spin wave-based logic circuits and architectures

14.5.1   Logic circuits

Since the first proposal on spin wave logic [13], the idea of using spin waves in logic circuitry has evolved in different ways [15, 24–26]. A variety of possible spin wave-based devices can be classified within the following groups: volatile and non-volatile, Boolean and non-Boolean, single-frequency and multi-frequency circuits. In this section we briefly describe the distinct features of these approaches and show some examples of magnonic architectures.

14.5.1.1   Volatile magnonic circuits

Volatile magnonic circuits provide functional output (i.e., inductive voltage) as long as external power is applied to the spin wave generating elements [13], or spin wave buses are combined with an electric circuit preserving the output voltage produced by the spin wave pulses [25] (e.g., as shown in Fig. 14.5). For example, the magnonic circuits described in [24] combine spin wave buses with micro-antennas. The circuit operates as long as the input antennas generate continuous spin waves. It is also possible to build a circuit combining spin wave buses with a bistable electric circuit, where the switching of the electric circuit is accomplished by the inductive voltage pulse [25]. In this case, there is no need for permanent spin wave generating elements, although external power is required to maintain the state of the electronic circuit.

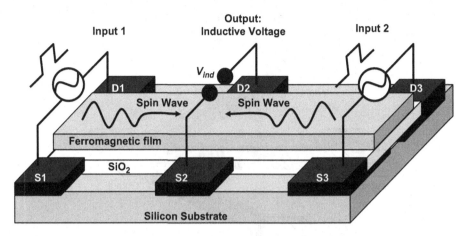

Fig. 14.5 Volatile spin wave device consisting of three spin-FETs connected via a spin wave bus. Three spin-FETs share the same gate made of a ferromagnetic material. Two spin waves excited by inputs 1 and 2 excite two spin waves with initial phases 0 or π corresponding to logic 0 and 1, respectively. The result of the interference is detected by the third spin-FET shown in the center. (Reprinted with permission from A. Khitun and K. Wang, "Nano scale computational architectures with spin wave bus." *Superlattices & Microstructures*, **38**, 184 (2005).)

14.5.1.2   Non-volatile magnonic logic

Non-volatile magnonic logic circuits are able to preserve the result of computation without external power applied (i.e., the circuit shown in Fig. 14.6). The storing of information is in the magnetic state of the output ME cell, where logic 0 and 1 are encoded into the two states of magnetization of the magnetostrictive material [15]. In general, the magnetic field produced by a spin wave is quite weak to reverse the magnetization of a large volume ferromagnetic required for reliable data storage (thermal stability >40). The switching is accomplished via the help of magneto-electric coupling, where an electric field applied to the ME cell rotates its magnetization in a metastable state, and the incoming spin wave defines the direction of relaxation [15]. Non-volatile magnonic logic devices have been recognized as a promising approach to post-CMOS circuitry for radical power consumption minimization [27].

14.5.1.3   Boolean magnonic circuits

Boolean magnonic circuits are aimed to provide the same basic set of logic gates (AND, OR, NOT) for general computing as currently provided by conventional transistor-based circuitry. The advantage of using waves (i.e., spin waves) is the ability to exploit the waveguides as passive logic elements for controlling the phase of the propagating wave. Waveguides of the same length but different width or composition introduce a different phase change to the propagating spin waves, which offers an additional degree of freedom for logic circuit construction. In addition, the utilization of spin wave interference is efficient for building high fan-in devices, which is a significant advantage over transistor-based circuits [17]. Overall, magnonic Boolean logic circuits can

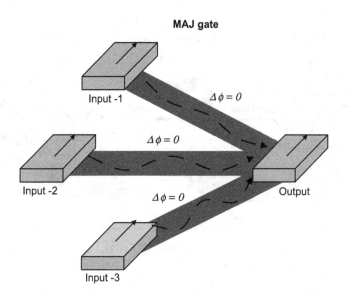

MAJ gate

Fig. 14.6 Non-volatile majority (MAJ) logic gate comprising multiferroic elements and spin wave buses. Three input and one output multiferroic cells are connected via spin wave buses. The magnetization state of the output element is controlled by the spin waves emitted by the input cells. External power is required only at the time of switching. (Reprinted with permission from A. Khitun and K. Wang, "Non-volatile magnonic logic circuits engineering." *Journal of Applied Physics*, **110**, 034306 (2011). Copyright 2013, American Institute of Physics.)

be constructed with fewer elements than required for their CMOS counterparts [28]. This advantage is more important for complex logic circuits. For example, a magnonic full adder circuit can be built with just five ME cells, while the conventional design requires at least 25 transistors [15].

14.5.1.4   Non-Boolean magnonic circuits

Non-Boolean magnonic circuits constitute a novel direction for magnonic circuit development aimed to complement scaled CMOS in special task data processing. In contrast to the Boolean logic gates used for conventional data processing, non-Boolean circuits are designed for one or several specific logic operations. Data search and image processing are examples of widely used tasks which require significant resources from a conventional processor. Parallel data processing of a large number of bits can be accomplished by utilizing multi-wave interference, where each wave (i.e., the phase of the wave) represents one bit of data. Examples of non-Boolean magnonic logic circuits designed for pattern recognition, finding the period of a given function, and magnonic holographic memory are described in [29]. The operation of these circuits is based on the massive use of spin wave interference within a magnetic template. This approach is similar to the methods developed for "all optical computing" [30], although the practical implementation of the magnonic circuits may be more feasible for integration on the silicon platform.

14.5.1.5   Multi-frequency magnonic logic circuits

Multi-frequency magnonic logic circuits use more than one operating frequency for data transmission and processing. Wave superposition allows us to send, process, and detect a number of waves propagating simultaneously within the same structure. The same circuit, made up of waveguides, junctions, and phase shifters as shown in Fig. 14.7, can operate over a range of frequencies $\{f_1, f_2, \ldots, f_n\}$, defined by the bandwidth of the waveguides, and the size and composition of the junctions and phase shifters. Each of the frequencies $\{f_1, f_2, \ldots, f_n\}$ is considered to be an independent information channel, where logic 0 and 1 are encoded into the phase of the propagating spin wave. An example of a multi-frequency circuit is described in [26]. The multi-frequency approach is an extension that can be applied to all types of magnonic circuits described above.

14.5.2   Architectures with spin wave buses

The combination of spin wave buses and multi-functional multiferroic elements provides a novel and interesting route to the implementation of complex computational architectures such as cellular non-linear network (CNN) [31] and holographic computing [32].

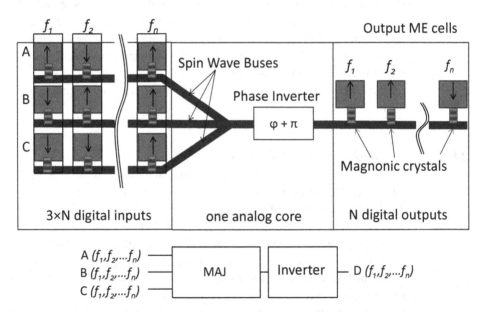

Fig. 14.7 Schematic view of the multi-frequency magnonic circuit. There are multiple ME cells on each of the input and output nodes aimed to excite and detect spin waves on the specific frequency (e.g., $f_1, f_2, \ldots, f_n$). The cells are connected to the spin wave buses via the magnonic crystals serving as the frequency filters. Within the spin wave buses, spin waves of different frequencies superpose, propagate, and receive a π-phase shift independently of each other. Logic 0 and 1 are encoded into the phases of the propagating spin waves on each frequency. The output ME cells recognize the result of computation (the phase of the transmitted wave) on one of the operating frequencies. (Reprinted with permission from A. Khitun, "Multi-frequency magnonic logic circuits for parallel data processing." *Journal of Applied Physics*, **111** (2012). Copyright 2013, American Institute of Physics.)

14.5.2.1 CNN with spin wave bus

The schematic of the magnonic CNN is shown in Fig. 14.8 [33]. The network consists of magneto-electric cells integrated onto a common ferromagnetic film/spin wave bus. The magneto-electric cell is an artificial two-phase multiferroic structure comprising piezoelectric and ferromagnetic materials similar to the ME cell in Fig. 14.7. One bit of information is assigned to the cell's magnetic polarization, which is controlled by the applied voltage. The information exchange among the cells is via the spin waves propagating in the spin wave bus. Each cell changes its state as a combined effect: magneto-electric coupling and the interaction with the spin waves. The distinct feature of the network with a spin wave bus is the ability to control the inter-cell communication using an external global parameter – the magnetic field. This makes it possible to realize different image processing functions on the same template without rewiring or reconfiguration. In Fig. 14.8(b), there are shown the examples of image processing functions dilation and erosion accomplished at two different magnetic bias fields. More complex image processing functions such as vertical and horizontal line detection, inversion, and edge detection can also be accomplished on one template by the proper choice of the strength and direction of the external magnetic field. It is important to note that none of the ME cells in the network has an individual contact, or a bias wire. The addressing of an individual cell is via the interference of two spin waves generated by the micro strips located at the edges of the structure as illustrated in Fig. 14.9, which offers an original solution to the interconnect problem inherent in most of the proposed nano-CNNs. Instead of a large number of wires or a crossbar structure, nano-cells can be addressed via wave interference produced by just two micro-antennas.

14.5.2.2 Magnonic holographic devices

Holographic memory is another promising architecture to benefit from spin wave utilization. Holographic techniques have been extensively developed in optics and the unique capabilities of the holographic approach for data storage and processing have been well described in a number of works [34, 35]. The concept of holography is based on the use of wave interference and diffraction, which can also be implemented in spin wave devices [29]. The schematics of the magnonic holographic device are shown in Fig. 14.9(a). It has multiple input/output ports located at the edges of the structure and a core consisting of a rectangular grid of ferromagnetic waveguides. For simplicity, just four ports are shown on each side, although the maximum number of nodes may exceed tens of hundreds. The I/O ports are magneto-electric (ME) cells designed to convert input electric signals into spin waves or spin waves into electric signals. The core of the structure consists of a two-dimensional grid of ferromagnetic waveguides (i.e., NiFe strips) designed to transmit spin waves between the input and output ports. An elementary mesh of the grid is a ferromagnetic cross-junction similar to the one shown in Fig. 14.2(a). The propagation of the spin wave through the junction depends on the junction magnetization discussed earlier. Each of the junctions may have several states for magnetization as illustrated in Fig. 14.9(b), so the whole

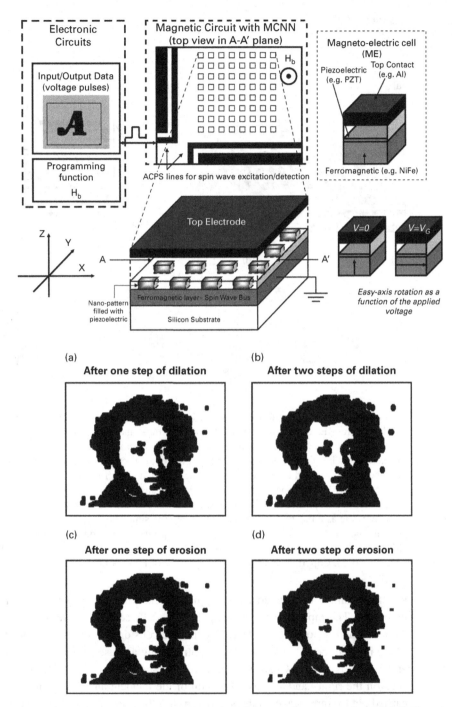

Fig. 14.8 (a) Schematic view of the magnonic cellular non-linear network (MCNN). There is an array of ME cells on the common ferromagnetic film/spin wave bus. Each cell is a bistable magnetic element. The interaction between the cells is via spin waves propagating through the spin wave bus. The read-in and read-out operations are accomplished by the edge micro-antennas. (b) Results of numerical modeling illustrating image processing with MCNN. The black and white pixels correspond to the two magnetic states of the ME cells. (Reprinted with permission from A. Khitun, M. Bao and K. Wang, "Magnetic cellular nonlinear network with spin wave bus for image processing." *Superlattices & Microstructures*, **47**(3) (2010).)

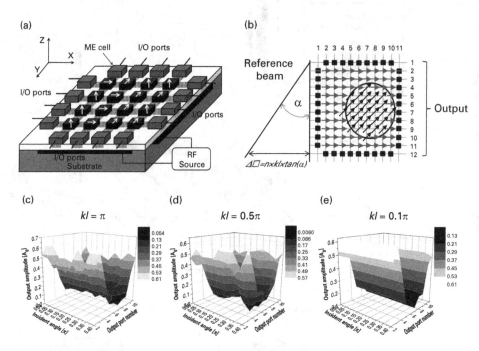

Fig. 14.9 (a) The schematics of the magnonic holographic memory. I/O ports at the edges of the device are ME cells aimed to convert input electric signals into spin waves and, vice versa. The core of the structure is a two-dimensional grid of ferromagnetic waveguides connected via magnetic cross junctions aimed to transmit spin waves between the input and output ports. (b) The input beam is generated by the ME cells on the left-hand side of the structure, and the output is detected by the ME cells on the right-hand side. The angle of illumination is controlled by the phase shift of the spin wave emitting cells. (c) The maps showing the output from the same template as a function of the incident angle. The simulations were carried out for three wavenumbers k: $kl = \pi$, $kl = 0.5\pi$, $kl = 0.01\pi$. (Reprinted with permission from A. Khitun, "Magnonic holographic devices for special type data processing." *Journal of Applied Physics*, **113**, 16 (2013). Copyright 2013, American Institute of Physics.)

structure can be considered as a magnonic scattering matrix similar to an optical hologram. The input beam is generated by the ME cells on the left-hand side of the structure, and the output is detected by the ME cells on the right-hand side. The angle of the incident beam α is controlled by introducing a phase shift among the spin wave emitting cells $\Delta\varphi = j \cdot kl \cdot \tan(\alpha)$. Figure 14.9(c) shows the output detected by the ME cells on the right-hand side as a function of the incident angle. The simulations were carried out for three wavenumbers k: $kl = \pi$, $kl = 0.5\pi$, $kl = 0.01\pi$. As one can see from Fig. 14.9(c), the output varies as a function of the incident angle. The angle dependence of the output disappears in the long wavelength limit $kl = 0.01\pi$, where the wavelength of the illuminating beam is much longer than the size of the junction. These results demonstrate the capabilities of magnonic holography for recording multiple images in the same structure. According to the estimates [29], magnonic holographic devices can provide up 1 Tb/cm^2 data storage density and provide data processing rates exceeding 10^{18} bits/s/cm^2.

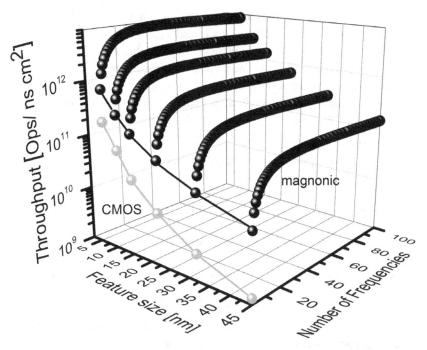

Fig. 14.10 Numerical estimates on the functional throughput for the full adder circuit built of CMOS and magnonic multi-frequency circuits. The estimates for the CMOS circuit are based on the 32 nm CMOS technology. The estimates for smaller feature size are extrapolated: the area per circuit scales as ×0.5 per generation, and the time delay scales as ×0.7 per generation. (Reprinted with permission from A. Khitun, "Multi-frequency magnonic logic circuits for parallel data processing." *Journal of Applied Physics*, **111** (2012). Copyright 2013, American Institute of Physics.)

14.6   Comparison with CMOS

The principle of operation of magnonic circuits is different from conventional CMOS technology and the design rules adopted in CMOS are not applicable to wave-based circuits. In order to compare magnonic devices with CMOS, we present estimates of phase-based non-volatile circuits [15], including area, time delay, energy per operation, and functional throughput.

The area of magnonic logic circuit is defined by several parameters: the size of the ME cell ($F \times F$); the number of ME cells per circuit N_{ME}; and the length L_{swb} and width W_{swb} of the spin wave buses. These parameters are related to each other via the same physical quantity – the wavelength of the spin wave. Theoretically, the feature size F of the ME cell can be much smaller than the wavelength λ of the information-carrying spin waves. On the other hand, the length of the ME cells should be about the wavelength, $F \approx \lambda$, for efficient spin wave excitation via the magneto-electric coupling. The width of the spin wave bus W_{swb} is also related to the wavelength λ via the dispersion law. However, the width of the spin wave bus can be much smaller than the wavelength.

In our estimates, we assume the feature size of the ME cell F to be equal to the wavelength λ ($F \approx \lambda$), $W_{swb} \ll L_{swb}$, and L_{swb} to be one or one and a half of the wavelength, depending on the particular logic circuit (e.g., $L_{swb} = \lambda$ for a buffer gate, $L_{swb} = \lambda/2$ for an inverter). The number of ME cells per circuit depends on the circuit functionality. At the moment, there is no empirical rule to estimate the size of the magnonic logic circuits based on the number of ME cells. Below, we present estimates for the area A of some logic circuits [15]:

- Buffer: $A = F \times (2F + \lambda) \approx 2\lambda^2$
- Inverter: $A = F \times (2F + \lambda/2) \approx 2.5\lambda^2$
- AND gate: $A = F \times (3F + \lambda + \lambda) \approx 3\lambda^2$
- MAJ gate/MOD2 gate: $A = (3F + 2\lambda) \times (2F + \lambda) \approx 15\lambda^2$
- Full adder circuit: $A = (3F + 2\lambda) \times (3F + 2\lambda) \approx 25\lambda^2$

Time delay per circuit is a sum of the following: the time required to excite spin waves by the input ME cells t_{ext}, propagation time for spin waves from the input to the output cells t_{prop}, and the time of magnetization relaxation in the output ME cells t_{relax}:

$$t_{delay} = t_{ext} + t_{prop} + t_{relax}$$

The minimum time delay for spin wave excitation is limited by the RC delay of the electric part, where R is the resistance of metallic interconnects, and C is the capacitance of the ME cell. Typically, the RC delay is much shorter than the time required for spin wave to propagate between the excitation and the detection ports [17]. The propagation time can be estimated by dividing the length of the spin wave bus connecting the most distant input and the output cells by the spin wave group velocity v_g, $t_{prop} = L_{swb}/v_g$. The group velocity depends on the material and geometry of the bus as well as the specific spin wave mode. The typical group velocity for magnetostatic spin waves propagating in conducting ferromagnetic materials (e.g., NiFe) is about 10^6 cm/s [7, 8].

We want to emphasize the difference between the volatile and non-volatile magnonic circuits in terms of the operational speed. The speed of operation of the volatile magnonic circuits is limited only by the spin wave excitation time, the length of the circuit, and the spin wave group velocity, while the non-volatile circuits require an additional time for the output bistable ME cell switching. The relaxation time of the output ME cell depends on the material properties of the magnetostrictive material (e.g., the damping parameter α). Realistically, the minimum time delay required for magnetization reversal t_{relax} of a bistable nanomagnet (thermal stability >40) is about 100 ps, which may be much longer than the propagation time (e.g., 100 nm/10^6 cm/s = 10 ps).

The energy per operation in the magnonic logic circuits depends on the number of ME cells per circuit and the energy required for magnetization rotation in each cell. ME cells are the only active (i.e., power-consuming) elements while the other components (e.g., spin wave buses, junctions, phase shifters) are passive elements. Input ME cells serve for spin wave generation while the output ME cells restore the spin wave signal. In both cases, the energy for spin wave generation/restoration comes from the electric domain via the magneto-electric coupling in the multiferroic element. For example, in the synthetic multiferroic comprising piezoelectric (e.g., PZT) and magnetostrictive

Table 14.1 Comparison between magnomic and conventional full adder circuits

	45 nm CMOS	32 nm CMOS	$\lambda= 45$ nm	$\lambda= 32$ nm
Area	$6.4 \, \mu m^2$	$3.2 \, \mu m^2$	$0.05 \, \mu m^2$	$0.026 \, \mu m^2$
Time delay	12 ps	10 ps	13.5 ps/0.1 ns	9.6 ps/0.1 ns
Functional throughput	1.3×10^9 Ops/[ns cm^2]	3.1×10^9 Ops/ [ns cm^2]	1.48×10^{11} Ops/[ns cm^2]	4.0×10^{11} Ops/[ns cm^2]
Energy per operation	12 fJ	10 fJ	24 aJ	15 aJ
Static power	>70 nW	>70 nW	-	-

material (e.g., Ni), an electric field applied across the piezoelectric produces stress, which in turn affects the anisotropy of the magnetostrictive material. According to the experimental data [36], the electric field required for a magnetization rotation of 90 degrees in Ni/PZT synthetic multiferroic is about 1.2 MV/m. This promises a very low (of the order of attojoules) energy per switch achievable in nanometer-scale ME cells (e.g., 24 aJ for 100 nm × 100 nm ME cell with 0.8 μm PZT) [15]. Thus, the maximum power dissipation density per 1 μm^2 area circuit operating at 1 GHz frequency can be estimated as 7.2 W/cm^2. In multi-frequency circuits, the addition of an extra operating frequency would linearly increase the power dissipation in the circuit [26].

The comparison between the magnonic and CMOS-based logic devices should be done at the circuit level by comparing the overall circuit parameters such as the number of functions per area per time, time delay per operation, and energy required for logic function. In Table 14.1, we summarize the estimates for magnonic full adder circuit and compare them with the parameters of the CMOS-based circuit. The data for the full adder circuit made on 45 nm and 32 nm CMOS technology is based on the ITRS projections [37] and available data on current technology [38]. The data for the magnonic circuits are based on the design described in [15] and the above estimates. The magnonic circuit predicts a significant (~100×) advantage in minimizing the circuit area due to the lower number of elements required per circuit (e.g., five ME cells versus 25–30 CMOSs). At the same time, magnonic logic circuits would be slower than their CMOS counterparts. In Table 14.1, we show two numbers for time delay, corresponding to volatile and non-volatile circuits. The time delay of the volatile circuit is mainly defined by the spin wave group velocity, while the time delay of the non-volatile circuit is restricted by the relaxation time of the output ME cell. The most prominent (~1000×) advantage over CMOS circuitry is expected in the minimization of power consumption. Besides the great reduction in active power, there is no static power consumption in magnonic logic circuits based on non-volatile magnetic cells. The overall functional throughput is about 100 times higher for magnonic logic circuits due to the smaller circuit area.

We also estimate the additional functional throughput enhancement due to the use of multiple frequencies. Figure 14.10 shows the estimates on the functional throughput in Ops/ns cm^2 for the full adder circuit built of scaled CMOS and magnonic multi-frequency circuits. The estimates for the CMOS full adder circuit are based on the data for the 32 nm CMOS technology (area = 3.2 μm^2, time delay = 10 ps from [38]). The estimates for further generations are extrapolated by using the following empirical rule: the area per circuit scales as $\times 0.5$ per generation, and the time delay scales as $\times 0.7$ per generation. The estimates for the magnonic circuit are based on the multi-frequency model presented in [26]. The plot in Fig. 14.10 shows the relative functional throughput enhancement as a function of the number of frequencies N (independent information channels). It should be noted that the introduction of an additional frequency channel is associated with the additional area and time delay introduced by each new input/output port. There is an optimum number of operational channels providing maximum functional throughput, which varies between different logic circuits.

14.7   Summary

Magnonic logic devices are one of the alternative approaches to post-CMOS logic circuits, promising a significant functional throughput enhancement. The essence of this approach is the use of wave-based phenomena for achieving logic functionality. Coding information into the phase of the propagating spin waves makes it possible to utilize the waveguides as passive logic elements and reduce the number of elements per circuit. The ability to use multiple frequencies as independent information channels opens a new dimension for functional throughput enhancement and may provide a route to their long-term development. However, there are many questions remaining to be answered before magnonic logic circuits can find any practical application. Most of these concerns are associated with circuit stability and immunity to structural imperfections. At the moment, all the demonstrated prototypes utilize spin waves of micrometer-scale wavelength, which makes them immune to variations in waveguide structure. It is not clear if scaling to the deep sub-micrometer range will significantly affect the signal-to-noise ratio as well as the speed of propagation.

Despite the number of technical issues, magnonic logic devices offer a new route to functional throughput enhancement with a substantial performance pay off. It is most probable that magnonic logic devices will be used to complement existing logic circuitry in special task data processing.

References

[1] C. Herring & C. Kittel, "On the theory of spin waves in ferromagnetic media." *Physical Review*, **81**, 869–880 (1951).

[2] R. W. Damon & J. R. Eshbach, "Magnetostatic modes of a ferromagnet slab." *Journal of Physics and Chemistry of Solids*, **19**, 308–320 (1961).

[3] P. Kabos, W. D. Wilber, C. E. Patton, & P. Grunberg, "Brillouin light scattering study of magnon branch crossover in thin iron films." *Physical Review B*, **29**, 6396–6398 (1984).

[4] C. Mathieu *et al.*, "Lateral quantization of spin waves in micron size magnetic wires." *Physical Review Letters*, **81**, 3968–3971 (1998).

[5] M. P. Kostylev *et al.*, "Dipole-exchange propagating spin-wave modes in metallic ferromagnetic stripes." *Physics Reviews B*, **76**, 054422 (2007).

[6] Z. K. Wang *et al.*, "Observation of frequency band gaps in a one-dimensional nanostructured magnonic crystal." *Applied Physics Letters*, **94**, 083112 (2009).

[7] T. J. Silva, C. S. Lee, T. M. Crawford, & C. T. Rogers, "Inductive measurement of ultrafast magnetization dynamics in thin-film permalloy." *Journal of Applied Physics*, **85**, 7849–7862 (1999).

[8] M. Covington, T. M. Crawford, & G. J. Parker, "Time-resolved measurement of propagating spin waves in ferromagnetic thin films." *Physical Review Letters* **89**, 237202 (2002).

[9] M. Bailleul, D. Olligs, C. Fermon, & S. Demokritov, "Spin waves propagation and confinement in conducting films at the micrometer scale." *Europhysics Letters*, **56**, 741 (2001).

[10] M. P. Kostylev, A. A. Serga, T. Schneider, B. Leven, & B. Hillebrands, "Spin-wave logical gates." *Applied Physics Letters*, **87**, 153501 (2005).

[11] T. Schneider *et al.*, "Realization of spin-wave logic gates." *Applied Physics Letters*, **92**, 022505 (2008).

[12] K.-S. Lee & S.-K. Kim, "Conceptual design of spin wave logic gates based on a Mach–Zehnder-type spin wave interferometer for universal logic functions." *Journal of Applied Physics*, **104**, 053909 (2008).

[13] A. Khitun & K. Wang, "Nano scale computational architectures with spin wave bus." *Superlattices & Microstructures*, **38**, 184–200 (2005).

[14] I. Krivorotov, *Western Institute of Nanoelectronics, Annual Review Abstract* **3**(1) (2012). Available at www.win-nano.org/.

[15] A. Khitun & K. L. Wang, "Non-volatile magnonic logic circuits engineering." *J. Applied Physics*, **110**, 034306 (2011).

[16] Y. Wu *et al.*, "A three-terminal spin-wave device for logic applications." *Journal of Nanoelectronics and Optoelectronics*, **4**, 394–397 (2009).

[17] P. Shabadi *et al.*, "Towards logic functions as the device." In *Proceedings of the Nanoscale Architectures (NANOARCH), 2010 IEEE/ACM International Symposium*, pp. 11–16 (2010).

[18] L. Berger, "Emission of spin waves by a magnetic multilayer traversed by a current." *Physical Review B*, **54**, 9353–9358 (1996).

[19] J. A. Katine, F. J. Albert, R. A. Buhrman, E. B. Myers, & D. C. Ralph, "Current-driven magnetization reversal and spin-wave excitations in Co /Cu /Co pillars." *Physical Review Letters*, **84**, 3149–3152 (2000).

[20] S. Kaka *et al.*, "Mutual phase-locking of microwave spin torque nano-oscillators." In *IEEE International Magnetics Conference, 2006*, **2**(01), pp. 1–2 (2006).

[21] F. B. Mancoff, N. D. Rizzo, B. N. Engel, & S. Tehrani, "Phase-locking in double-point-contact spin-transfer devices." *Nature*, **437**, 393–395 (2005).

[22] A. Kozhanov, "Spin wave topology and imaging." *Annual Report to the Western Institute of Nanoelectronics* (2011). Available at www.win-nano.org/.

[23] S. Cherepov *et al.*, "Electric-field-induced spin wave generation using multiferroic magneto-electric cells." In *Proceedings of the 56th Conference on Magnetism and Magnetic Materials (MMM 2011), DB-03* (2011).

[24] A. Khitun, M. Bao, & K. L. Wang, "Spin wave magnetic nanofabric: a new approach to spin-based logic circuitry." *IEEE Transactions on Magnetics*, **44**, 2141–2152 (2008).

[25] A. Khitun *et al.*, "Inductively coupled circuits with spin wave bus for information processing." *Journal of Nanoelectronics and Optoelectronics*, **3**, 24–34 (2008).

[26] A. Khitun, "Multi-frequency magnonic logic circuits for parallel data processing." *Journal of Applied Physics*, **111**, 054307 (2012).

[27] "Process integration and device structure." *International Technology Roadmap for Semiconductors (ITRS)* (2011). Available at: www.itrs.net.

[28] P. Shabadi *et al.*, "Spin wave functions nanofabric update." In *Proceedings of the IEEE/ACM International Symposium on Nanoscale Architectures (NANOARCH-11)*, pp. 107–113 (2011).

[29] A. Khitun, "Magnonic holographic devices for special type data processing." *J. Applied Physics*, **113**, 164503 (2013).

[30] S. H. Lee, Ed., *Optical Information Processing Fundamentals* (Berlin: Springer, 1981).

[31] L. O. Chua & L. Yang, "Cellular neural networks: theory." *IEEE Transactions on Circuits & Systems*, **35**, 1257–1272 (1988).

[32] P. Ambs, "Optical computing: a 60-year adventure." *Advances in Optical Technologies*, vol. 2010, Article ID 372652, 15 pages (2010). doi: 10.1155/2010/372652.

[33] A. Khitun, B. Mingqiang, & K. L. Wang, "Magnetic cellular nonlinear network with spin wave bus." In *2010 12th International Workshop on Cellular Nanoscale Networks and their Applications (CNNA 2010)*, pp. 1–5 (2010).

[34] D. Gabor, "A new microscopic principle." *Nature*, **161**, 777–778 (1948).

[35] P. Hariharan, ed., *Optical Holography: Principles, Techniques and Applications* , 2nd edn. (Cambridge: Cambridge University Press, 1996).

[36] T. K. Chung, S. Keller, & G. P. Carman, "Electric-field-induced reversible magnetic single-domain evolution in a magnetoelectric thin film." *Applied Physics Letters*, **94**, 132501 (2009).

[37] Yearly report. *International Technology Roadmap for Semiconductors (ITRS)* (2007). Available at: www.itrs.net..

[38] A. Chen, private communication (2010).

Section V

Interconnect considerations

15 Interconnect considerations

Shaloo Rakheja, Ahmet Ceyhan, and Azad Naeemi

15.1 Introduction

The exponential growth of the electronics industry has been guided by the continual reduction in feature size of microchips manufactured using silicon-based CMOS technology. This reduction in feature size, commonly known as dimensional scaling, has enabled significant improvements in transistor performance and power – a higher transistor density for improved functionality, complexity, and performance of microchips; and a reduction in the cost per function. These benefits have enabled the semiconductor industry to offer a wide range of new products at every technology generation.

The research pipeline of the semiconductor industry involves increasingly radical potential solutions to carry technology advancement through dimensional scaling to beyond the 10-year visibility limit. Many logic devices that require innovations in materials, use of heterogeneous technologies, and the exploitation of alternative state variables and non-binary computation schemes are under investigation to extend Moore's law to beyond the year 2020. These logic devices differ in structure and operating principles, and include various physical quantities that may be used for encoding information, such as charge, electric dipole, magnetic dipole (spin), orbital state, mechanical position, light intensity, etc.

Besides smaller and faster transistors, the semiconductor industry requires fast and dense interconnects to manufacture high-performance microchips. The evolution of integrated circuits from an embedded system of only a few components to large-scale systems with billions of devices transformed the interconnection problem into one of the major threats to continue improving the performances of microchips at each new technology node [1]. Interconnects impose major limits on the performance of integrated circuits because of the delay they add to critical paths, the energy they dissipate, the noise and jitter they induce, and the degrading metal and dielectric reliability due to vulnerability to electromigration (EM) and time-dependent dielectric breakdown (TDDB), respectively. All of these limitations worsen with dimensional scaling.

This chapter focuses on the interconnect challenges and opportunities associated with emerging charge- and non-charge-based devices. In Section 15.2, we describe the problems associated with interconnecting devices in charge-based systems. Section 15.3 investigates the potential benefits offered by carbon-based emerging interconnect technologies over the conventional Cu/low-k technology for various voltage-controlled, charge-based devices. Section 15.4 describes the physical transport mechanisms for alternative state variables and introduces how information can be conveyed using electron spin. In Section 15.5, compact models for spin relaxation length (SRL) in silicon and gallium arsenide for a wide range of doping concentration and operating temperature are presented. In Section 15.6, the compact models are used to quantify spin injection and transport efficiency (SITE) in both conventional and non-local spin valve (NLSV) devices with semiconducting channels. The role of electric field in modifying the spin transport characteristics in conventional spin valves in presented. In Section 15.7, the performance and energy-per-bit of semiconducting spintronic interconnects are compared with those of their CMOS counterparts at the end of the silicon technology roadmap (minimum feature size of 7.5 nm). Section 15.8 concludes the chapter by providing an outlook on the future of nanoelectronics.

15.2   The interconnect problem

15.2.1   The routing problem

Ignoring every other problem caused by interconnects, simply routing the tremendous number of wires on a microchip in the same footprint is a growing concern and a complicated problem. The International Technology Roadmap for Semiconductors (ITRS) projects 12 wiring levels in 2012; 14 in 2020; and 16 in 2026 [2]. Short interconnects that carry signals between transistors that are relatively close to each other, within a certain functional block, are routed at local interconnect levels with fine pitches for high density because their total resistances can be tolerable. As interconnects get longer, they are made wider and thicker to reduce the associated resistance; hence delay. Therefore, this multilevel interconnection architecture is a partial solution for both the routing problem and the interconnect latency problem, which will be discussed later [3]. This approach requires a substantial amount of effort to be devoted to both the design and process optimizations to ensure manufacturability, while meeting performance constraints [4]. As a consequence, the cost for manufacturing electronic equipment becomes dominated by the cost associated with routing interconnects and increases with each additional metal layer [5]. Adding a new metal level has become a problem that has to be considered thoroughly, weighing the extra cost against the benefit in performance. Furthermore, if size effects are not mitigated and scaling the barrier thickness continues to be a challenge, the number of metal levels may have to be increased even more aggressively [6].

15.2.2   Interconnect problem from the resistance–capacitance perspective

The current CMOS technology is essentially a charge-based technology. The presence or absence of mobile electron charge in the inversion layer of the transistor is used to distinguish between digital logic states of "0" and "1." Information between transistors in the CMOS technology is communicated through voltage-diffusion via electrical interconnects typically implemented in copper.

Historically, the delay of short local and intermediate level interconnects have been much smaller compared with the delay of switches and their lengths have scaled with technology [7]. The delay of short interconnects is determined by the output resistance of transistors and interconnect capacitance. The length of long global interconnects, however, did not scale with technology scaling since they ran across the chip. The delay of repeated global interconnects remained constant, resulting in an increasing delay trend compared with gate delays. Therefore, global interconnects were thought to be the more serious interconnect problem [8–10].

The intrinsic delay of signal propagation through a distributed RC electrical interconnect is proportional to $L^2/r_w c_w$, where L is the interconnect length, and r_w and c_w are the resistance and capacitance per unit length of the interconnect. Rewriting this expression as $\rho \varepsilon L^2/HT$, it can be seen that the interconnect delay can be reduced by: (1) reducing metal resistivity, ρ, using new materials; (2) scaling insulator permittivity, ε; (3) reducing interconnect length, L, using novel architectures; and (4) reverse scaling interconnect height, H, and insulator thickness, T [11]. A variety of solutions have materialized to mitigate the global interconnect problem in the last decade. Some of these include: switching to the Cu/low-k technology to introduce a lower $\rho \varepsilon$ product [12, 13], transitioning to many-core architectures [14], and 3-D integration to reduce the maximum global interconnect length [15], and reverse scaling interconnect height. Another potential solution to the global interconnect scaling problem is changing the physical means of interconnection by introducing on-chip optical interconnects [16, 17]. Although some of these solutions have in turn introduced other problems, such as router power dissipation in many-core architectures [18], it is undeniable that the nature of the global interconnect problem has changed as a result of these advances.

Besides, there is a radical change in local-level interconnect behavior at sub-20 nm technology nodes. In addition to the reduction in cross-sectional interconnect dimensions due to the aggressive dimensional scaling, the effective cross-sectional dimensions of Cu interconnects are further reduced because of an increasing fraction of the trench being occupied by the barrier/liner/Cu seed tri-layer thickness. Furthermore, at such small dimensions, the resistivity of Cu interconnects significantly increases as a consequence of size effects that include electron scatterings at the sidewalls and grain boundaries, and line edge roughness (LER). Interconnect size effect parameters, namely the specularity parameter, p, which determines the fraction of electrons that scatter specularly at the wire surfaces, and the reflectivity parameter, R, which determines the fraction of electrons that are scattered backwards at the grain boundaries determine the severity of this resistivity increase [19–27]. As a consequence, the resistance per unit length associated with Cu interconnects increases rapidly such that the resistances of even the local-level interconnects become significant. The delay of local interconnects can no longer be determined by just the output resistance of transistors and interconnect capacitance. It is now known that both the local/intermediate and global interconnects make significant contributions to the overall circuit delay [28]. Potential solutions, including the use of atomic layer deposition instead of sputter deposition for better control and the use of self-forming barrier layers, may reduce the

Table 15.1 Interconnect parameter projections extracted from the 2012 update of ITRS. Calculated metrics are indicated with the * sign

	2015	2020	2025
M1 half pitch (nm)	21	12	7
Aspect ratio	1.9	2	2.2
Cu resistivity ($\mu\Omega$-cm)	6.61	9.74	15.02
Barrier/cladding thickness for Cu M1 wiring (nm)	1.9	1.1	0.6
*Resistance per unit length for M1 wires, r ($\Omega/\mu m$)	101	434	1750
Capacitance per unit length for M1 wires, c (pF/cm)	1.8–2	1.6–1.8	1.5–1.8
NMOS intrinsic delay, $\tau = CV/I$ (Multi-gate, MG) (ps)	0.32	0.19	0.12
NMOS dynamic power indicator per device width, $E = CV^2$ (fJ/μm)	0.42	0.25	0.15
*Distributed RC delay of 1mm M1 wire, $\tau_{int} = 0.4rcL^2$ (ps)	7676	29512	115500
*Length at which $\tau_{int} = \tau$ (μm)	6.5	2.5	1
*M1 wire dynamic power indicator per length, $E_{int} = C_{int}V^2$ (fJ/μm)	0.1216	0.079	0.057
*Length at which $E_{int} = E$ assuming minimum width NMOS (in unit of minimum device width)	3.45	3.16	2.63

liner thickness and improve reliability [29], but the resistivity increase that is inherent to Cu will eventually require a material-based solution to mitigate the latency issues at the local/intermediate levels.

In fact, the only plausible solution to the interconnect latency problem at the local levels has been the introduction of progressively lower-k materials in various technology generations in the last decade. The difficulties associated with introducing new ultra-low-k dielectric materials with dielectric constant values that are below 2 will put more burden on optimizing layout and design for shortening local interconnects in the future. Some projections of the ITRS are tabulated in Table 15.1 to emphasize the severity of the interconnect problem. In Fig. 15.1, RC delay and energy-delay product (EDP) for ten-gate-pitch long, minimum-size interconnects and inverters are compared. Two sets of size effect parameters covering a pessimistic and a more optimistic case are considered in this comparison; their values are taken from the aforementioned experiments in the literature. Predictive technology model results for inverter characteristics are also plotted [30].

Furthermore, as the interconnect dimensions scale and the density of interconnects increases, the aggregate interconnect length and, hence, the total interconnect capacitance on a microchip also increase. As a consequence, interconnect dynamic power dissipation becomes a larger fraction of the total power dissipation on a microchip at each technology generation. An interconnect power analysis study performed on a microprocessor designed for power efficiency, consisting of 77 million transistors, fabricated using 130 nm technology in 2004, revealed that interconnects account for 50% of the total dynamic power dissipation. This study also revealed that both local- and global-level interconnects are equally important in terms of power dissipation, each accounting for 25% of the total dynamic power of the chip [31].

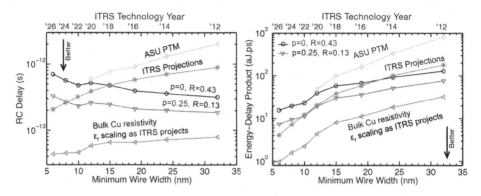

Fig. 15.1 RC delay (left) and EDP (right) comparison between ten-gate-pitch long Cu/low-k interconnect and minimum size inverter at sub-20 nm technology nodes. Both pessimistic ($p = 0$, $R = 0.43$) and more optimistic ($p = 0.25$, $R = 0.13$) size effect parameters are considered for interconnects. Both ITRS projections and predictive technology models (ASU PTM) for finFETs are considered for inverters. The large gaps between intrinsic interconnect and inverter RC delay and EDP narrow and disappear at future technology generations.

15.3   Interconnect options for emerging charge-based device technologies

We have established in the previous section that the Cu/low-k interconnect technology faces various major challenges and innovative solutions are urgently needed. There are rising opportunities for carbon-based emerging interconnect technologies as dimensional scaling continues. Any device technology that offers advantages in performance, power dissipation, or ease in dimensional scaling will have to be complemented with an interconnect technology that offers similar trades to avoid major bottlenecks due to interconnects. In this section, we investigate the interactions of interconnects with various voltage-controlled, charge-based devices similar to the conventional Si/CMOS transistor, including high-performance and ultra-low-power options. For that purpose, we pair finFETs, MOSFET-like carbon nanotube FETs (CNFETs), homojunction III-V tunnel FETs (TFETs), and sub-threshold CMOS devices with conventional Cu/low-k and emerging interconnect options, and compare the relative performances at the circuit level. The emerging interconnect options that we consider are single-wall carbon nanotube (SWNT) bundles, one or a few parallel SWNTs, and mono- and multi-layer graphene nanoribbon (GNR) interconnects. Since various device technologies offer very different characteristics in terms of the output current, input capacitance, sub-threshold swing, etc., the constraints that they put on interconnects and the best interconnect option for each device are different as well. This difference stems from the fact that the impacts of technology parameters of various interconnect technologies on the speed and energy advantages of a circuit may differ depending on the transistors used. Emerging CNT and GNR interconnects are mostly more resistive than Cu/low-k, but they offer lower capacitances. The intrinsic properties of these options are compared with those of Cu/low-k interconnects in Fig. 15.2. Many carbon-based designs offer comparable or better performance than Cu/low-k in terms of both RC delay and EDP

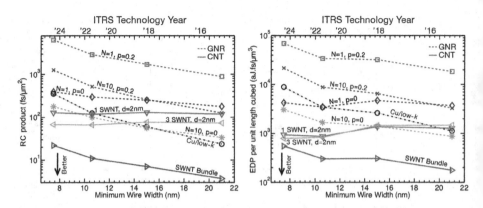

Fig. 15.2 Comparison of the RC product per unit length squared and EDP per unit length cubed associated with Cu interconnects, SWNT bundles, SWNT interconnects considering various number and diameter of tubes in a single layer, and mono- and multi-layer GNR interconnects with perfect and rough (20% edge scattering probability, $p = 0.2$ [32]) edges. Emerging carbon-based interconnects can potentially outperform Cu/low-k in both RC product and EDP.

beyond the year 2020. Despite major technological progress in fabricating such interconnects, there are many major challenges that must be overcome before they can become commercially viable options.

Predictive technology models (PTMs) for finFET devices are developed jointly by the Arizona State University PTM Group and ARM [30] based on the Berkeley short-channel IGFET common gate model (BSIM-CMG). The development of the model parameters for the BSIM-CMG model is performed using the scaling theory of multi-gate devices, physical models, and ITRS projections. Compared to planar bulk CMOS devices, finFET devices have significantly improved short channel effects (SCE) due to better electrostatics, can carry more current, and offer improved area efficiency. The channel width is quantized; hence, the number of fins has to be optimized for drive current choices.

CNFET devices are alternative solutions to performance enhancement of transistors in the "atomic dimension limit" beyond 2020. A compact model developed by Deng and Wong [33] is calibrated to meet reasonable on-current requirements while controlling the threshold voltage to keep leakage current at a reasonable value; and used for simulating MOSFET-like CNFET gates at the 16 nm technology node to predict the circuit- and system-level properties of CNFET devices. This model includes various non-idealities, such as the quantum confinement effects on both circumferential and axial directions, elastic scattering in the channel region, the resistive source/drain, the Schottky-barrier resistance, the acoustical and phonon scattering in the channel region, the screening effect by the parallel CNTs in CNFETs that contain multiple parallel CNTs under the same gate, and parasitic gate capacitances.

Conduction in a TFET occurs through band-to-band tunneling between the source and the channel barrier. The gate voltage is used to shift the bands in energy and change the probability of tunneling of carriers. Various materials and structures are proposed including homojunction III-V TFETs, heterojunction III-V TFETs, and GNR TFETs

[34]. TFETs with single gate, double gate, or gate-all-around (GAA) architectures are also studied. The principle of operation for all these devices is the same, but they differ in parameters such as supply voltage and drive current. We focus on InAs nanowire-based GAA TFET devices, and use physics-based compact models to calculate reasonably accurate current–voltage characteristics and gate/parasitic capacitances. InAs nanowires are considered due to their direct bandgap that eliminates the necessity for phonon assistance in tunneling. InAs is a promising material for realizing TFETs thanks to their small bandgap and light hole and electron effective masses [35], which both increase the on-current of the TFET device. Both p-type and n-type TFETs are realized by assuming n^{++}-i-p^+ and p^{++}-i-n^+ structures, respectively.

The delay and EDP performances of the device–interconnect pairs are calculated assuming a driver connected to a receiver through an interconnect of varied length, considering a typical fan-out of 3. To perform a fair comparison, we assume a CNFET inverter that is $5\times$ the minimum size as the driver; and the number of fins in a finFET and the number of nanowires in a TFET are calculated such that the total width of the devices is the same as the CNFET. The fin pitch and nanowire pitch are assumed to be equal and as given in [30] for each technology node.

In finFET circuits, per unit length values of both the interconnect resistance and capacitance have a significant impact on the circuit delay. Therefore, to outperform Cu/low-k interconnects in finFET circuits, either of these parameters has to be reduced significantly while avoiding a significant opposite change in the other parameter. Individual SWNT interconnects with 2 nm diameter have much smaller capacitance per unit length compared to copper interconnects, but they are too resistive to be used in high-performance circuits at the 16 nm technology node. On the other hand, bundles of SWNT interconnects have significantly lower resistance per unit length values compared to Cu interconnects at similar capacitance values. Therefore, as illustrated in Fig. 15.3, the best interconnect option for a finFET circuit in terms of circuit delay is SWNTs manufactured in horizontal bundles. Multi-layer GNR interconnects may outperform Cu interconnects if the edges are perfectly smooth, with a probability of electrons backscattering at the edges equal to 0.

Even though it is not possible to outperform Cu interconnects in terms of circuit delay with a few parallel SWNT interconnects at the 16 nm technology node due to their high resistance per unit length, it is possible to benefit from their smaller capacitance per unit length compared to Cu interconnects, which translates into a lower power dissipation. As Fig. 15.3 demonstrates, a mono-layer of SWNTs as dense as 250 SWNTs/μm can offer ~2× better EDP performance than Cu at ~100 gate pitches. It is important to note that as technology scales, the impact of size effects on Cu interconnect becomes more pronounced. Since the resistance per unit length of an individual SWNT does not change with scaling as long as its diameter is the same and that of GNRs with perfect edges increase at a slower rate compared to Cu interconnects, more opportunities arise for using these carbon-based interconnect technologies at highly scaled technology nodes. This fact is illustrated in Fig. 15.3, where it can be seen that SWNTs can offer much larger gains in both circuit delay and EDP at the 7 nm technology node. For this to be possible, however, the density of SWNTs has to increase significantly since the

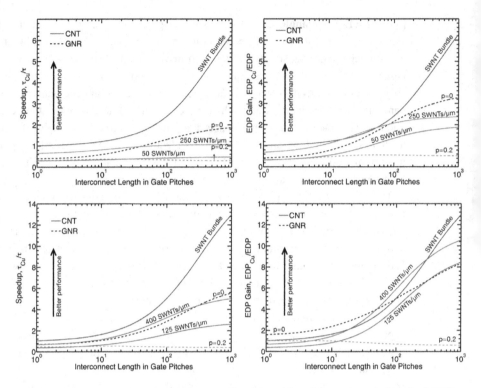

Fig. 15.3 Speedup and EDP gain advantages offered by various interconnect designs in finFET circuits at the 16 nm (top) and 7 nm (bottom) technology nodes. Multiple interconnect options using emerging carbon-based interconnects can potentially outperform Cu/low-k at the 16 nm technology node finFET circuits in terms of EDP. Even more opportunities arise at the 7 nm technology node as a consequence of both the rapid increase in Cu resistivity and low capacitance of carbon-based interconnects.

minimum interconnect dimensions where tubes have to be placed are much smaller at future technology nodes. Assuming perfectly reliable connections, a density of at least 125 SWNTs/µm is required to have a connection between the driver and the receiver at the 7 nm technology node.

The conclusions that can be drawn from simulations using CNFETs are very similar to finFET circuits as plotted in Fig. 15.4. Interconnect resistance and capacitance are equally effective in determining the circuit delay and bundles of SWNTs can offer the best delay performance due to their smaller interconnect resistance per unit length compared to Cu. CNFET devices offer the highest output current among the device types that are considered in this work. As a consequence, CNFETs are affected more severely from the changes in interconnect resistance per unit length.

TFETs have very different interconnect requirements than finFETs and CNFETs. Due to the lower on-current offered by TFETs, the output resistance of TFET devices is much larger than both of these device technologies. As a result, interconnect resistance is not as crucial, but not negligible, in TFET circuits as it is in finFET and CNFET circuits. Reducing interconnect capacitance per unit length is more beneficial in reducing the

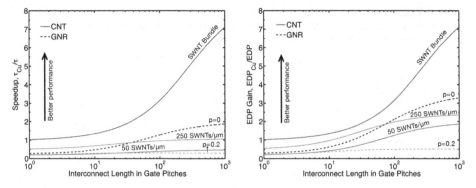

Fig. 15.4 Speedup and EDP gain advantages offered by various interconnect options in CNFET circuits at the 16 nm technology node. Due to the low output resistance of CNFETs, interconnect resistance becomes the determining factor in circuit delay. Bundles are the best option targeting circuit delay, but multiple alternative options may outperform Cu/low-k in terms of EDP in CNFET circuits due to their low capacitance.

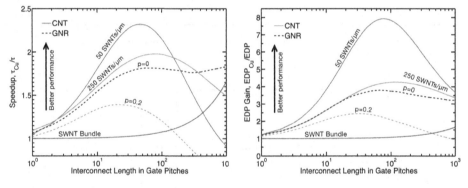

Fig. 15.5 Speedup and EDP gain advantages offered by various interconnect designs in TFET circuits at the 16 nm technology node. The best interconnect option in TFET circuits is a function of length.

circuit delay in TFET circuits than reducing the interconnect resistance per unit length. Clearly, reduced interconnect capacitance means lower interconnect power dissipation as well. Figure 15.5 demonstrates that the best circuit delay can be obtained by using a low-density mono-layer of SWNTs because they offer the smallest interconnect capacitance per unit length. However, the diameter of the tubes in the mono-layer has a non-negligible impact on the speedup as shown in Fig. 15.5 due to the different resistance per unit length values. If tubes with a diameter of 2 nm are used, the resistance per unit length can be reduced compared to 1 nm diameter tubes and a better speedup can be achieved. In short, moderately resistive low-capacitance interconnect technology options must be considered for obtaining the best performance in delay in TFET circuits.

Considering planar bulk MOSFET devices at the 16 nm technology node and operating them at 20% of the nominal supply voltage value, it is possible to reduce the power dissipation of a circuit significantly. For these circuits operating at sub-threshold voltage

Table 15.2 Comparison table summarizing the simulation results. The number of checkmarks indicates the importance of each interconnect parameter. The best interconnect options are sorted from top to bottom based on the performance improvements they can offer with respect to Cu/low-k wires

Device type	Device resistance	Device capacitance	Interconnect resistance impact		Interconnect capacitance impact		Promising interconnect option (targeting EDP)	
			Short	Long	Short	Long	Short	Long
FinFET	Reference	Reference	✓	✓✓✓	✓✓✓	✓✓✓	Bundle	Bundle GNR* SWNT*
CNFET	Low	Low	✓✓	✓✓✓	✓✓✓	✓✓✓	Bundle	Bundle GNR* SWNT*
TFET	High	Low		✓✓	✓✓✓✓	✓✓✓	SWNT GNR	SWNT SWNT* GNR
Sub-Vt	Very high	Reference		✓✓	✓✓✓✓	✓✓✓✓	SWNT SWNT* GNR*	SWNT SWNT* GNR*

* Indicates smooth edges in case of GNR interconnect and high-density in case of mono-layer SWNT interconnect.

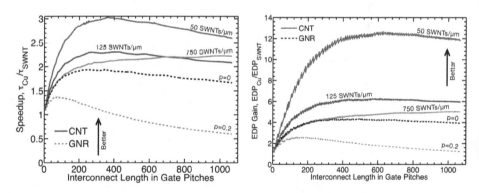

Fig. 15.6 Speedup and EDP gain advantages offered by various interconnect designs in sub-threshold operation at the 16 nm technology node. Due to the low output currents in sub-threshold operation, interconnect capacitance becomes the dominating factor in determining circuit delay.

values, devices can be assumed highly resistive. Therefore, the resistance associated with interconnects is a secondary concern to the capacitance. Since wire resistance does not have a significant impact on circuit performance, only SWNTs with 1 nm diameter are considered in plotting Fig. 15.6. Note that as the density of tubes is increased, the associated capacitance per unit length increases and the maximum speedup is lowered.

Table 15.2 quantifies how much interconnect resistance and capacitance per unit length impact circuit performance for various device types at short and long interconnect lengths. Also, the first three best interconnect options that maximize EDP performance at short and long interconnect lengths for each device are tabulated.

15.4   Interconnect considerations for spin circuits

Alternative state variable devices (called "post-CMOS devices" hereafter) are envisaged on the idea of using state variables other than electron charge to store and manipulate information [36, 37]. To date, several potential candidates have emerged as the state variable of the future. Some noteworthy examples include the electron spin, pseudospin in graphene, excitons, phonons, domain walls, and photons [38] (see Fig. 15.7).

    Non-voltaic information tokens may be communicated with various transport mechanisms involving either particle transport or in a wave-based fashion [38]. Amongst the particle-based transport mechanisms, there are diffusion, drift, and ballistic transport modes, while amongst the wave-based mechanisms there are spin waves and electromagnetic waves as shown in Table 15.3. These transport mechanisms may or may not involve net charge transfer. Diffusion is a passive process that can be used even for those state variables that have no net charge associated with their carriers. For

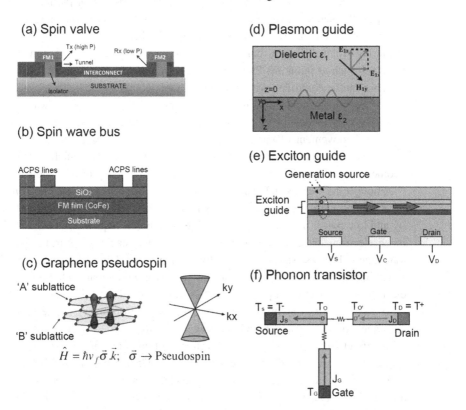

Fig. 15.7 Some examples of emerging devices based on alternative state variables. (Figure taken from [39].)

Table 15.3 A map of the physical transport mechanism and the corresponding state variable for which the transport mechanism can potentially be used to establish communication between the on-chip driver and the receiver

Transport mode	Information token	Transit time of carriers through the interconnect
Diffusion	Spin, pseudospin in graphene, temperature (phonons), direct and indirect excitons	L^2/D, where D is the diffusion coefficient of the information token
Drift	Indirect excitons, spin, pseudospin	L/v_d, where v_d is the drift velocity of the information token and is a function of the applied electric field across the interconnect
Ballistic	Spin, pseudospin in graphene, temperature (phonons)	L/v_{ball}, where v_{ball} is the ballistic velocity and will be equal to unidirectional thermal velocity of the information token
Spin waves	Spin	L/v_{SWB}, where v_{SWB} is the propagation velocity of spin waves through the interconnect medium
Electromagnetic waves	Photons, plasmons	L/v_p, where v_p is the propagation velocity of electromagnetic waves through the interconnect medium

diffusion, only a concentration gradient of information-bearing particles is required to transport the signal between the on-chip driver and the receiver. Ballistic transport is usually observed in low-dimensional materials that have a limited phase space for scattering. Like diffusion, ballistic transport can also be used for non-charged state variables such as phonons. Particle drift, on the other hand, works only when the state variable can be controlled with the help of an external electric field. As such, drift only works for spins, pseudospins in graphene, and indirect excitons. In wave-based interconnects, communication of information is supported in a wave-based fashion without the actual movement of particles. The delay of wave-based interconnects is a strong function of the propagation velocity of the information token, which depends upon the geometry and design of the interconnect, and the frequency of operation.

Amongst the various state variables currently under investigation, electron spin is the most studied with potential advantages in terms of its robustness, enhanced functionality and low power dissipation [40, 41]. Electron spin is a quantum mechanical property of an electron that can be quantized in one of two stable states in the presence of a magnetic field; these states are typically labeled as $+1/2$ spin and $-1/2$ spin. Since spin is inevitably associated with electron charge, it is easier to imagine hybrid circuits where electron spin devices can provide for "nonconventional extended CMOS" [42].

Recent interest in spintronics was sparked by the discovery of the giant magnetoresistance (GMR) effect by Gruenberg *et al.* in 1985 [43]. The discovery of the GMR effect led to renewed research efforts in tunnel magnetoresistance (TMR) [44], originally discovered at low temperatures in 1975 by Julliere [45]. The spin valve effect forms the underlying principle of the GMR and the TMR effects. A spin valve is a 2-terminal magnetoelectronic device consisting of a non-magnetic layer sandwiched between two ferromagnetic layers. In a typical GMR device, a non-magnetic metal spacer is inserted between two ferromagnetic electrodes either in a current-in-plane

(CIP) or current-perpendicular-to-plane (CPP) geometry. On the other hand, for a typical magnetic tunnel junction (MTJ) exhibiting the TMR effect, an insulator is inserted between the ferromagnetic electrodes. Both GMR and MTJ devices are characterized using a magnetoresistance ratio (MR) as the figure of merit. The MR ratio is given as $(G^P - G^{AP})/G^{AP}$, where G^P and G^{AP} are the electrical conductances of the parallel and antiparallel configurations. Although the initial MTJs were made with AlO_x tunnel barriers, it was the development of MgO tunnel barriers that led to the dramatic progress of MTJs [9, 46]. The application of GMR effect to implement magnetic random access memories (MRAMs) is discussed in various review articles [47–49]. However, directly using the 2-terminal spin valves to implement digital logic may not be feasible since these are passive devices. As such, their applications are limited to non-volatile storage devices and to provide re-configurability in electronic devices to enhance their functionality.

The basic requirements of a logic gate are: gain to restore attenuating signals, high drivability for sufficient fan-out, and directionality of information propagation from input to output and not vice versa. Since the first proposal of a spin transistor by Datta and Das [50] and Johnson [51, 52], some interesting proposals on spin logic gates have been put forth. Some noteworthy proposals on spin logic are all-spin logic (ASL) [53], spin wave bus (SWB) logic [54–56], and spin hall effect (SHE) logic [57, 58].

In an ASL device, information is stored in the magnetization of magnets that communicate with each other in a multi-magnet network using pure spin currents. These pure spin currents can be detected at the receiver magnet through the spin torque effect. Spin torque refers to the phenomena in which a spin polarized current deposits its spin angular momentum as it passes through a small magnetic conductor [59]. As a result, the magnetization of the magnetic conductor will undergo precession or may even switch its direction. ASL devices are amenable to scaling since the amount of spin current required to switch the magnetization of the magnet decreases upon the reduction of the size of the magnet.

Recently, it was demonstrated that spin current generated by the spin Hall effect (SHE) in heavy metals like tantalum and tungsten can be used to apply spin torque to an adjacent magnet to drive energy-efficient reversal [60]. This advantage of SHE over spin torque effect is the charge-to-spin amplification. That is, the spin current generated (in units of $\hbar / 2$) to apply torque on the receiving magnet can be an order of magnitude or more greater than the applied charge current (in units of e).

A spin wave is a collective oscillation of electron spins around the direction of magnetization. In a spin wave, information is encoded into the amplitude or the phase of the spin wave. The underlying idea of the spin wave is to establish communication between inductively coupled devices using a magnetic flux rather than a current flow through wires. Ferromagnetic interconnects such as NiFe or CoFe can be used as conduits of spin waves.

To compare any novel logic technology with CMOS logic, it is imperative to quantify the physical limits of novel interconnects needed to communicate information in post-CMOS logic. It is important that the interconnects in novel logic transmit information in the same domain as the alternative state variable. Otherwise, the amount of energy and

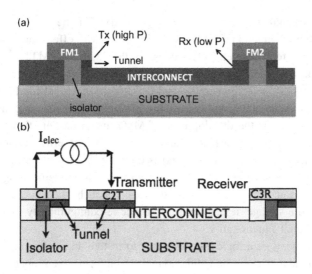

Fig. 15.8 (a) A conventional spin valve. Electrical and spin currents flow along the same path. (b) A non-local spin valve (NLSV) device. A pure spin diffusion current flows through the interconnect to establish communication between the transmitter and the receiver.

circuit-area overhead needed for signal conversion back and forth between the electrical and novel state variable domains will be prohibitive for the new technology. The work presented in this chapter is based on a spin valve device that is considered as the prototype of a switching element in the spin domain. Spin valves are typically used in experiments to demonstrate successful spin injection in various non-magnetic materials and to examine the role of interface materials on spin injection and transport. Both conventional and non-local spin valves (NLSV) devices are shown in Figs. 15.8(a) and (b). A conventional spin valve uses particle drift to transmit information encoded in the electron spin, while an NLSV uses pure spin diffusion current to communicate between the injecting driver and the receiver nanomagnets.

Interconnects in a spin valve device can be implemented using a variety of materials – metals like copper and aluminum, semiconductors like silicon and gallium arsenide, and even the novel carbon-based material, graphene. Semiconductors are particularly attractive for spintronics because of the mature semiconductor-process technology and the flexibility offered by semiconductors to tune their electrical and spintronic properties via impurity doping. Unlike charge, spin is not a conserved quantity. That is, spin information can decay as it propagates through a physical medium because of the spontaneous spin flipping mechanisms present in the interconnect. The spin relaxation length (SRL[1]) in semiconductors is highly influenced by electric field through the semiconductor, lattice temperature, and the doping concentration. This in turn influences the energy-per-bit of spintronic interconnects. If SRL is much longer than the spin interconnect length, spin information can be preserved over

---

[1] For a non-magnetic material, spin-relaxation length, $L_s = (D\tau_s)^{0.5}$, where D is the electron diffusion coefficient, and τ_s is the spin relaxation time.

long distances without the need for spin repeaters.[2] This can lead to substantial savings in energy and area overhead associated with spin repeaters.

15.5   Spin relaxation mechanisms

Spin relaxation refers to the process by which a non-equilibrium population of electron spins is brought to its equilibrium value in a material. If an electron suddenly changes its spin orientation then it is referred to as "spin flip." However, if a spin population changes gradually with time then it is referred to as "spin relaxation." The main cause of spin relaxation is the spin orbit coupling (SOC) that arises due to various intrinsic and extrinsic effects in materials. It is SOC that in conjunction with momentum relaxation of electrons gives rise to the relaxation of electron spins.

15.5.1   Spin relaxation in silicon

The spin relaxation time (SRT) in silicon governed by the Elliott–Yafet mechanism [61] can be given by Mattheissen's rule as

$$\frac{1}{\tau_s} = \frac{1}{\tau_s^{ph}} + \frac{1}{\tau_s^{imp}}, \tag{15.1}$$

where τ_s^{ph} and τ_s^{imp} denote the phonon- and impurity-dominated spin relaxation times in silicon. The phonon-dominated SRT in silicon is given as [39]

$$\frac{1}{\tau_s^{ph}} = \frac{1}{\tau_0 \left(\frac{T}{300K}\right)^{-\theta}}, \tag{15.2}$$

where T is the operating temperature, τ_0 is the phonon-dominated SRT in silicon at room temperature, and θ denotes the roll-off of SRT in silicon with temperature. The values of τ_0 and θ are extracted from calibration with experimental data. It is found that $\theta = 3$ provides the best match with experimental data [62]. The value of $\theta = 3$ also matches with theoretical predictions from [63, 64]. τ_0 is extracted as 7.7 ns, which is also in good agreement with the theoretically predicted value in [64].

The spin orbit interaction due to impurities in silicon becomes especially important at high doping concentrations. Hence, the complete model for SRT in silicon must account for the impurity-dominated spin relaxation. It is shown in [65–68] that τ_s^{imp} for non-degenerately doped silicon is given by

$$\frac{1}{\tau_s^{imp}} = \frac{\alpha_0 \left(\frac{T}{300\,K}\right)}{\mu^{imp}(T, N_d)}, \tag{15.3}$$

where μ^{imp} is the impurity-dominated mobility in silicon and N_d is the doping concentration. Here, α_0 is a fitting parameter whose value is determined from calibration with

---

[2] A spin repeater is essentially a nanomagnet.

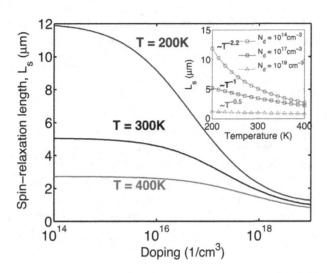

Fig. 15.9 Spin relaxation length versus doping concentration in Si at various temperatures. A long spin relaxation length is desirable for spintronic interconnects. Inset plot shows the impact of temperature on spin relaxation length for various doping levels. (Figure taken from [39].)

experiments. It is found that $\alpha_0 = 5 \times 10^{10}\, \text{cm}^2/\text{vs}^2$ provides an excellent match with experimental data [39].

The SRL versus doping in silicon is shown in Fig. 15.9 at various temperatures. The room temperature SRL in Si degrades from $5\,\mu\text{m}$ at $N_d = 10^{14}\,\text{cm}^{-3}$ to $1\,\mu\text{m}$ at $N_d = 10^{19}\,\text{cm}^{-3}$. At 200 K, the degradation in the SRL is from $12\,\mu\text{m}$ to $1.2\,\mu\text{m}$ as the doping increases from $10^{14}\,\text{cm}^{-3}$ to $10^{19}\,\text{cm}^{-3}$, while at 400 K the degradation in SRL is from $2.72\,\mu\text{m}$ to $9.82\,\mu\text{m}$ for the same change in the doping density. Hence, the SRL degrades more rapidly with an increase in doping concentration at low temperatures. The inset plot of Fig. 15.9 shows the temperature dependence of SRL for various doping levels. It can be clearly seen that the roll-off of SRL with temperature is a strong function of the doping concentration in silicon.

15.5.2 Spin relaxation in gallium arsenide

Due to the lack of bulk inversion symmetry, the dominant spin orbit interaction in GaAs is the Dresselhaus spin orbit interaction. Due to the presence of Dresselhaus spin orbit coupling (DSOC), there exists an intrinsic momentum-dependent magnetic field, $B_i(k)$, in the crystal around which the electron spins undergo precession with a Larmor frequency, $\Omega(k) = e/m_c B_i(k)$. Here, e is the electron charge and m_c is the effective mass of electrons in the crystal. The momentum-dependent precession of electrons in conjunction with momentum relaxation (characteristic time τ_p) leads to spin de-phasing. There are two limiting cases of spin de-phasing due to DSOC: (a) $|\Omega|\tau_p \ll 1$ ("strong-scattering limit") and (b) $|\Omega|\tau_p \gg 1$ ("weak-scattering limit"). The strong-scattering limit is also identified as the "motional-narrowing regime" of the D'yakonov–Perel (DyP) spin relaxation mechanism [69]. For the motional-narrowing regime of DyP mechanism, the spin relaxation time is inversely proportional to the momentum-relaxation time. In 1988, Pikus and coworkers [70] obtained the spin relaxation rate in bulk GaAs as

$$\frac{1}{\tau_s(E_k)} = \frac{32}{105} \frac{\gamma_3^{-1}\tau_p(E_k)\alpha^2 E_k^3}{\hbar^2 E_g}. \tag{15.4}$$

In Eq. (15.4), α denotes the strength of SOC in GaAs and its value is typically between 0.063–0.07; γ_3 depends on the dominant scattering mechanism and is between unity and six for most scattering mechanisms [69]; E_k is the momentum-dependent energy of the electron; and τ_p is the momentum-relaxation time of electrons. The SRT in bulk GaAs is shown as a function of doping concentration at 300 K in Fig. 15.10(a). Also shown on this plot are the experimental data points from Bungay and coworkers [71] and Kimel and coworkers [72]. The experimental data points fit very well on the analytical curve for SRT.

The SRL is independent of doping in both degenerately and non-degenerately (NDG) doped GaAs. However, it depends on operating temperature in non-degenerately doped GaAs. The SRL in NDG GaAs can be expressed as

$$L_s = L_0 \left(\frac{T}{300\,\text{K}}\right)^{-1}, \tag{15.5}$$

where L_0 is the SRL in bulk GaAs at 300 K and is ~0.5 µm. The SRL in GaAs is plotted in Fig. 15.10(b) as a function of temperature for a doping concentration of $10^{16}\,\text{cm}^{-3}$. The inset plot in Fig. 15.10(b) shows that for NDG levels in GaAs, the SRL is independent of the doping concentration and is approximately equal to 0.5 µm.

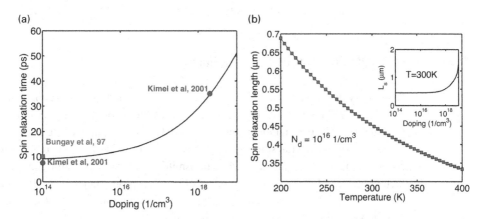

Fig. 15.10 (a) SRT versus doping concentration in bulk GaAs at 300K. Experimental data points from Bungay and coworkers [71] and Kimel and coworkers [72] are also shown. (b) The temperature dependence of spin relaxation length in bulk GaAs for non-degenerate doping levels. The inset plot shows SRL versus doping concentration in bulk GaAs at 300 K. (Parts (a) and (b) taken from [39].)

Table 15.4 Spin relaxation length and time in various materials. W/T denotes the width-to-thickness ratio of the interconnect. For graphene, n_0 denotes the carrier concentration

Material	Spin relaxation length (L.T.)	Spin relaxation length (R.T.)	Reference
Cu [$W/T = 100\,\text{nm}/54\,\text{nm}$]	546 nm	147.9 nm	[73]
Cu [$W/T = 220\,\text{nm}/320\,\text{nm}$]	1000 nm	400 nm	[74]
Cu [$W/T = 150\text{--}200\,\text{nm}/100\,\text{nm}$]	1000 nm	400 nm	[75]
Al [$W/T = 100\,\text{nm}/15\,\text{nm}$]	660 nm	350 nm	[76]
Al [$W/T = 150\,\text{nm}/50\,\text{nm}$]	1200 nm	600 nm	[77]
Graphene (SiO_2, $n_0 = 10^{12}\,\text{cm}^{-2}$)	–	0.8 µm	[78]
Graphene (SiO_2, $n_0 \approx 0$)	–	1.4 µm	[79]
Graphene (SiO_2, $n_0 = 5 \times 10^{12}\,\text{cm}^{-2}$)	–	7 µm	[80]

15.5.3   Spin relaxation in metals and graphene

A survey of experimentally measured SRL in Cu, Al, and graphene is given in Table 15.4. It can be seen from this table that the SRL in NDG Si and GaAs is more than that in metals, but is lower than that in graphene at R.T. The SRL in metals degrades as the operating temperature increases because of the phonon-induced spin relaxation. Further, the SRL in metals is shorter than that in wide graphene. This is to be expected because the intrinsic SOC in graphene is quite low owing to its low atomic number ($Z = 6$). Even though the theoretically predicted SRL in graphene is $\approx 20\,\mu\text{m}$ [81], experiments have demonstrated only few µm SRL. It is believed that the low values of SRL in graphene are a result of the interaction of graphene with adatoms that hybridize with the carbon atoms of graphene and induce a localized SOC that leads to a rapid spin relaxation [82]. A summary of various experimental techniques for measuring spin relaxation and an exhaustive survey of SRTs in metallic channels are presented in [83].

15.6   Spin injection and transport efficiency

To quantify the performance of spin valves, a metric called "spin injection and transport efficiency" (SITE) is defined. SITE is defined as the amount of spin polarized current reaching the receiver nanomagnet normalized to the input electrical current. SITE incorporates losses in the spin signal encountered during injection from the ferromagnet into the semiconductor and also while transmitting the spin signal through the semiconductor channel. Hence, SITE is a product of spin injection efficiency (SIE) and transport efficiency (TE). To ensure correct functionality of spintronic logic built with spin valves as basic switching elements, it is desired to maximize SITE.

The cross-sectional view of a conventional spin valve is shown in Fig. 15.11. The spin resistances of various materials are also given. For a conventional spin valve, the spin resistance of the semiconductor depends upon the upstream transport length, L_u, which is a function of electric field across the semiconductor and is typically shorter

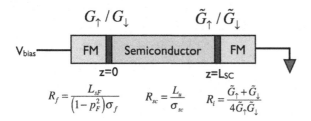

$$R_f = \frac{L_{sF}}{(1-p_F^2)\sigma_f} \qquad R_{sc} = \frac{L_u}{\sigma_{sc}} \qquad R_i = \frac{\tilde{G}_\uparrow + \tilde{G}_\downarrow}{4\tilde{G}_\uparrow \tilde{G}_\downarrow}$$

Fig. 15.11 Cross-sectional view of a conventional spin valve. R_f, R_{SC}, and R_i denote the spin resistances of the ferromagnet, semiconductor interconnect, and interface barrier, respectively. L_{sF} denotes the spin relaxation length in the ferromagnet, p_F is the spin polarization, and s_F is the conductivity of the ferromagnet. For the semiconductor, L_u denotes the upstream transport length.

than the SRL [84]. The interface is characterized by a voltage- and temperature-independent up-spin and a down-spin conductance, $G\uparrow$ and $G\downarrow$, respectively. This is a good approximation in the low bias regime where $eV \ll k_B T$ [84].

SIE depends upon the relative spin dependent resistances of various materials in the spin valve. When $R_i \gg R_{SC} \gg R_f$, the interface barrier helps to increase SIE. On the other hand, when R_{SC} is greater than the other resistances, then electric field may play a role in improving both SIE and TE. Furthermore, when doping is increased in the semiconducting channel, σ_{SC} increases, which helps to mask the "conductivity mismatch" problem between the injecting ferromagnet and the semiconducting channel improving SIE. Transport efficiency, on the other hand, depends upon the ratio L_{SC}/L_d, where L_{SC} is the length of the semiconductor sandwiched between the ferromagnets, and L_d is the downstream transport length. Like the upstream transport length, L_d is also a function of the electric field across the semiconductor interconnect. However, L_d in Si can be several 100s of μm at low doping densities, indicating that spins once injected into Si can be transported over a 100 μm thick Si slab. For $L_{SC}/L_d \ll 1$, TE is unity. Both L_u and L_d degrade at high doping densities in Si but are independent of doping in GaAs.

Figure 15.12(a) shows the SITE in a conventional spin valve with Si and GaAs channels as a function of doping in the channel. The length of the semiconductor sandwiched between the injecting and receiving ferromagnets is assumed to be 1 μm. All other simulation parameters are noted in the figure caption. The improvement in SITE is significant when doping increases for low doping levels. However, the improvement in SITE with doping gradually reduces for higher doping levels. The SITE of spin valves with GaAs channels is more than that with Si channels. This is primarily because of the higher conductivity of GaAs, which improves the injection efficiency. The transport efficiency for a 1 μm channel is nearly the same for both Si and GaAs channels because the modified spin transport length, L_d, is much greater than the channel length, L_{SC}. Further, electric field can be used to enhance SITE particularly when R_{SC} is large at low doping concentration in the semiconductor channel.

In Fig. 15.12(b), the SITE in the NLSV structure with a Si channel is shown as a function of the doping concentration. The SITE exhibits a maximum for a certain doping concentration. This is because as the doping increases, the resistivity of the material reduces, which can help to lower the "conductivity mismatch" between the

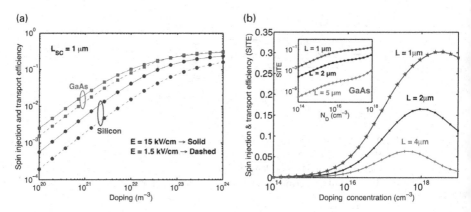

Fig. 15.12 (a) SITE as a function of doping concentration in conventional spin valves. The conductivity polarization of injecting ferromagnet is taken to be 0.5, while that of the receiving ferromagnet is -0.5. $L_{sF} = 10$ nm, $\sigma_F = 1.3 \times 10^8$ $(\Omega m)^{-1}$, $G\uparrow = 6 \times 10^9$ $(\Omega m^2)^{-1}$, $G\downarrow = 3 \times 10^9$ $(\Omega m^2)^{-1}$ for both injecting and receiving ferromagnets. A large value of SITE is desirable in spintronic devices to lower their energy dissipation. (Figure taken from [39].) (b) SITE versus doping concentration for a non-local spin valve with a Si channel at 300 K. L denotes the length of the semiconducting channel. The inset shows a similar plot for GaAs. $G\uparrow = 7.5 \times 10^8$ $(\Omega m^2)^{-1}$, $G\downarrow = 2.25 \times 10^9$ $(\Omega m^2)^{-1}$. The contact area for each of C1T, C2T, and C3R for the NLSV geometry shown in Fig. 15.12(b) is assumed to be 130 nm². For Si interconnects, there occurs an optimal doping concentration that maximizes SITE. For GaAs, SITE continually improves with an increase in doping concentration.

channel and the injecting ferromagnet. However, a reduction in SRL with an increase in doping leads to a greater signal loss within the channel itself. The doping concentration that maximizes SITE is a function of the interconnect length. For interconnect lengths longer than 1 μm, the doping concentration that maximizes SITE is less than 3×10^{18} cm^{-3}. The inset plot of Fig. 15.12(b) shows that SITE in the NLSV structure with GaAs channel improves with an increase in doping concentration. This is because the SRL in GaAs is independent of N_d because of the dominant spin relaxation mechanism being the DyP mechanism. Hence, the loss in the spin signal incurred within the GaAs interconnect is independent of N_d. Hence, an increase in N_d leads to a reduction in the conductivity mismatch between the interconnect and the ferromagnet, which helps to improve SITE consistently with an increase in N_d.

A comparison of SITE at 300 K as a function of channel length in conventional and NLSV structures is shown in Fig. 15.13. For high-field transport in conventional spin valves, L_d is sufficiently large for both Si and GaAs such that, for L_{sc} up to 10 μm, the SITE is relatively unaffected by L_{SC}. However, when electric field is reduced, L_d reduces, and the transport efficiency begins to degrade with an increase in L_{SC}, leading to a reduction in SITE with L_{SC}. Even though the roll-off of SITE with L_{SC} is greater in GaAs channels at low electric fields, the absolute value of SITE in conventional spin valves with GaAs channels is better than that with Si channels. Further, the SITE in conventional spin valves is better than that in NLSV structures for the same doping and same channel length.

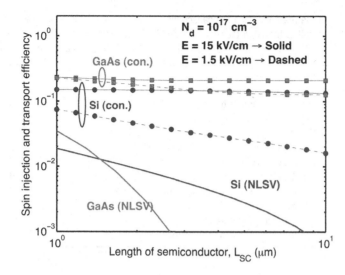

Fig. 15.13 SITE versus interconnect length for both Si and GaAs channels in conventional and NLSV structures at 300 K. Material and device parameters are the same as that in Fig. 15.12(a). SITE degrades with an increase in the length of the semiconductor between the nanomagnets in the spin device. The degradation in SITE is more pronounced for NLSV devices. However, for conventional spin valves, the presence of an electric field helps to improve SITE over that of NLSV devices.

15.7   Comparison of electrical and semiconducting spintronic interconnects

The performance and the energy dissipation of spintronic and CMOS interconnects are compared at the end of the silicon-technology roadmap (minimum feature size of 7.5 nm). For this analysis, the nanomagnet delay and energy consumption is not considered for the spintronic circuit. However, for CMOS circuits, the overhead of devices is also included in the analysis. This gives an upper bound on the performance of spintronic interconnects.

The transport of electron spins in conventional spin valves is governed by particle drift. Hence, the delay of interconnects in conventional spin valves is given as $t_{con} = L_{SC}/v_d$, where v_d is the drift velocity, μ is the mobility of the electron spins, and L_{SC} is the interconnect length. The low-field drift velocity for bulk semiconducting interconnects is given as $v_d = \mu \Delta V / L_{SC}$, where ΔV is the voltage drop across the interconnect. For interconnects in NLSV devices, particle transport is governed by diffusion. Hence, the delay of interconnects in NLSV is given as $t_{NLSV} = L_{SC}{}^2/(2D)$, where D is the diffusion coefficient of electron spins in the interconnect.

Figure 15.14(a) shows the delay of both NLSV spintronic and CMOS interconnects as a function of interconnect length in gate pitches at the 2024 ITRS technology roadmap. The delay of CMOS interconnects is plotted at two driver sizes: channel width-to-length (W/L) ratio equal to unity and five (called 5× driver). The parameters used for CMOS interconnect delay are listed in Table 15.5. The doping concentration is fixed at $10^{17}\,cm^{-3}$ for both silicon and GaAs interconnects in Fig. 15.14(a). As the

Table 15.5 Values of the resistance and the capacitance of devices and interconnects to evaluate the delay of the CMOS circuit at the 2024 technology node [2]. The delay of the CMOS circuit is given as $t_{CMOS} = 0.69R_s(C_s + C_L) + 0.69(R_sc_w + r_wC_L)L + 0.38r_wc_wL^2$, where L is the interconnect length

Symbol	Meaning	Value
W	Width taken equal to ½-M1 pitch	7.5 nm
V_{dd}	Supply voltage	0.75 V
I_{DSAT}	Saturation current of minimum-sized NFET	2170 μA/μm
R_s	On-resistance of minimum-sized NFET	37 kΩ
C_s	Parasitic capacitance of minimum-sized inverter	6.3 aF
C_L	Load capacitance of minimum-sized inverter	6.3 aF
c_w	Per unit length capacitance of local-level M1	1.2 pF/cm
AR	Aspect ratio = height/width	2.1
H	Height of the interconnect	15 nm
r_w	Per unit length resistance of local-level M1	2.7×10^7 Ω/cm

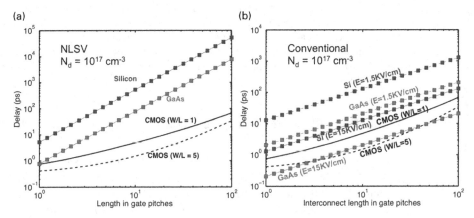

Fig. 15.14 (a) Delay versus length for NLSV spintronic interconnects at the 7.5 nm ITRS technology node. The performance of spintronic interconnects in an NLSV setup is much lower than that of their CMOS counterparts. The disparity in the performance of NLSV spintronic and CMOS interconnects grows with an increase in the interconnect length. (b) Delay versus length for spintronic interconnects based on conventional spin valves at the 7.5 nm ITRS technology node. The delay of conventional spintronic interconnects grows linearly with interconnect length, so the disparity in the performance of CMOS and conventional spintronic interconnects does not degrade with interconnect length for short, local interconnects. The performance of GaAs spintronic interconnects in conventional spin valves is comparable to that of CMOS interconnects driven by a 5× driver.

doping concentration increases, the electrical transport parameters - electron diffusion coefficient and mobility - in semiconductors degrade due to the enhanced impurity-dominated scattering of carriers. Hence, the delay of NLSV spintronic interconnects would degrade if the doping concentration were increased [85]. Both Si and GaAs NLSV interconnects are significantly slower than their conventional CMOS counterparts even for short interconnect lengths. It must be noted that the delay of NLSV spintronic interconnects increases quadratically with the interconnect length, while the delay of CMOS interconnects grows linearly with interconnect length for short, local interconnects considered here. Hence, the disparity in delays of NLSV and CMOS circuits grows with interconnect length.

The delay of conventional spintronic interconnects is plotted in Fig. 15.14(b) as a function of interconnect length in gate pitches. Unlike the NLSV spintronic interconnects, the delay of conventional spintronic interconnects increases linearly with interconnect length. The delay of conventional spintronic interconnects can be increased by increasing the drift velocity of carriers via electric field.[3] The delay of GaAs conventional interconnects at a high electric field of 15 kV/cm can be comparable to that of CMOS interconnects driven by a 5× driver. However, Si spintronic interconnects are slower than CMOS interconnects largely due to the lower mobility of carriers in Si compared to that in GaAs at the same doping concentration.

It can be seen from Fig. 15.14(a) and (b) that spintronic interconnects particularly in the NLSV configuration are fairly sluggish compared to their CMOS counterparts even when the overhead associated with nanomagnets is ignored. While the delay of the native spintronic interconnect is not affected by spin relaxation, a degradation in spin concentration caused by spin relaxation in the interconnect will necessitate insertion of spin repeaters that may add to the overall delay of the signal path.

The energy-per-bit of spintronic interconnects depends on the Joule heating occurring along the path of the current flow, and is given as

$$E_{\text{spin}} = I_{\text{elec}}^2 R \Delta t, \tag{15.6}$$

where I_{elec} is the electrical current used for pumping electron spins in the interconnect, R is the resistance of the interconnect for the electric current flow, and Δt is the transit time of electron spins through the interconnect. For signal detection at the receiver in the spin valve, $I_{\text{spin,thres}} > \eta I_{\text{elec}}$, where η is the SITE of the spin valve and $I_{\text{spin,thres}}$ is the minimum spin current required to cause magnetization reversal of the receiver nanomagnet via the spin torque effect. The minimum energy dissipation in spintronic interconnects can be expressed in terms of the receiver threshold as

$$E_{\text{spin,min}} = \left(\frac{I_{\text{spin,thres}}}{\eta} \right)^2 R \Delta t. \tag{15.7}$$

For conventional spintronic interconnects, R is the sum of the semiconductor resistance and the interface resistances; for NLSV spintronic interconnects, R is only the sum of the interface resistances. Interestingly, the energy dissipation of spin valves can be optimized by changing the interface properties. An increase in the interface resistance may lead to an improvement in SITE (η) that will help to reduce the electric current

---

[3] This comment is true only when the mobility of carriers does not degrade at high electric fields across the conventional spin valve devices. To accurately estimate the performance benefits of increasing electric field in conventional spin valves, an electric field-dependent expression for mobility must be used. Typically, at very high electric fields, the drift velocity of carriers will saturate at the receiver terminal in long-channel semiconducting interconnects.

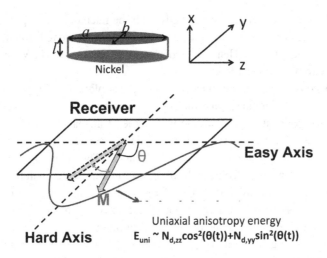

Fig. 15.15 The top figure shows the elliptical nanomagnet with the coordinate axes. The major-axis dimension of the nanomagnet is denoted as a, the minor-axis dimension is denoted as b, while the thickness of the nanomagnet is denoted as l. The bottom figure is the shape anisotropy energy landscape of the nanomagnet. (Figure taken from [86].)

requirement for the circuit, but at the same time the overall resistance of the spin valve will increase. Furthermore, a reduction in $I_{spin,thres}$ can lead to substantial savings in the energy dissipation of spintronic interconnects. $I_{spin,thres}$ is a function of the nanomagnet material properties, its volume, and the switching scheme utilized to switch the nanomagnet. The energy-landscape of the nanomagnet is illustrated in Fig. 15.15. In a full spin torque assisted switching (STS), the receiver nanomagnet is assumed to lie along its easy axis and the incoming spin current toggles the magnetic state of the receiver nanomagnet. Even though this scheme is less prone to thermal errors, the amount of spin current and the time for magnetization reversal are huge [86, 87]. Alternatively, an external clock may be used to align the nanomagnet along its hard axis, which is only a marginally stable state of the nanomagnet. This supports the Bennett clocking scheme for the nanomagnet-based logic. A Bennett clock is implemented by forcing the magnet to orient along its hard axis temporarily prior to propagating information signal through it [88, 89]. In the mixed-mode switching (MMS) scheme, magnetization reversal is accomplished in three distinct phases. In Phase 1, the nanomagnet is aligned along its hard axis using Bennett clocking. In Phase 2, the spin current through the nanomagnet is switched on and the nanomagnet magnetization rotates from $(180 - v)°$ to $v°$, where $v°$ is slightly smaller than $90°$. In the last phase, the spin torque current is switched off and the intrinsic damping of the nanomagnet brings the magnetization to its stable orientation. There are a few innovative ways to rotate the magnet from its easy axis to the hard axis and provide a metastable quiescent point. These possible schemes are [53, 87]:

(1)   Use of voltage-generated stress (VGS) in a multiferroic magnet consisting of a magnetostrictive layer (nickel) and a piezoelectric layer to rotate the nanomagnet from $(\pi - v)$ to $(\pi - \varphi)$, where $\varphi < \pi/2$.

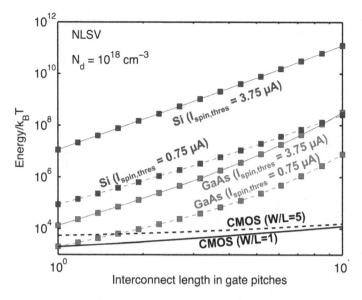

Fig. 15.16 Energy dissipation of NLSV spintronic and CMOS interconnects at the 7.5 nm ITRS technology node. The energy dissipation of NLSV spintronic interconnects increases more rapidly with interconnect length. Hence, even short, local interconnects consume more energy than their CMOS counterparts. A lower spin threshold current for the receiver helps to reduce the energy dissipation of spintronic interconnects.

(2)   Applying a voltage to a fixed magnetic layer in contact with the output layer through a spacer region; the spacer region accumulates spins in the direction of the fixed layer and helps to exert spin torque on the output nanomagnet.

In this work, the nanomagnet is assumed to be made of nickel with a switching threshold of 3.75 μA (unless otherwise stated). This corresponds to a uniaxial anisotropy of $15k_BT$ for a volume of $1.62 \times 10^4 \, \text{nm}^3$.[4]

The energy-per-bit of NLSV spintronic and CMOS interconnects is plotted in Fig. 15.16 as a function of interconnect length in gate pitches at the 2024 ITRS technology roadmap. Due to the limited spin transport length in NLSV devices, the energy dissipation of both Si and GaAs spintronic interconnects is much higher than that of CMOS interconnects. The energy dissipation in NLSV spin valves increases in a super-linear fashion with interconnect length. The energy dissipation of GaAs interconnects with $I_{\text{spin,thres}} = 0.75$ μA is comparable to that of CMOS interconnects ($W/L = 5$) only for short interconnects of up to 2–3 gate pitches.

Figure 15.17 shows the energy-per-bit of conventional spintronic and CMOS interconnects as a function of interconnect length in gate pitches at the 2024 ITRS technology roadmap. Two values of interface conductance are chosen: $G_c = 9 \times 10^9$ $(\Omega\text{m}^2)^{-1}$ (tunneling barrier) and 9×10^{11} $(\Omega\text{m}^2)^{-1}$ (relatively transparent). The area of

---

[4] For an elliptical nickel nanomagnet with major axis equal to 82.5 nm, minor axis of 50 nm, this corresponds to a switching current density of $1.15 \times 10^9 \, \text{J/m}^2$.

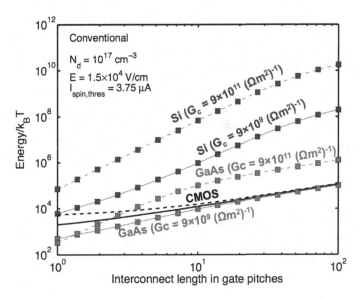

Fig. 15.17 Energy dissipation of spintronic interconnects in a conventional spin valve and CMOS interconnects at the 7.5 nm ITRS technology node. The energy dissipation of Si spintronic interconnects in a conventional spin valve is more than that of CMOS interconnects even for interconnects as short as one gate pitch at the 7.5 nm technology node. GaAs spintronic interconnects for conventional valves can offer lower energy dissipation than their CMOS counterparts for interconnects shorter than few gate pitches.

the magnets is taken to be $3230\,\text{nm}^2$ and the switching spin threshold current is taken to be $3.75\,\mu\text{A}$. The energy dissipation of GaAs conventional spin interconnects with tunneling barriers is lower than that of CMOS interconnects. However, for a relatively transparent barrier, the energy dissipation increases more rapidly with interconnect length and quickly exceeds that of CMOS interconnects for interconnects longer than 2–3 gate pitches. The energy dissipation of Si interconnects in conventional spin valves is significantly higher than that of CMOS interconnects. This is largely due to a severe degradation in the SITE of Si interconnects at longer interconnect lengths.

A comparison of the performance and energy dissipation of conventional and NLSV configurations of spin devices reveals that the conventional spin valve with GaAs benefits greatly from (i) superior electron transport characteristics (high electron diffusivity, mobility, and lower electrical resistivity), (ii) doping-independent spin relaxation length, and (iii) an enhancement in SITE due to the presence of an external electric field. Hence, GaAs-based conventional spin valves may be comparable both in energy and performance to CMOS circuits. However, the results presented here only provide an upper bound on the performance metrics since the overhead associated with spin devices has been ignored. In principle, designing a fully optimized spin circuit requires a multivariable optimization with respect to the nanomagnet material and dimensions, interconnect material and dimensions, doping concentration for semiconducting interconnects, and the interface resistance–area product. The results presented here serve to highlight the advantages, challenges, and limits of semiconductor-based spin valve circuits.

15.8 Conclusion and outlook

As the CMOS technology advances towards ultra-scaled dimensions, interconnects become increasingly problematic from the resistance–capacitance, reliability, and routability perspectives. Various new device structures are being proposed to improve the performance of transistors at each technology node. To fully utilize the performance benefit offered by making better transistors, however, an equally beneficial interconnect technology has to be invented. Due to the significant increase in the resistivity of Cu interconnects at ultra-scaled dimensions, new opportunities arise for ultra-low-capacitance carbon-based interconnect technologies. These options may offer speed and energy gain in circuits with conventional and emerging charge-based FET technologies. The optimal interconnect technology parameters for various FETs may differ due to the differences in the output resistances and input capacitances of these devices.

Post-CMOS devices are envisaged on the idea of using state variables other than electron charge to store and manipulate information. To this date, various potential candidates have emerged as the state variable of the future. Some noteworthy examples include the electron charge, pseudospin in graphene, excitons, phonons, domain walls, and photons. Amongst these state variables, electron spin is the most studied with potential advantages in terms of its robustness and non-volatility. Since electrons inherently possess both charge and spin, it is possible to conceive a logic system where spin is used to enhance the functionality provided only by the electron charge. The interconnections form an integral component of any logic system as they provide the physical medium to carry information between the devices in the system. Through the quantitative analysis presented here, we have highlighted some of the challenges and limitations associated with spintronic interconnects in a spin valve device. One of the ways to improve the performance while reducing energy dissipation of the current spin valve devices is to design novel circuits with a reduced complexity. This can cut down on interconnect lengths and help overcome the throughput limitations imposed by slow particle transport through the spintronic interconnects.

Smarter architectures that have massive concurrency or parallelism of operation built into them may be required to harness the full potential of these logic options. A promising architecture for spintronic systems must meet the following requirements: (i) implement massive concurrency to mask the low computation and communication speeds; (ii) implement highly localized computation to avoid long interconnects which require repeaters or spin-to-electrical signal convertors; and (iii) take advantage of the non-volatile nature of spin-based devices. An example of such an architecture is the array processor in which the arrays operate either in parallel or in pipeline. Each processor is connected only to its nearest neighbor [90]. Array processors with spin logic can be utilized towards specific applications such as matrix manipulation, video filtering, data processing, etc. A recent proposal on using electron spin computation to implement non-Boolean, analog mode, majority evaluation is the neuromorphic architecture using lateral spin valves [91]. This architecture can be used for analog data sensing, data conversion, cognitive computing, associative memory, and analog and digital signal processing.

References

[1] M. T. Bohr, "Interconnect scaling – the real limiter to high performance ULSI." *Electron Devices Meeting (IEDM), 1995 IEEE International*, pp. 241–244 (1995).

[2] *International Technology Roadmap for Semiconductors (ITRS)* (2012). Available at: www. itrs.net.

[3] H. B. Bakoglu & J. D. Meindl, "Optimal interconnection networks for ULSI." *IEEE Transactions Electron Devices*, **32**(5), 903–909 (1985).

[4] I. Young & K. Raol, "A comprehensive metric for evaluating interconnect performance." *Interconnect Technology Conference, 2001 IEEE International*, pp. 119–121 (2001).

[5] J. Baliga, "Chips go vertical [3D IC interconnection]." *IEEE Spectrum*, **41**(3), 43–47 (2004).

[6] A. Ceyha & A. Naeemi, "Multilevel interconnect networks for the end of the roadmap: conventional Cu/low-*k* and emerging carbon based interconnects." In *Interconnect Technology Conference, IEEE International*, pp. 1–3 (2011).

[7] H. B. Bakoglu, *Circuits, Interconnections and Packaging for VLSI* (Reading, MA, Addison-Wesley, 1990).

[8] J. A. Davis, R. Venkatesan *et al.*, "Interconnect limits on gigascale integration (GSI) in the 21st century." *Proceedings of the IEEE*, **89**(3), 305–324 (2001).

[9] J. Mathon & A. Umerski, "Theory of tunneling magnetoresistance of an epitaxial Fe/MgO/Fe(001) junction." *Physical Review B*, **63**(22) (2001).

[10] J. D. Meindl, R. Venkatesan *et al.*, "Interconnecting device opportunities for gigascale integration (GSI)." In *Electron Devices Meeting, IEEE International*, pp. 23.21.21–23.21.24 (2001).

[11] J. D. Meindl, J. A. Davis *et al.*, "Interconnect opportunities for gigascale integration." *IBM Journal of Research and Development*, **46**(2–3), 245–263 (2002).

[12] D. Edelstein, J. Heidenreich *et al.* "Full copper wiring in a sub-0.25mum CMOS ULSI technology." In *Electron Devices Meeting, 1997 IEEE International*, pp. 773–776 (1997).

[13] M. Bohr, "The new era of scaling in an SoC world." In *Solid-State Conference, 2009 IEEE International, Digest of Technical Papers*, pp. 23–28 (2009).

[14] S. Borkar, "A thousand core chips." In *Proceedings of the 44th Annual Design Automation Conference (DAC)* (2007).

[15] W. R. Davis, J. Wilson *et al.*, "Demystifying 3D ICs: the pros and cons of going vertical." *IEEE Design and Test of Computers*, **22**(6), 498–510 (2005).

[16] R. G. Beausoleil, P. J. Kuekes *et al.*, "Nanoelectronic and nanophotonic interconnect." *Proceedings of the IEEE*, **96**(2), 230–247 (2008).

[17] A. V. Krishnamoorthy, R. Ho *et al.* "Computer systems based on silicon photonic interconnects." *Proceedings of the IEEE*, **97**(7), 1337–1361 (2009).

[18] A. Balakrishnan, "*Analysis and Optimization of Global Interconnects for Many-core Architectures.*" Department of Electrical and Computer Engineering. Atlanta, GA, Georgia Institute of Technology. MS (2010).

[19] J. F. Guillaumond, L. Arnaud *et al.*, "Analysis of resistivity in nano-interconnect: full range (4.2–300 K) temperature characterization." In *Interconnect Technology Conference, 2003 International*, pp. 132–134 (2003).

[20] W. F. A. Besling, M. Broekaart *et al.*, "Line resistance behavior in narrow lines patterned by a TiN hard mask spacer for 45 nm node interconnects." *Microelectronic Engineering*, **76**(1–4), 167–174 (2004).

[21] W. Steinhoegl, G. Schindler *et al.*, "Impact of line edge roughness on the resistivity of nanometer-scale interconnects." *Microelectronic Engineering*, **76**(1–4), 126–130 (2004).

[22] W. Steinhoegl, G. Schindler *et al.*, "Unraveling the mysteries behind size effects in metallization systems." *Semiconductor International*, **28**, 34–38 (2005).

[23] W. Steinhoegl, G. Schindler *et al.*, "Comprehensive study of the resistivity of copper wires with lateral dimensions of 100 nm and smaller." *Journal of Applied Physics*, **97**(2), 023701–023707 (2005).

[24] H.-C. Chen, H.-W. Chen *et al.*, "Resistance increase in metal-nanowires." In *VLSI Technology, Systems, and Applications, International Symposium on*, pp. 1–2 (2006).

[25] J. J. Plombon, E. Andideh *et al.*, "Influence of phonon, geometry, impurity, and grain size on copper line resistivity." *Applied Physics Letters*, **89**(11), 113123–113124 (2006).

[26] M. Shimada, M. Moriyama *et al.*, "Electrical resistivity of polycrystalline Cu interconnects with nano-scale linewidth." *Journal of Vacuum Science & Technology B: Microelectronics and Nanometer Structures*, **24**(1), 190–194 (2006).

[27] H. Kitada, T. Suzuki *et al.*, "The influence of the size effect of copper interconnects on RC delay variability beyond 45nm technology." In *Interconnect Technology Conference, International*, pp. 10–12 (2007).

[28] N. S. Nagaraj, W. R. Hunter *et al.*, "Impact of interconnect technology scaling on SoC design methodologies." In *International Interconnect Technology Conference*, pp. 71–73 (2005).

[29] A. Kaloyeros, E. T. Eisenbraun *et al.*, "Zero thickness diffusion barriers and metallization liners for nanoscale device applications." *Chemical Engineering Communications*, **198**(11), 1453–1481 (2011).

[30] S. Sinha, G. Yeric *et al.*, "Exploring sub-20-nm finFET design with predictive technology models." In *Design Automation Conference (DAC)*, pp. 283–288 (2012).

[31] N. Magen, A. Kolodny *et al.*, "Interconnect-power dissipation in a microprocessor." In *International Workshop on System Level Interconnect Prediction (SLIP)* (2004).

[32] X. Wang, Y. Ouyang *et al.*, "Room-temperature all-semiconducting sub-10-nm graphene nanoribbon field-effect transistors." *Physics Review Letters*, **100**(20), 206803 (2008).

[33] J. Deng & H.-S. P. Wong, "A compact SPICE model for carbon nanotube field-effect transistors including non-idealities and its application part 1: model of the intrinsic channel region." *IEEE Transactions Electron Devices*, **54**(12), 3186–3194 (2007).

[34] A. C. Seabaugh & Q. Zhang, "Low-voltage tunnel transistors for beyond CMOS logic." *Proceedings of the IEEE* **98**(12), 2095–2110 (2008).

[35] M. Luisier, & G. Klimeck, "Atomistic full-band design study of InAs band-to-band tunneling field-effect transistors." *IEEE Electron Devices Letters*, **30**(6), 602–604 (2009).

[36] V. V. Zhirnov, R. K. Cavin, III *et al.*, "Limits to binary logic switch scaling – a gedanken model." *Proceedings of the IEEE* **91**(11), 1934–1939 (2003).

[37] K. Galatsis, A. Khitun *et al.*, "Alternate state variables for emerging nanoelectronic devices." *IEEE Transactions on Nanotechnology*, **8**(1), 66–75 (2009).

[38] S. Rakheja & A. Naeemi, "Interconnects for novel state variables: physical limits and device and circuit implications." *IEEE Transactions on Electron Devices*, **57**(10) (2010).

[39] S. Rakheja & A. Naeemi, "Communicating novel computational state variables: post-CMOS logic." *IEEE Nanotechnology*, **7**, 15–23 (2013).

[40] D. Nikonov, & G. Bourianoff, "Operation and modeling of semiconductor spintronics computing devices." *Journal of Superconductivity and Novel Magnetism*, **21**(8), 479–493 (2008).

[41] M. Cahay & S. Bandyopadhyay, "An electron's spin – part I." *Potentials, IEEE*, **28**(3), 31–35 (2009).

[42] S. Sughara, "Spin-transistor electronics: an overview and outlook." *Proceedings of the IEEE*, **98**(12), 2124–2154 (2010).

[43] P. Gruenberg, R. Schreiber *et al.*, "Layered magnetic structures: evidence for antiferromagnetic coupling of Fe layers across Cr interlayers." *Physical Review Letters*, **57** (1986).

[44] J. Moodera, L. Kinder *et al.*, "Large magnetoresistance at room temperature in ferromagnetic thin film tunnel junctions." *Physical Review Letters*, **74**, 3273–3276 (1995).

[45] M. Julliere, "Tunneling between ferromagnetic films." *Physics Letters A*, **54**, 225–226 (1975).

[46] S. Yuasa, T. Nagahama *et al.*, "Giant room-temperature magnetoresistance in single crystal Fe/Mgo/Fe magnetic tunnel junctions." *Nature Materials* **3**(12), 868–871 (2004).

[47] J. M. Daughton, "Magnetic tunneling applied to memory." *Journal of Applied Physics*, **81**(8), 3758–3763 (1997).

[48] S. Tehrani, J. M. Slaughter *et al.*, "Magnetoresistive random access memory using magnetic tunnel junctions." *Proceedings of the IEEE*, **91**(5), 703–714 (2003).

[49] Z. Jian-Gang, "Magnetoresistive random access memory: the path to competitiveness and scalability." *Proceedings of the IEEE*, **96**(11), 1786–1798 (2008).

[50] S. Datta, & B. Das, "Electronic analog of the electro-optic modulator." *Applied Physics Letters*, **56**(7), 665–667 (1990).

[51] M. Johnson, "Bipolar spin switch." *Science*, **260**(5106), 320–323 (1993).

[52] M. Johnson, "The bipolar spin transistor." *Nanotechnology*, **7**(4), 390–396 (1996).

[53] B. Behin-Aein, D. Datta *et al.*, "Proposal for an all-spin logic device with built in memory." *Nature Nanotechnology*, **5**, 266–270 (2010).

[54] A. Khitun, M. Bao *et al.*, "Spin wave logic circuit on silicon platform." In *Information Technology: New Generations, ITNG 2008, Fifth International Conference on* (2008).

[55] A. Khitun, & K. L. Wang, "Nano scale computational architectures with spin wave bus." *Superlattices and Microstructures*, **38**(3), 184–200 (2005).

[56] A. Khitun, D. E. Nikonov *et al.*, "Feasibility study of logic circuits with a spin wave bus." *Nanotechnology*, **18**(46) (2007).

[57] L. Liu, O. J. Lee *et al.*, "Magnetic switching by spin torque from the spin Hall effect." arXiv:1110.6846 (2012).

[58] L. Liu, C.-F. Pai *et al.*, "Spin torque switching with the giant spin Hall effect of tantalum." *Science*, **336** (2012).

[59] J. Z. Sun, M. C. Gaidis *et al.*, "A three-terminal spin-torque-driven magnetic switch." *Applied Physics Letters*, **95**, 083506 (2009).

[60] C.-F. Pai, L. Liu, *et al.*, "Spin transfer torque devices utilizing the giant spin Hall effect of tungsten." *Applied Physics Letters*, **101** (2012).

[61] R. J. Elliott, "Theory of the effect of spin-orbit coupling on magnetic resonance in some semiconductors." *Physical Review*, **96**(2), 14 (1954).

[62] D. J. Lepine, "Spin resonance of localized and delocalized electrons in phosphorus-doped silicon between 20K and 300K." *Physical Review B*, **2** (1970).

[63] J. Cheng, M. Wu *et al.*, "Theory of the spin relaxation of conduction electrons in silicon." *Physical Review Letters*, **104**(1) (2010).

[64] O. D. Restrepo & W. Windl, "Full first-principles theory of spin relaxation in group-IV materials." *Physics Review Letters*, **109**, 166604 (2012).

[65] H. Kodera, "Effect of doping on the electron spin resonance in phosphorus doped silicon." *Journal of the Physical Society of Japan*, **19**(6) (1964).

[66] H. Kodera, "Effect of doping on the electron spin resonance in phosphorus doped silicon. II." *Journal of the Physical Soceity of Japan*, **21**(6) (1966).

[67] H. Kodera, "Effect of doping on the electron spin resonance in phosphorus doped silicon. III. Absorption intensity." *Journal of the Physical Society of Japan*, **26**, 377–380 (1969).

[68] H. Kodera, "Dyson effect in the electron spin resonance of phosphorus-doped silicon." *Journal of the Physical Society of Japan*, **28**, 89–98 (1970).

[69] J. Fabian, A. Matos-Abiague *et al.*, "Semiconductor spintronics." *Acta Physica Slovaca*, **57**(4 & 5) (2007).

[70] G. E. Pikus & A. N. Titkov, Chapter 3, in *Optical Orientation*, ed. F. Meier and B. P. Zakharchenya (Amsterdam: North-Holland, 1984).

[71] A. Bungay, S. Popov *et al.*, "Direct measurement of carrier spin relaxation times in opaque solids using the specular inverse Faraday effect." *Physics Letters A*, **234** (1997).

[72] A. Kimel, F. Bentivegna *et al.*, "Room-temperature ultrafast carrier and spin dynamics in GaAs probed by the photoinduced magneto-optical Kerr effect." *Physical Review B*, **63**(23) (2001).

[73] S. Garzon, I. Zutic *et al.*, "Temperature-dependent asymmetry of the nonlocal spin-injection resistance: evidence for spin nonconserving interface scattering." *Physical Review Letters*, **94**(17) (2005).

[74] T. Yang, T. Kimura *et al.*, "Giant spin-accumulation signal and pure spin-current-induced reversible magnetization switching." *Nature Physics*, **4**(11), 851–854 (2008).

[75] H. Zou, X. J. Wang *et al.*, "Reduction of spin-flip scattering in metallic nonlocal spin valves." *Journal of Vacuum Science & Technology B* **28**(6) (2010).

[76] N. Poli, M. Urech *et al.*, "Spin-flip scattering at Al surfaces." *Journal of Applied Physics*, **99**(8) (2006).

[77] F. J. Jedema, M. S. Nijboer *et al.*, "Spin injection and spin accumulation in all-metal mesoscopic spin valves." *Physical Review B*, **2003**(67) (2003).

[78] N. Tombros, S. Tanabe *et al.*, "Anisotropic spin relaxation in graphene." *Physical Review Letters*, **101**(4), 4 (2008).

[79] M. Popinciuc, C. J. Zsa *et al.*, "Electronic spin transport in graphene field effect transistors." *Physics Reviews B*, **80**, 214427 (2009).

[80] M. Wojtaszek, I. J. Vera-Marun *et al.*, "Enhancement of spin relaxation time in hydrogenated graphene spin-valve devices." *Physical Review B* **87**(8), 5 (2013).

[81] D. Huertas-Hernando, "Spin relaxation times in disordered graphene." *The European Physical Journal*, **148**(1), 177–181 (2007).

[82] P. Zhang & M. W. Wu, "Electron spin relaxation in graphene with random rashba field: comparison of d'Yakonov–Perel and Elliott–Yafet-like mechanisms." *New Journal of Physics* **14**(March) (2012).

[83] J. Bass, & W. P. Pratt Jr, "Spin-diffusion lengths in metals and alloys, and spin-flipping at metal/metal interfaces: an experimentalist's critical review." *Journal of Physics: Condensed Matter*, **19** (2007).

[84] Z. Yu & M. Flatte, "Spin diffusion and injection in semiconductor." *Physical Review B* **66**, 14 (2002).

[85] S. Rakheja & A. Naeemi, "Roles of doping, temperature, and electric field on spin transport through semiconducting channels in spin valves." *IEEE Transactions on Nanotechnology*, **12**(5), 796–805 (2013).

[86] S. Rakheja & A. Naeemi, "Interconnect analysis in spin-torque devices: Performance modeling, optimal repeater insertion, and circuit-size limits." In *Quality Electronic Design, 2012 13th International Symposium on*, pp. 283–290 (2012).

[87] K. Roy, S. Bandyopadhyay *et al.*, "Hybrid spintronics and straintronics: A magnetic technology for ultra low energy computing and signal processing." *Applied Physics Letters*, **99**(6), 3 (2011).

[88] C. H. Bennett, "The thermodynamics of computation – a review." *International Journal of Theoretical Physics*, **21**(12), 905–940 (1982).

[89] M. S. Fashami, K. Roy *et al.*, "Magnetization dynamics, Bennett clocking and associated energy dissipation in multiferroic logic." *Nanotechnology*, **22**, 155201–155210 (2011).

[90] S. Kung, "VLSI array processors." *ASSP Magazine, IEEE*, **2**(3), 4–22 (1985).

[91] M. Sharad, C. Augustine *et al.*, "Boolean and non-Boolean computation with spin devices." In *Electron Devices Meeting, 2012 IEEE International*, (2012).

Index

Page numbers in **bold** refer to figures.